Hydraulic Fracturing and Microseismicity for Applications of CO_2 Storage

Hydraulic fracturing (HF) has become essential for extracting oil and gas from unconventional reservoirs and heat energy from hot dry rock masses. However, a paradox remains in the fracturing mode of HF: stress conditions predicted by elasticity theory suggest tensile fracturing, while shear fracturing is dominant in HF-related microseismicity (MS). This book resolves this paradox by reviewing the results of various laboratory and small-scale field experiments conducted by the author over nearly 40 years.

It explores the use of carbon dioxide (CO_2) in HF as a method for geologic sequestration and demonstrates that CO_2 can induce three-dimensional, sinuous cracks with numerous secondary branches, creating ideal pathways for heat, oil, and gas recovery. The author provides a detailed explanation of the fracturing mechanisms based on rock and fluid properties, particularly the use of supercritical (SC) CO_2 as a fracturing fluid. Readers will benefit from a deeper understanding of HF processes and their applications for resource extraction and environmental management.

Hydraulic Fracturing and Microseismicity for Applications of CO_2 Storage will serve as a reference for researchers, engineers, and students who are involved or interested in the areas of geomechanics, geophysical exploration, seismology, and geology for petroleum and geothermal engineering, as well as for academics and professionals working in the fields of general rock mechanics, geophysics, and civil engineering.

ABOUT THE SERIES

The Geomechanics Research Series was set up in the early 1990s and focused on geomechanics not only from a geotechnical engineering point of view but also emphasized geosciences/geophysics theory and applications. Volumes published thus far have covered the topics Fractals in Rock Mechanics, True Triaxial Testing of Rocks, Experimental Rock Mechanics, and Mechanical Behaviour of Rocks under High Pressure Conditions. Now re-launched under the title Geomechanics Research, we would be delighted to consider new book proposals for publishing in the series, please contact the series editor for further information.

OBJECTIVES

The objective of this series is to provide knowledge on geomechanics to help engineers, researchers, and students including beginners in the field. Each book in the series takes readers on a journey from foundational concepts to breakthrough discoveries in a specialized area of geomechanics, showcasing the career-defining research and insights of leading experts in the field.

MAIN FIELDS COVERED BY THE BOOK SERIES

This series covers the field of geomechanical engineering related to metal and coal mining, petroleum exploitation in conventional and unconventional reservoirs, geothermal energy extraction, and civil engineering for dam foundations, slopes, underground facilities such as powerhouses, tunnels, high level radioactive nuclear disposal and others. It also covers the field of geosciences related to seismology, volcanology, geology, tectonophysics among other areas.

BOOKS PUBLISHED IN THE SERIES

Volume 1. Rock Mass Response to Mining Activities
Inferring Large-Scale Rock Mass Failure, 2018
Tadeusz Szwedzicki

Volume 2. Hydraulic Fracturing and Microseismicity for Applications of CO_2 Storage
The Role of Rock Properties and Fluid Viscosity, 2026
Tsuyoshi Ishida

Hydraulic Fracturing and Microseismicity for Applications of CO_2 Storage

The Role of Rock Properties and Fluid Viscosity

Tsuyoshi Ishida

CRC Press
Taylor & Francis Group
Boca Raton London New York Leiden

CRC Press is an imprint of the
Taylor & Francis Group, an **informa** business
A BALKEMA BOOK

Designed cover image: Getty Images

First published 2026
by CRC Press/Balkema
4 Park Square, Milton Park, Abingdon, Oxon, OX14 4RN

and by CRC Press/Balkema
2385 NW Executive Center Drive, Suite 320, Boca Raton FL 33431

CRC Press/Balkema is an imprint of the Taylor & Francis Group, an informa business

British Library Cataloguing-in-Publication Data
A catalogue record for this book is available from the British Library

ISBN: 9780367608439 (hbk)
ISBN: 9781032253152 (pbk)
ISBN: 9781003282631 (ebk)

DOI: 10.1201/9781003282631

Typeset in Times New Roman
by codeMantra

Contents

Foreword

The marriage of hydraulic fracturing (HF) and acoustic emission (AE) is a natural one, and the content of this book, *Hydraulic Fracturing and Microseismicity for Applications of CO_2 Storage: The Role of Rock Property and Fluid Viscosity*, highlights fundamental knowledge and innovative research in both HF and AE. The problem of an internally pressurized cylindrical cavity (borehole) within a stressed medium is reviewed for both dry and fluid-saturated rock, along with the tension cutoff failure criterion. It is clear from theory and experiments that macroscopic crack growth is in the same direction as the maximum principal stress (compression positive), and several laboratory and field investigations are discussed. Timely studies on the use of CO_2 as the fracturing fluid are included.

The use of HF as an *in-situ* stress measurement method dates back to the early 1960s, when Professor Charles Fairhurst of the University of Minnesota and his research team designed and built equipment capable of determining the stress state in a rock mass at depths of several hundred meters. After careful study of various techniques, HF was determined to be the most promising procedure, where the fluid pressure developed in the borehole during fracture of the rock (or opening of existing fractures) can be related to the pre-existing stress state in the rock mass. Several graduate students and visiting scholars participated in research on HF and stress measurements. One of those researchers was Professor Yoshiaki Mizuta, who, upon his return to Japan, influenced many other scholars, including Professor Tsuyoshi Ishida.

Professor Ishida worked in the Central Research Institute of Electric Power Industry and performed rock mechanics research related to construction of a number of underground facilities, including large powerhouses. He earned his PhD degree from Kyoto University for the dissertation entitled *Study on In-situ Stress State in Rock.* In the application of HF as a stress measurement method, the location and orientation of the crack plane are required. Thus, Professor Ishida realized that he needed a technique well suited for identifying fractures, and that technique is AE, also known as microseismicity.

Rock is generally thought of as a solid material, but what is sometimes not fully appreciated is that rock is damaged: small cracks and pores are common features found, in varying amounts, in all rock, even those of high strength. As the stress state in a rock mass is varied and further damage is generated, energy is released in the form of stress waves called acoustic emission that travel through the rock, are captured by AE sensors, and are recorded through a high-speed data acquisition system. The location of each AE event can be obtained from the first arrival of the wave at different sensors, and if a macrocrack propagates, the distribution of the locations can be used to determine the crack plane. In addition, more advanced quantitative analysis can be performed to evaluate the source mechanisms – opening, closing, and/or sliding of microcracks – as well as orientations. Thus, AE is a powerful tool for the characterization of fracture over a range of applications, from laboratory testing to monitoring an underground mine to observing crack growth during HF.

This book presents AE as the elastodynamics problem of a point source of displacement discontinuity, a microcrack, represented by a specific combination of force dipoles known as the moment tensor. A convenient analysis of microcrack modes called focal mechanism involves relating polarities of the P-wave and the normal displacement calculated from the far-field, point-source solution. Tensile opening of the microcrack causes positive first motions over the entire far-field region, whereas closing gives all negative first motions; pure mode II shear sliding causes an equal distribution of positive and negative polarities. A mixed-mode condition of sliding with opening requires a detailed study of the point-source solution. Nonetheless, advantages of the polarity analysis are the simplicity of the calibration and the insensitivity of polarity to coupling.

Other topics of this book include the use of the distinct element method to model fracture, detailed laboratory experiments examining the effect of fluid viscosity, and the explanation of the apparent inconsistency between the macroscopic fracture of meter length propagating as a mode I opening crack and microcracks of millimeter length that display mixed-mode characteristics because of the local tortuosity. *Hydraulic Fracturing and Microseismicity for Applications of CO_2 Storage* is a welcome addition to the contemporary geomechanics literature.

Joseph F. Labuz
University of Minnesota
USA

Preface

Recently, hydraulic fracturing (HF) has become crucial for recovering oil and gas from unconventional reservoirs and extracting heat energy from hot dry rock masses. On the other hand, the issue of global warming, caused by the release of anthropogenic carbon dioxide (CO_2), must be addressed seriously. HF using CO_2 instead of water has been proposed as a means of geologic sequestration. However, there remains an unsolved inconsistency or paradox in the crack displacement mode of HF: the stress conditions derived from elasticity theory predict tensile fracturing in the direction of the maximum compressive stress, while shear fracturing is the dominant focal mechanism observed in acoustic emission (AE) and microseismicity (MS) associated with HF in laboratory experiments and actual production fields.

This book provides foundational knowledge and understanding of the fracturing mechanism in HF, considering rock and fluid properties, including pre-existing cracks and the viscosity of fracturing fluids. With a review of relevant elastostatic and elastodynamic solutions, this book focuses on two areas: (i) explaining the paradox of HF fracturing modes, and (ii) demonstrating the utility of using supercritical (SC) CO_2 as a fracturing fluid.

This book clarifies the HF paradox by reviewing the results of various laboratory and small-scale field experiments in which the author has been involved for nearly 40 years. Indeed, macroscopic tensile fracturing is induced in the direction of the maximum compressive stress in intact rock. However, locally, shear fracturing is observed through AE and MS due to the tortuosity of the fracture at a smaller scale. Local heterogeneities, such as pre-existing cracks, divert the direction of the HF crack extension from the direction of the global maximum compressive stress. This concept helps to explain why focal mechanisms are dominated by shear events in HF production fields, where various discontinuities interact with the HF crack extension.

This book discusses the use of SC-CO_2 as a fracturing fluid, focusing on how fluid viscosity affects crack characteristics and the fracture mode of HF. CO_2 transitions SC state due to the phase changes deep underground, where HF is applied. SC-CO_2 has low viscosity and induces three-dimensional, sinuous cracks with numerous secondary branches that feature shear fractures – ideal pathways for the recovery or extraction process. Additionally, small-scale field experiments show that HF using SC-CO_2 stimulates a significantly larger region and improves injectivity much more than HF using water.

Acknowledgments

This book discusses the results of various laboratory and small-scale field experiments on hydraulic fracturing (HF) that the author has been involved in for nearly 40 years. First of all, the author would like to express his gratitude to Prof. Yoshiaki Mizuta of Yamaguchi University, who organized and led the laboratory HF experiments and facilitated various international collaborative HF research projects, including those with the University of Minnesota and the University of Oklahoma. The author also sincerely appreciates Dr. Tadashi Kanagawa for sharing invaluable knowledge, skills, and opportunities to initiate Acoustic Emission Monitoring in both laboratory and field settings. Additionally, he is deeply grateful to Dr. Yoshihiro Sawada and Dr. Shunji Sasaki for their guidance in conducting small-scale field HF experiments, as well as their support in monitoring and analyzing microseismicity during HF experiments at the Central Research Institute of Electric Power Industry.

His HF research described in this book has been supported by the knowledge, ideas, data, and skills generously provided by many researchers, engineers, and students. Among them, the author sincerely appreciates the following researchers for their help. (Affiliations' names are listed chronologically as they were when they supported the author's HF research.)

Central Research Institute of Electric Power Industry:
Dr. Tadashi Kanagawa, Dr. Yoshihiro Sawada, Dr. Shunji Sasaki, Dr. Naoto Yoshioka, Dr. Yuji Kanaori, and Dr. Hideshi Kaieda.

National Institute for Resources and Environment:
Dr. Michio Kuriyagawa, Dr. Isao Matsunaga, Dr. Hideo Kobayashi, and Dr. Tsutomu Yamaguchi.

Yamaguchi University:
Prof. Yoshiaki Mizuta, Prof. Osam Sano, and Dr. Qu Chen.

University of Minnesota:
Prof. Joseph F. Labuz and Prof. Emmanuel Detournay.

University of Oklahoma:
Prof. Jean-Claude Roegiers.

Kyoto University:
Dr. Youqing Chen, Dr. Makoto Naoi, Dr. Hiroyuki Shimizu, Dr. Ziad Bennour, and Dr. Kazuhei Aoyagi.

Japan Oil, Gas and Metals National Corporation (JOGMEC):
Dr. Yamamoto Koji, Mr. Sunao Takagi, Dr. Takashi Akai, Mr. Isao Kurosawa, Mr. Yoshihiro Ito, Ms. Yu Nagano, Mr. Yutaro Arima, Mr. Sigenobu Onozuka, Mr. Kazuhito Oseto, Dr. Kotaro Sekine, Mr. Hiroyuki Tanaka, Mr. Shigehiro Kitamura, Mr. Kokei Tamura, Mr. Hyodo Daisuke, Mr. Tsutau Takeuchi, Mr. Akira Fujimoto, and Mr. Ryohei Inoue.

3D Geoscience, Inc.:
Dr. Yoshiki Nakayama, Mr. Masaharu Itamoto, Mr. Kazumichi Kuwabara, and Takeo Tanno.

Lazoc Inc.:
Mr. Yuichi Machijima, Dr. Masahiro Tanaka, and Mr. Hirokazu Fujii.

Japan Atomic Energy Agency (JAEA):
Dr. Toshinori Sato and Mr. Hiroya Matsui.

Obayashi Corporation:
Dr. Kenichiro Suzuki.

OYO Corporation:
Dr. Tatsuya Yokoyama, Mr. Kouji Ogawa, and Dr. Takashi Danjo.

In conclusion, the author would like to express special thanks for their support of his wife, Yoko, and his children, Go and Yu.

About the Author

Tsuyoshi Ishida has been involved in Acoustic Emission (AE) and Microseismicity (MS) monitoring for various laboratory and small-scale field experiments on hydraulic fracturing (HF) for nearly 40 years. He obtained his B.E. (1977) and M.S. (1979) degrees in the Department of Mineral Science and Technology from Kyoto University. From 1980 to 1991, he worked at the Central Research Institute of Electric Power Industry, where he conducted rock mechanics research related to the construction of various power generation facilities, such as large underground powerhouse caverns, dams, nuclear power stations, and feasibility studies on high-level radioactive nuclear waste disposal. During this research, he began monitoring AE and MS associated with rock fracturing. He received his Ph.D. (1989) from Kyoto University for his dissertation titled "Study on In-situ Stress State in Rock".

From 1991 to 2006, he was a faculty member at Yamaguchi University, where he initiated laboratory HF experiments with AE monitoring. During this period, he also stayed at the University of Minnesota from April 1993 to March 1994 and at the University of Oklahoma from October 1997 to January 1998 as Visiting Associate Professor. In 2002, he was promoted to Professor at Yamaguchi University, and in 2006, he moved to Kyoto University as a professor. At Kyoto University, he conducted laboratory and field experiments on HF using water and carbon dioxide for geothermal heat extraction, shale gas/oil exploitation, and underground carbon dioxide sequestration. He retired from Kyoto University in March 2019 upon reaching the mandatory retirement age. From April 2019 to March 2022, he served as a visiting researcher at the Tono Research Institute of Earthquake Science (TRIES), Association for the Development of Earthquake Prediction (ADEP). He is now Professor Emeritus of Kyoto University, and serves as an adviser for OYO Corporation, a geological consulting company from April 2019.

Chapter 1

Introduction

1.1 Brief historical review and paradox of hydraulic fracturing

According to Montgomery and Smith (2010), the first hydraulic fracturing (HF) treatment specifically designed to stimulate well production was conducted by Stanolind Oil in 1947 at the Hugoton gas field in Grant County of southwestern Kansas, USA. Following this experiment, HF treatment became more widely adopted in the industry starting in 1948. In 1949, a patent for the process was issued, with an exclusive license granted to Halliburton Oil Well Cementing Company. That same year, the company performed the first two commercial fracturing treatments in Stephens County, Oklahoma, and Archer County, Texas.

After the commercial introduction of HF, it has been applied to generate man-made cracks for stimulating oil and gas reservoirs and extracting geothermal energy from hot dry rock (HDR). Additionally, HF has been used in some cases to recover rock salt by injecting water into the rock salt reservoir and melting it (Maisons et al., 1997), as well as to induce caving as part of routine operations in longwall mining at a colliery (Jeffrey et al., 2001). On the other hand, HF has often been used to measure in situ rock stress conditions, which can be determined from pressure and flow rate data (e.g., Haimson, 1978; Cornet & Valette, 1984; Mizuta et al., 1987).

Recently, the development of horizontal drilling technology and HF treatments has made it possible to exploit oil and gas from unconventional shale reservoirs, with HF being the essential method for this. As the issue of global warming due to greenhouse gases is being taken more seriously, the use of carbon dioxide (CO_2) instead of water in HF has been proposed for shale oil and gas extraction as well as for geothermal energy extraction from HDR. If CO_2 can be used as a fracturing fluid for these operations, the underground sequestration of CO_2 could be developed and adopted more eagerly, as it would offer economic benefits that help offset the costs associated with CO_2 capture, transportation, and underground injection.

On the other hand, the first theoretical investigation into the fracturing mechanism of HF was conducted by Hubbert and Willis (1957). They demonstrated that the cracks resulting from HF are created by tensile fracture and extend along the direction of maximum compressive stress, assuming that the fracturing hole is aligned with one principal stress direction and that the surrounding rock is isotropic, homogeneous, impermeable, and elastic. This idea is based on traditional elastic theory and has been supported by many researchers (e.g., Zoback et al., 1977; Haimson, 1978; Schmitt and Zoback, 1993). However, regarding the fracturing mechanism of HF, many researchers (e.g., Talebi and Cornet, 1987; Cornet, 1992; Horálek et al., 2010) have reported that shear fracturing dominates focal mechanisms of Acoustic Emission (AE) or Micro-Seismicity (MS) observed during field operations of HF. This discrepancy between the mode of HF and its focal mechanism is referred to as the "paradox of HF" by Maxwell and Cipolla (2011).

DOI: 10.1201/9781003282631-1

1.2 Outline of this book

This chapter explains why and how the paradox arises. In Section 1.3, the stress conditions around an HF hole, derived from traditional elastic theory, are discussed, showing that HF is induced by tensile fracturing. In Section 1.4, the analysis and data demonstrating that HF is induced by shear fracturing are clarified, along with the method used to obtain the focal mechanism for MS and AE. With this background knowledge of the stress condition around the HF hole and the focal mechanism of MS and AE, the book addresses how to solve the paradox through the results of various laboratory and small-scale field experiments in which the author has been involved for over 40 years. This book highlights the tendency for tensile-dominant fractures to form near the HF hole immediately after BD (breakdown) and at the edges of regions where AE events have occurred previously. In other words, these fractures are induced by crack initiation in intact rock. In contrast, the book also identifies the tendency for shear-dominant fractures to form during crack extension, with cracks linking together, i.e., the interaction of pre-existing inherent cracks. These tendencies help explain why shear events dominate in HF in actual production fields, where various discontinuities are present. Crack extensions in these fields are likely to interact with these discontinuities. The two tendencies regarding fracture modes offer a solution to the paradox: while the elastic theory predicts tensile fractures, shear events dominate in AE monitoring in actual production fields.

This book also discusses how the viscosity of fracturing fluid affects the fracturing mechanism of HF. Specifically, concerning CO_2 capture, utilization, and storage (CCUS) projects, the effect of CO_2's fracturing fluid viscosity is examined. CO_2, in a supercritical (SC) state under high temperature and pressure deep underground, has very low viscosity. The focus is on the applicability of using CO_2 for HF in the exploitation of oil, gas, and heat energy from HDR. Laboratory experiments show that HF using CO_2 tends to induce three-dimensional sinuous cracks with many secondary branches through shear fracturing. These cracks appear to be desirable pathways for the exploitation processes. Additionally, small-scale field experiments indicate that HF using CO_2 stimulates a larger region compared to HF using water, with injectivity improvements from HF using CO_2 being significantly greater than those achieved with water.

Chapter 2 describes laboratory HF experiments on granite specimens with varying grain sizes. The results from located AE sources and the focal mechanism of the AE events indicate the tendency for tensile-dominant fractures to form in intact, homogeneous rock, unaffected by inherent pre-existing cracks. In contrast, shear-dominant fractures occur during crack extension, with cracks linking together—i.e., the interaction of inherent pre-existing cracks. These tendencies are confirmed by the results of HF experiments on impermeable acrylic resin, as well as a comparison of crack extensions normal to and along the rift plane in granite specimens, which are discussed later in this chapter. Chapter 3 elucidates the effects of the injected fluid's viscosity and the injected fluid itself on HF, based on the results of laboratory experiments that used water and viscous oil as a fracturing fluid, in comparison to sleeve fracturing (dry fracturing without injected fluid). When fracturing fluids are used, a crack extends rapidly and continuously once it starts, as the fluid invades the crack and applies pressure at the crack tip. In contrast, during fracturing using an expanding urethane sleeve without any fracturing fluid, the crack does not extend unless there is an additional increase in the applied load on the urethane sleeve. Regarding the effect of fracturing fluid viscosity, viscous oil tends to create thicker and more planar cracks with fewer branches and wider apertures compared to water. This difference in crack characteristics is consistent with the focal mechanism of AE events: the tensile-type mechanism is dominant during viscous oil injection, whereas the shear-type mechanism prevails during water injection. Additionally, to further clarify the effect of fluid viscosity, we introduce an experimental observation: when a viscous fluid is used as a fracturing fluid, a new crack is created in a specimen with a pre-existing crack, effectively "ignoring" pre-existing crack.

Chapter 4 presents the results of numerical simulations using the distinct element method to examine the influence of grain size distribution and fluid viscosity, as observed in the experiments described in Chapters 2 and 3. The simulation results revealed that shear cracks, which emit a large amount of energy, form to connect pre-existing microcracks. In contrast, tensile cracks emit a relatively small amount of energy, as the tensile strength of rock is lower than its shear strength. Consequently, small AE events are easily buried in noise and are difficult to measure experimentally. The simulation results suggest that this is why shear-type AE events, which have large energy, are predominantly observed in AE monitoring, even though the tensile cracks are primarily generated in the simulation. On the other hand, shear fracturing occurs along the grain boundaries, diagonally across the direction of maximum confining stress in the inhomogeneous rock model, where particle radii are widely distributed and large particles are present. In contrast, in the homogeneous rock model, where all particles are small, the fractures during HF tend to develop straight in the direction of maximum confining stress due to tensile fracturing. These simulation results are consistent with the experimental findings presented in Chapters 2 and 3, providing valuable insight into resolving the paradox between the elastic theory of fracturing and the focal mechanism of AE.

Chapter 5 presents the results of laboratory HF experiments using CO_2, water, and viscous oil. Since CO_2 becomes SC at high temperatures and pressures, similar to those at great depths where HF is typically conducted in the field, CO_2 was injected in the SC state, which has a much lower viscosity than water. For comparison, a liquid (L) CO_2 was also injected. The results showed that HF using SC-CO_2 tends to create three-dimensional sinuous cracks with many secondary branches, which appear to be desirable pathways for heat recovery from HDR and for oil and gas extraction from shale reservoirs. Regarding the fracturing mechanism, the ratios of the P-wave first motion polarity of AE events suggested that a low-viscosity fracturing fluid like SC-CO_2 tends to induce shear-dominant fractures, while more viscous fluids tend to induce tensile-dominant fractures. The results, derived from AE data, are also consistent with the observations of HF cracks under microscopy.

Chapter 6 presents a small-scale field experiment aimed at extending HF cracks by increasing the flow rate of injected water, mimicking the refracturing process commonly used in shale oil and gas fields. The experiment took place at a site located in a gallery 500 m below the surface in the Mizunami underground laboratory, which is dedicated to research and development for the geological disposal of high-level radioactive waste (HLW) by the Japan Atomic Energy Agency (JAEA). An 86-mm diameter HF hole was drilled downward from the gallery floor, and four parallel AE-monitoring holes, each 66 mm in diameter, were drilled 1 m horizontally from the HF hole. To inject water, a 36-mm diameter pilot hole was drilled at the center of the bottom of the 86-mm diameter HF hole. After sealing the upper section of the pilot hole with O-rings attached to a packer unit, water was pressurized and injected into a 160-mm long section at a depth of 5.34–5.49 m. Initially, water was injected at a constant flow rate of 10 mL/min to achieve the first BD. After that, the flow rate was increased to 30 mL/min to induce the second BD, simulating the refracturing process. The locations and ratios of the P-wave first motion polarity of AE events indicated that most of the recorded AE events were shear-dominant. However, tensile-dominant AE events occurred near the HF hole immediately after the first BD. After the second BD, tensile-dominant AE events were observed primarily at the edges of regions where AE events had already been induced. These results suggest that tensile-dominant AE events occurred in the intact rock near the HF hole during the initiation stage of HF in the first BD. Furthermore, increasing the injection flow rate was effective in generating new cracks during the refracturing process. The new crack expansion associated with the second BD was induced by tensile fracturing at the edges of the regions where fractures had already been initiated, as indicated by the AE events.

Chapter 7 presents the results of small-scale field HF experiments to inject SC-CO_2 using holes drilled in a similar arrangement to that of Chapter 6. In a laboratory experiment, examining the effects of macroscopic pre-existing cracks, which are often encountered in field operations, is challenging because specimens typically do not contain any macroscopic cracks. Additionally, it is difficult to evaluate the size of crack extensions induced by BD, as induced cracks immediately extend to the specimen surfaces. To address these limitations and compare the effects of pre-existing cracks and induced crack size between SC-CO_2 fracturing and conventional water fracturing, small-scale field HF experiments were conducted at a site with a hot rock mass at temperatures high enough to bring CO_2 into the SC state. The AE source distribution during HF suggested that CO_2 can easily intrude and enhance AE occurrence along a pre-existing crack. CO_2 fracturing also stimulated a much larger region and improved injectivity to a much greater extent than conventional water fracturing under the same flow rate. Regarding the fracturing mechanism, the ratios of the *P*-wave first motion polarity of AE events suggested that CO_2 fracturing induces many compression-dominant AE events, which were not observed in the water fracturing experiments. The compression-dominant events were likely caused by crack closure due to the degassing of injected CO_2. Although CO_2 injection induces many AE events, the compressive-dominant events are unlikely to trigger natural earthquakes, as they do not create new cracks.

Chapter 8 summarizes two key topics that this book focuses on: the first is solving the paradox of the fracturing mode of HF, and the second is clarifying the characteristics of HF using CO_2 as a fracturing fluid in comparison to conventional HF that uses water. This chapter outlines how these two topics are described, discussed, and concluded throughout this book.

1.3 Stress condition around an HF hole indicating tensile fracturing

This section explains the stress conditions around an HF hole derived from traditional elastic theory. The stress conditions suggest that HF is induced by tensile fracturing, which is inconsistent with shear fracturing demonstrated by the focal mechanism of monitored MS and AE, resulting in the paradox of HF.

Figure 1.1a shows the horizontal section of the vertical HF hole. In this figure, suppose that one of the principal stresses acts in the vertical direction, while the two horizontal principal compressive stresses, Q and P (with $Q<P$; compression is positive), act in the X and Y directions, respectively, from sufficiently far from the HF hole. In this condition, according to the solution by Kirsch (1898), the tangential compressive stress on the HF hole wall reaches a maximum compressive stress, $3P–Q$, at the two points A_1 and A_2, where the X coordinates represent the maximum and minimum values, respectively. The minimum compressive stress, $3Q–P$, occurs at the two points B_1 and B_2, where the Y coordinates represent the maximum and minimum values, respectively. Next, when fluid pressure, W, is applied in the HF hole as shown in Figure 1.1b, according to the solution by Lamé (1852), the tangential stress at any point on the HF hole wall becomes the tensile stress, $-W$. Thus, when the fluid pressure, W, is applied in the HF hole subjected to the two horizontal principal compressive stresses, Q and P ($Q<P$) from

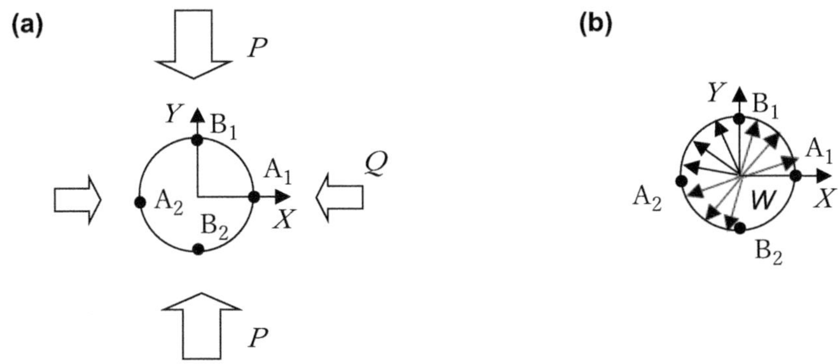

Figure 1.1 The horizontal section of the vertical HF hole. (a) The two horizontal principal compressive stresses, *Q* and *P*, act in the *X* and *Y* directions, respectively, from sufficiently far from the HF hole. (b) Injected fluid pressure, *W*, is applied in the HF hole.

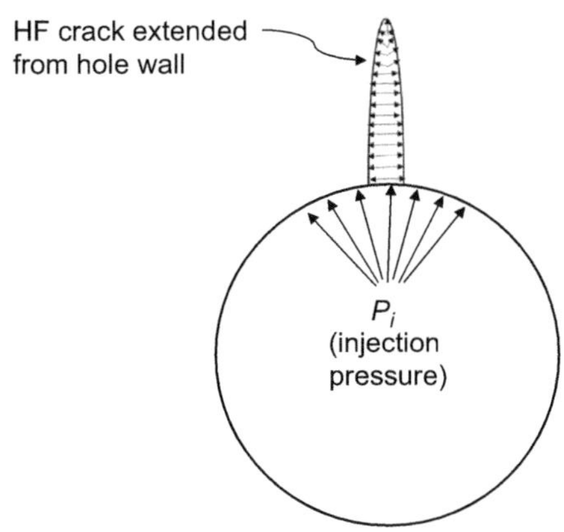

Figure 1.2 Conceptual illustration of the fluid pressure application after an HF crack initiates and extends from the wall of the HF hole.

sufficiently far from the HF hole, by superimposing Figure 1.1b on Figure 1.1a, the maximum tensile stress, $|-W+3Q-P|$, is induced at the two points B_1 and B_2. As the fluid pressure, W, increases and $|-W+3Q-P|$ becomes larger than the tensile strength |St|, HF is induced, and the HF crack extends in $+Y$ and/or $-Y$ directions. Consequently, the stress condition derived by Kirsch (1898) and Lamé (1852), following traditional elastic theory, indicates that HF is induced by tensile fracturing.

This theory can explain the fracturing mechanism only for initiating the crack on the hole wall. Once the HF crack extends, stress around the hole is disturbed by the crack, making it difficult to calculate the stress condition around the crack using conventional elastic theory. However, after the crack extends, the injection fluid penetrates the crack, and the fluid pressure is applied to both the crack surfaces and the injection hole wall, as shown in Figure 1.2. If the rock around the hole is impermeable and contains no microcracks, the fluid pressure applied to the crack surfaces wedges them open and extends the crack by inducing tensile fracturing at the crack tip.

1.4 Focal mechanism of MS and AE induced by HF indicating shear fracturing

This section explains the analysis and data that demonstrate HF induced by shear fracturing. This first subsection shows the *P*-wave radiation patterns of MS and AE induced by shear fracturing. Next, typical displacements associated with tensile fracturing and crack closure are presented, along with the corresponding *P*-wave radiation patterns. Additionally, the ratio *R* of compressional polarity, which is frequently used in this book, is explained, based on the *P*-wave radiation patterns induced by shear, tensile, and crack-closing displacements.

1.4.1 P-wave radiation pattern induced by shear displacement and fault plane solution

Figure 1.3 shows shear displacements associated with shear fracturing and the *P*-wave radiation pattern they induce. When shear displacement occurs along a plane with a normal X_3, as shown in Figure 1.3a, compressional *P*-wave initial motions are radiated in the first and third quadrants along the X_2 and X_3 axes. In contrast, dilatational *P*-wave initial motions are radiated in the second and fourth quadrants, as shown in Figure 1.3c. Additionally, when a shear displacement occurs along a plane with a normal X_2, as shown in Figure 1.3b, compressional and dilatational *P*-wave initial motions are radiated in the same pattern as in Figure 1.3c (e.g., Aki and Richards, 1980). The two shear displacements shown in Figure 1.3a and b are induced under the same stress conditions, where the maximum compressive stress, *P*, and the minimum compressive stress, *T*, apply in the directions indicated in Figure 1.3c. In this context, the two pairs of shear displacements in Figure 1.3a and b, which correspond to the pairs of arrows around the origin in Figure 1.3c, are induced alternatively under the same stress conditions. In this case, AE sensors deployed in the first and third quadrants record compressional *P*-wave initial motions, while those in the second and fourth quadrants record dilatational motions. Conversely, by analyzing the distribution of AE sensors recording compressional and dilatational *P*-wave motions, we can determine the directions of the maximum and minimum compressive stresses (the *P*-axis and *T*-axis) that caused the shear fracture. We can also identify the directions of the two alternative

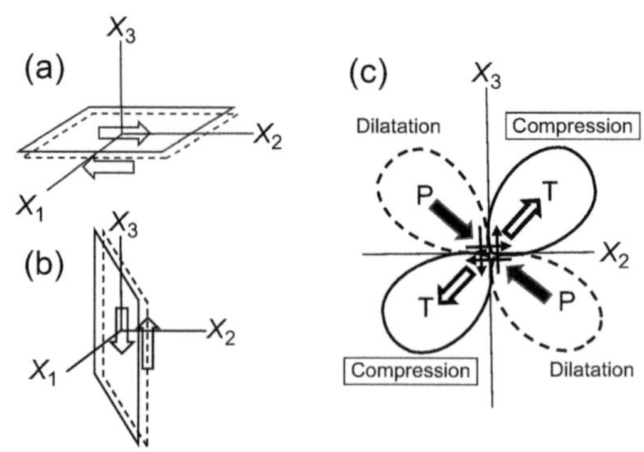

Figure 1.3 (a) and (b) Alternative shear displacements that are induced under the stress condition that the maximum and the minimum compressive stress, *P* and *T*, apply in the directions shown in Figure (c). *P*-wave radiation pattern induced by the alternative shear displacement shown in Figure (a) or (b).

share fracture planes; however, without additional information about the fracture, such as the fracture itself, we cannot determine which plane is the true fracture plane. This procedure and its results are called the "fault plane solution".

The fault plane solution is essentially a three-dimensional analysis using a stereo net, although Figure 1.3c illustrates its concept in two dimensions for simplicity. The three-dimensional solution can be obtained by analyzing the distribution of the polarities of the *P*-wave initial motions on a stereo net, with the azimuth (horizontal angles) and angle of emergence (vertical angles) from the source to respective sensors. Kasahara (1981) explains the actual procedure for using a stereo net in three dimensions for an earthquake, which helps understand the method.

As observed in a laboratory fracture test of a rock specimen, two conjugate fracture planes never intersect at a 90° angle due to the rock's internal friction angle, θ. Instead, they intersect at an angle of $90°-\theta$. Based on this fact, the directions of the P- and T-axes obtained through the method differ by an angle of $\theta/2$ from the actual directions of the maximum and the minimum compressive stresses (Yamakawa, 1971). Strictly speaking, attention should be paid to the difference. However, since the fault plane solution is primarily used to determine the general trend of the fracturing mechanism, the difference is not considered significant and, for this book, will be ignored.

Many researchers (e.g., Talebi and Cornet, 1987; Cornet, 1992; Horálek et al., 2010) have reported that most AE or MS associated with field operations of HF exhibit a focal mechanism of shear fracturing. Especially, the initial *P*-wave motions of these events show a quadrant radiation pattern as illustrated in Figure 1.3c, which is inconsistent with the tensile fracturing predicted by traditional elastic theory.

1.4.2 *P*-wave radiation patterns induced by tensile displacement and crack-closing displacement

Figure 1.4 illustrates a tensile displacement and the corresponding *P*-wave radiation pattern induced by it. When a tensile displacement occurs on a plane with a normal in the X_3 direction, as shown in Figure 1.4a, compressional *P*-wave initial motions are radiated in all directions from the source. Although an explosion source also radiates compressional *P*-wave initial motions in all directions from the source, it radiates equal amplitudes in all directions. In contrast, the tensile displacement induces the maximum amplitude in the direction of the displacement and the minimum amplitude in the directions normal to the displacement, as shown in Figure 1.4b (Sasaki et al., 1988; Walter and Brune, 1993).

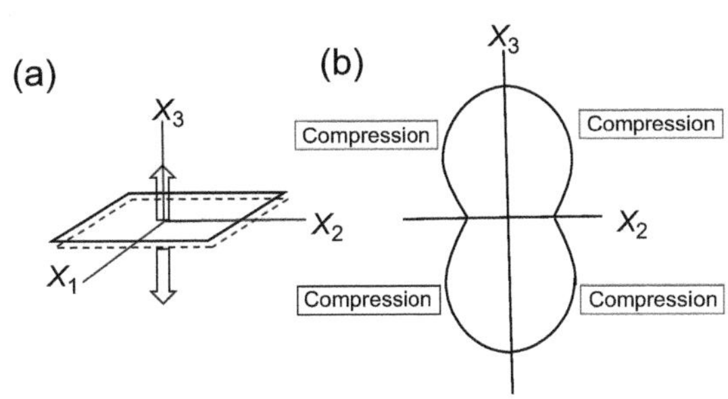

Figure 1.4 (a) Tensile displacements. (b) *P*-wave radiation pattern induced by the tensile displacement shown in Figure (a) (Sasaki et al., 1988).

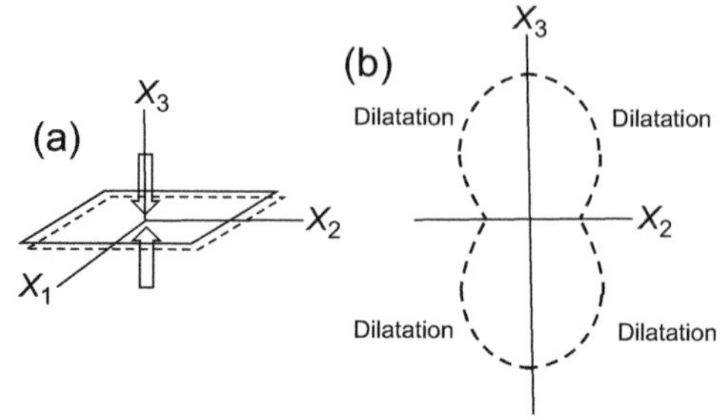

Figure 1.5 (a) Crack-closing displacement. (b) *P*-wave radiation pattern induced by the crack-closing displacement shown in Figure (a).

In contrast, Figure 1.5 shows a crack-closing displacement and the *P*-wave radiation pattern induced by it. When the displacement occurs on a plane with a normal in the X_3 direction, as shown in Figure 1.5a, dilatational *P*-wave initial motions are radiated in all directions from the source. Although an implosion source induced by pore collapse, cave-in of a cavity, rupture of a vacuum bulb, and similar events also radiates dilatational *P*-wave initial motions in all directions from the source, it radiates nearly the same amplitude in all directions. In contrast, the crack-closing displacement induces the maximum amplitude in the direction of the displacement and the minimum amplitude in the directions normal to the displacement, as shown in Figure 1.5b. This is the same as the pattern produced by the tensile displacement shown in Figure 1.4b, except that "compressional" is replaced with "dilatational".

The radiation pattern is examined in three dimensions using a stereo net for actual data. In this case, we cannot determine the displacement direction or the direction of the fracture plane without amplitude data recorded at the respective sensors. Consequently, when all polarities recorded at the sensors are compressional, we regard the fracture as tensile. Conversely, when all polarities are dilatational, we regard it as crack-closing.

1.4.3 *Ratio of compressional polarity, R, to infer fracturing mechanism*

From the relation between the typical displacements and the *P*-wave radiation patterns they induce, as shown in Figures 1.3–1.5, the percentage *R* (%) of compressional polarities relative to the total number of the polarities detected by sensors should be 100% for a tensile or explosion fracture, 50% for a pure shear fracture, and 0% for a compression or implosion fracture, assuming good coverage of multiple AE sensors around the source. However, the actual displacements associated with the fracture are not always normal (tensile fracture and/or crack-closing) or parallel (shear fracture) to the fracture plane. In reality, these displacements typically form an angle, θ, ranging from 0 to 90° to the normal of the fracture plane, as shown in Figure 1.6. As a result, although the percentage *R* (%) ranges from 0 to 100%, an AE event with *R* close to 100% can be considered to be induced by a tensile-dominant fracture, one close to 50% by a shear-dominant fracture, and one close to 0% by a crack-closing dominant fracture. Thus, the percentage *R* can be used as an index to estimate the fracture mode.

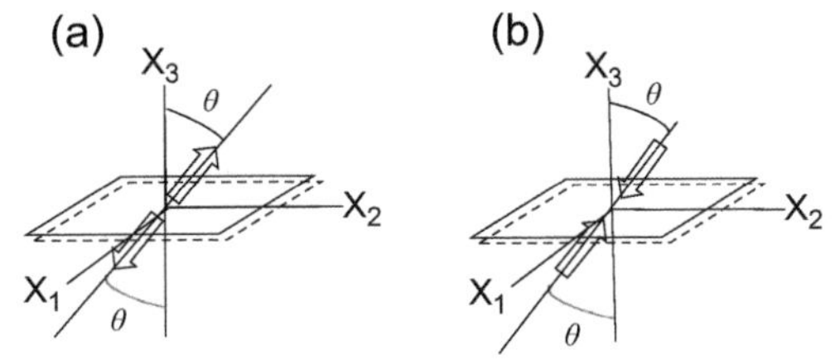

Figure 1.6 (a) Crack displacement inducing tensile-dominant fracturing. (b) Crack displacement inducing compression-dominant fracturing.

1.5 Conclusion of this chapter

Building on the knowledge of the stress conditions around the HF hole presented in Section 1.3 and the focal mechanism of MS and AE discussed in Section 1.4, this chapter addresses the paradox between tensile fracture, indicated by the stress condition around the HF hole, and shear fracture, indicated by the focal mechanism of MS and AE. This is explored through the results of various laboratory and small-scale field experiments.

The fracturing mechanisms of HF are primarily discussed based on the fault plane solution and the percentage R (%) of compressional polarities relative to the total number of polarities of P-wave initial motion detected by sensors. The author explains the AE monitoring methods used throughout this book, as these methods vary significantly depending on the period of the experiment due to the remarkable advancements in measuring instruments. For further information on AE monitoring methodology, please refer to the following three Suggested Methods from the International Society for Rock Mechanics: Ishida et al. (2017) for laboratory, Feng et al. (2019) for in situ small-scale field monitoring covering less than 1 m^3, and Xiao et al. (2016) for in situ large scale field monitoring covering hundreds of m^3. For the methodology of fault plane solutions in AE, refer to Chapter 3 of Kasahara (1981).

In this book, nearly all specimens subjected to laboratory and small-scale field experiments are granitic rocks; however, the author believes that the results can also be applied to sedimentary rocks, including various shales, sandstones, and others.

References

Aki, K. & Richards, P. G. (1980) *Quantitative Seismology – Theory and Method – Volume 1*, W. H. Freeman and Company, San Francisco, p. 557.

Cornet, F. H. (1992) Fracture processes induced by forced fluid percolation. In: Gasparini, P., Scarpa, R., Aki, K. (eds) *Volcanic Seismology, IAVCEI Proceedings in Volcanology*, vol 3. Springer, New York, pp. 407–431.

Cornet, F. H. &Valette, B. (1984) In situ stress determination from hydraulic injection test data. *J. Geophys. Res.*, 89(B13), 11527–11537.

Feng, X-T., Young, R. P., Reyes-Montes, J. M., Aydan, Ö., Ishida, T., Liu, J-P. & Liu H-J. (2019) ISRM suggested method for in situ acoustic emission monitoring of the fracturing process in rock masses. *Rock Mech. Rock Eng.*, 52(5), 1395–1414. doi: 10.1007/s00603-019-01774-z

Haimson, B. C. (1978) The hydrofracturing stress measuring method and recent field results. *Int. J. Rock Mech. Min. Sci. & Geomech. Abstr.*, 15, 167–178.

Horálek, J., Jechumtálová, Z., Dorbath, L., Šílený, J. (2010) Source mechanisms of micro-earthquakes induced in a fluid injection experiment at the HDR site Soultz-sous-Forêts (Alsace) in 2003 and their temporal and spatial variations. *Geophys. J. Int.*, 181, 1547–1565.

Hubbert, M. K. & Willis, D. G. (1957) Mechanics of hydraulic fracturing. *Petrol. Trans. Am. Soc. Min. Eng.*, 210, 153–168.

Ishida, T., Labuz, J. F., Manthei, G., Meredith, P. G., Nasseri, M. H. B., Shin, K., Yokoyama, T. & Zang, A. (2017) ISRM suggested method for laboratory acoustic emission monitoring. *Rock Mech. Rock Eng. Rock Mech. Rock Eng.*, 50(3), 665–674. doi: 10.1007/s00603-016-1165-z

Jeffrey, R. G., Zhang, X., Settari, A., Mills, K. W. & Detournay, E. (2001) Hydraulic fracturing to induce caving: fracture model development and comparison to field data. In: *Proceedings of the 38th U.S. Symposium on Rock Mechanics,* 7–10 July, Washington, D.C., USA. American Rock Mechanics Association. ARMA-01-0251.

Kasahara, K. (1981) *Earthquake Mechanics*. Cambridge University Press, Cambridge, p.248.

Kirsch, C. (1898) Die Theorie der Elastizität und die Bedürfnisse der Festigkeitslehre. *Z. VDI.*, 42, 797–807. (in German).

Lamé, G. (1852) *Leçons sur la théorie mathématique de l'élasticité des corps solides*. Gauthier-Villars, Paris, p. 335 (in French).

Maisons, C., Fortier, E. & Valette, M. (1997) Induced microseismicity and procedure for closure of brine production caverns. *Pure Appl. Geophys.*, 150(3/4), 585–603.

Maxwell, S.C. & Cipolla, C. (2011) What does microseismicity tell us about hydraulic fracturing? In: *SPE Annual Technical Conference and Exhibition,* Denver, Colorado, USA, SPE 146932.

Mizuta, Y., Sano, O., Ogino, S. and Katoh, H. (1987) Three dimensional stress determination by hydraulic fracturing for underground excavation design. *Int. J. Rock Mech. Min. Sci. & Geomech. Abstr.*, 24, 15–29.

Montgomery, C. T. & Smith, M. B. (2010) Hydraulic fracturing –History of an enduring technology. *J. Pet. Technol.*, December 2010, 26–41.

Sasaki, S., Matsunaga, I., Kobayashi, H. & Ishida, T. (1988) *Mechanism of Fracturing Inferred from Acoustic Emission.* CRIEPI (Central Research Institute of Electric Power Industry), Abiko, Chiba, Japan, U88034 (in Japanese with English abstracts).

Schmitt, D. R. & Zoback, M. D. (1993) Infiltration effects in the tensile rupture of thin walled cylinders of glass and granite: implications for the hydraulic fracturing breakdown equation. *Int. J. Rock. Mech. Min. Sci. Geomech. Abstr.*, 30, 289–303. https://doi. org/10.1016/0148-9062(93)92731-5

Talebi, S. & Cornet, F.H. (1987) Analysis of the microseismicity induced by a fluid injection in a granitic rock mass. *Geophys. Res. Lett.*, 14, 227–230.

Walter, W. R. & Brune, J. N. (1993) Spectra of seismic radiation from a tensile crack. *J. Geophys. Res.*, 98(B3), 4449–4459.

Xiao, Y.-X., Feng, X.-T., Hudson J. A., Chen, B.-R., Feng, G.-L. & Liu, J. P. (2016) ISRM suggested method for in situ microseismic monitoring of the fracturing process in rock masses. *Rock Mech. Rock Eng.,* 49, 843–869. doi: 10.1007/s00603-015-0859-y

Yamakawa, N. (1971) Stress fields in focal regions. *J. Phys. Earth*, 19(4), 347–355.

Zoback, M. D., Rummel, F., Jung, R., Raleigh, C.B. (1977) Laboratory hydraulic fracturing experiments in intact and pre-fractured rock. *Int. J. Rock. Mech. Min. Sci. Geomech. Abstr.*, 14, 49–58.

Chapter 2

Fracturing mechanism depending on rock properties

2.1 Introduction

The properties of rock mass should affect hydraulic fracturing (HF) mechanisms and the features of crack extension, in addition to the viscosity of the fracturing fluid, rock stress conditions, and other factors. In HF's field operations, crack extension and the associated mechanisms have been investigated by monitoring acoustic emission (AE) and/or microseismicity (MS). For example, in hot dry rock (HDR) geothermal projects, AE events were monitored in Cooper Basin, Australia (Baisch et al., 2006, 2009, 2015), Soultz HDR, France (Evans et al., 2005), Hijiori, Japan (Sasaki, 1997, 1998), and Ogachi, Japan (Kaieda et al., 1995). In oil and gas recovery, AE events were monitored in Cotton Valley sandstone, TX, USA (Rutledge et al., 2004) and the Barnett shale, TX, USA (Hummel and Shapiro, 2013). Under laboratory conditions, Lockner and Byerlee (1977) located AE hypocenters and inferred the fracturing mechanism from the directions of the induced cracks. However, they did not discuss the fracturing mechanism from the AE data. Majer and Doe (1986) recorded AE waveforms using an analog data recorder and used the data not only to locate the hypocenters but also to analyze the mechanism by examining the polarities of the P-wave initial motions. However, since the number of sensors was limited to five, they could not conduct more advanced analyses of AE data. Recently, with the advancement of AE monitoring equipment, Stanchits et al. (2014), Hampton et al. (2018), Naoi et al. (2018), Yamamoto et al. (2019), and others have monitored AE events with a large number of sensors in laboratory experiments and discussed the fracturing mechanism of HF.

In field monitoring, many researchers have reported that shear events are dominant (e.g., Talebi and Cornet, 1987), whereas elastic theory predicts that HF should induce tensile fractures (e.g., Hubbert and Willis, 1957). This paradox (e.g., Maxwell and Cipolla, 2011) suggests that the fracturing mode is still unclear, despite the publication of many excellent papers, including those mentioned above, that discuss the problem from various perspectives based on field and laboratory data.

This chapter focuses on how the texture property of rock affects HF mechanisms and features of crack extension based on the results of laboratory experiments. In Section 2.2, the results of laboratory HF experiments on granite specimens with different grain sizes are presented, highlighting differences in features of crack extension and focal mechanism of monitored AE events. In Section 2.3, the results of the HF experiment on an impermeable acrylic resin are discussed. Additionally, in Section 2.4, the differences in the focal mechanisms when cracks extend normally and along the rift plane of granite are also presented.

Since the term "rift plane" is frequently used in this chapter, a brief explanation is provided below. The rift plane is a plane along which microcracks are preferentially oriented in the

DOI: 10.1201/9781003282631-2

granitic rock. Dale (1923) and others investigated the properties of granitic rocks and found that these rocks generally have two sets of preferred orientations for microcracks. Sano et al. (1992) summarized the findings as follows: most of the microcracks are aligned within the "rift plane". A secondary crack orientation is referred to as the "grain plane". The "hardway plane" is defined as the plane perpendicular to both the rift plane and the grain plane. These three planes are named by quarrymen, who used them to split granitic rock more easily along the rift plane. Dale (1923) noted that an observer can detect the rift plane with the unaided eye after some experience. However, when P-wave velocities in the directions normal to these planes are labeled as V_R, V_G, and V_H, their order is $V_R < V_G < V_H$ (e.g., Sano et al., 1992). Consequently, by measuring P-wave velocity, we can identify the rift plane as a plane normal to the direction of the lowest velocity.

2.2 HF experiments on granite specimens with different grain sizes

Early experiments investigating the effects of rock properties on HF were conducted by Matsunaga et al. (1993). They examined fracturing mechanisms in granite, marble, andesite, and acrylic resin with injections of water and viscous oil. Although they obtained some interesting results, the findings lacked consistency due to the many variables involved. Following these experiments, in collaboration with Matsunaga et al., Chen et al. (1998), Ishida et al. (2000a,b), and Ishida (2001) conducted HF experiments using four 20 cm-cubic granitic rock specimens of different grain sizes to investigate the effects of rock properties. The experiments allowed for a consistent discussion of the effects of grain size on the magnitudes of breakdown (BD) pressure (defined as the peak pressure immediately before it drops suddenly), crack extension directions, and the focal mechanism of monitored AE events. In this section, the experimental results and discussions are presented based on the work of Chen et al. (1998), Ishida et al. (2000a,b), and Ishida (2001).

2.2.1 *Specimen, experimental setup, and AE monitoring method*

Here, the features of rock specimens and the experimental setup, including the geometry of the specimen and the loading conditions, are explained, and after that, the AE monitoring method is described.

2.2.1.1 *Specimen and experimental setup*

Four kinds of granitic rocks with different grain sizes were used in the HF laboratory experiments. Their photomicrographs are shown in Figure 2.1. The features of the rocks are as follows:

i (Symbol: LL; average grain size: 1.12 mm): Pink Granite produced in China, consisting of quartz, plagioclase, biotite, and potassium feldspar phenocryst with 1 to 4 cm grain size.
ii (L; 0.88 mm): Ashio-Sohri Granite produced in Japan, consisting of quartz, potassium feldspar, plagioclase, and biotite. Some continuous microcracks develop.
iii (M; 0.68 mm): Makabe-Kome Granodiorite produced in Japan, consisting of quartz, potassium feldspar, plagioclase, biotite, and monoclinic pyroxene. Weak alternations such as chloritization, carbonatization, sericitic alteration, and epidotization can be observed.
iv (S; 0.51 mm): Nukame Granodiorite produced in China, consisting of quartz, plagioclase, biotite, and a small amount of potassium feldspar. It is fresh and shows almost no alternation.

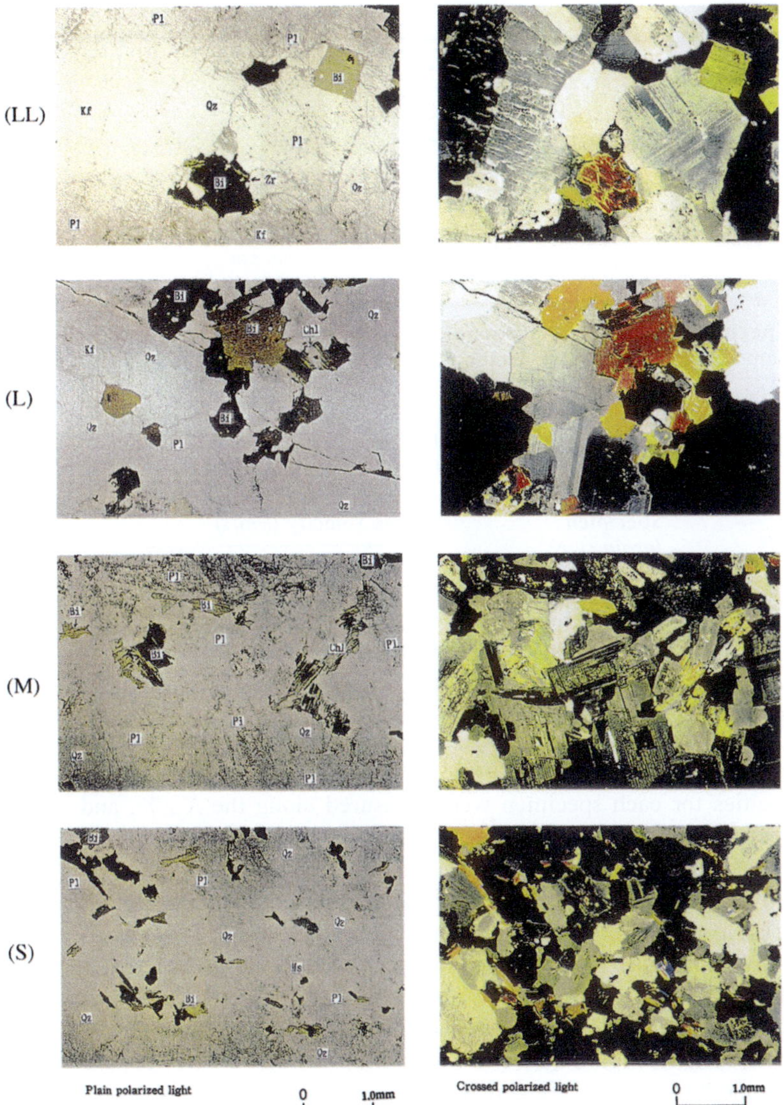

Figure 2.1 Photomicrographs of the specimens with plane polarized light (Left) and with crossed polarized light (Right). All the figures are the same size as indicated by the bar scales. From the top to the bottom, in the order of the grain size, (LL) Chinese Pink Granite, (L) Japanese Ashio-Sohri Granite, (M) Japanese Makabe-Kome Granodiorite, and (S) Chinese Nukame Granodiorite. The symbols in the photomicrographs are as follows: Pl (plagioclase), Bi (biotite), Qz (quartz), Kf (potassium feldspar), Chl (Chlorite), Zr (Zircon), and Ms (Muscovite) (Chen et al., 1998; Ishida et al., 2000a,b; Ishida, 2001).

Cubic specimens ($20 \times 20 \times 20\,cm^3$) of these granitic rocks were used in the experiments. Each contains a 1 cm diameter hole as shown in Figure 2.2. The rift plane of the granitic rock specimen was oriented to correspond to the *XY* plane in the chosen Cartesian coordinate system.

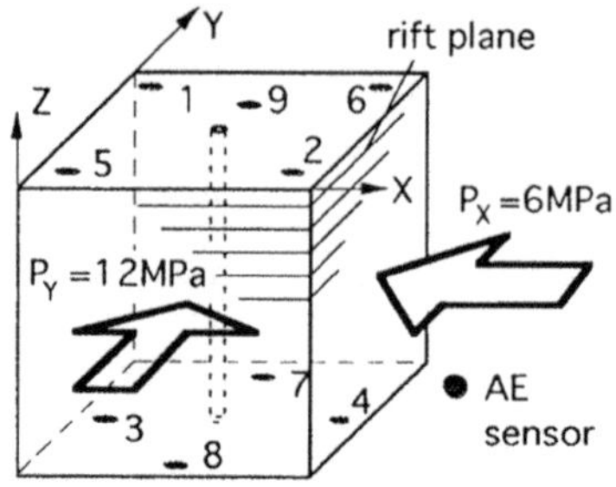

Figure 2.2 Specimen, loading condition, setting positions of AE sensors, and coordinate system (Chen et al., 1998; Ishida et al., 2000a,b; Ishida, 2001).

Table 2.1 P-wave velocities (Chen et al., 1998, Ishida et al., 2000a, Ishida, 2001)

Specimen (grain size)	*P-wave velocity (km/s)*		
	V_X	V_Y	V_Z
LL (1.12 mm)	4.83	4.85	4.24
L (0.88 mm)	4.48	4.37	3.59
M (0.68 mm)	5.18	5.01	4.94
S (0.51 mm)	4.77	4.63	4.22

P-wave velocities for each specimen were measured along the *X*-, *Y*-, and *Z*-directions, and measured velocities are shown in Table 2.1.

Two horizontal confining pressures, 6 MPa in the *X*-direction and 12 MPa in the *Y*-direction, were applied to the specimen through flat jacks inserted between a loading frame and the specimen. The loading system is shown in Figures 2.3 and 2.4. A mechanical-type double packer, as shown in Figure 2.5, was used to inject water. This packer applied fluid pressure to a 5.0 cm sealed-off section of the borehole. To inject water into the sealed-off section, oil pressure applied from a hydraulic pump was transferred to water pressure through a booster pump, as shown previously in Figure 2.3. The water injection was controlled, and flow rates measured by a turbine flow meter were kept at almost constant flow rates from 4.4 to 4.9 cm^3/min. The injecting pressure was measured by a pressure head set on the steel pipe connected to the packer.

2.2.1.2 *AE monitoring method*

To monitor AE events, nine cylindrical sensors, each with a 1.8 cm diameter, 1.85 cm length, and resonance frequency of 500 kHz (Type AE904E, NF Corporation), were attached to the upper and lower surfaces of the specimen, as shown previously in Figure 2.2. Because the experiments were conducted in the 1980s, the instruments to record AE waveforms were out of date compared to current state-of-the-art technology. Thus, two independent systems were employed, one for recording and one for reproducing the signals, as shown in Figure 2.6. In the recording system shown in Figure 2.6a, the detected AE signals were amplified by 80 dB

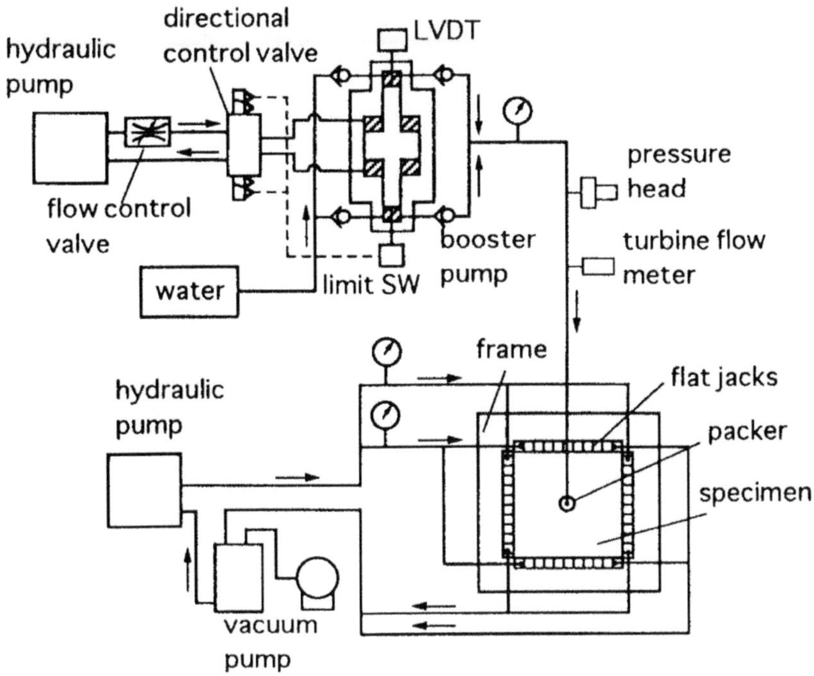

Figure 2.3 Illustration of the loading system (Chen et al., 1998).

in total through a pre-amplifier, a main amplifier, and a signal conditioner. When a signal from one of the three sensors (Nos. 1, 7, and 8 in Figure 2.1) exceeded a threshold level, the wave memory was triggered with a recording length of 2 kwords. After the signal was digitized using a sampling interval of 0.2 µs, an analog signal was produced again using a sampling interval of 5 µs and recorded on an analog data recorder (SR-51, TEAC Corporation). Through this process, the waveform was extended 25 times (5 µs/0.2 µs =25) along the time axis to fit the recordable frequency band from 0 Hz (direct current) to 40 kHz of the analog data recorder. In the reproducing system shown in Figure 2.6b, the waveform extended 128 times along the time axis was reproduced at a tape speed of 1.19 cm/s, which is 128 times slower than that of 152.4 cm/s at the recording. The waveform extended 3,200 times ($25 \times 128 = 3{,}200$) was A/D converted with a sampling time of 1 ms and a resolution in amplitude of 12 bits, and stored in a removable hard disk of 47 Mbytes. ISRM (International Society for Rock Mechanics) suggested method for laboratory AE monitoring (Ishida et al., 2017), which may help your understanding of the above descriptions.

2.2.2 *Experimental results*

2.2.2.1 *AE hypocenter distributions and pressure records*

AE hypocenters were located, considering the anisotropy of the P-wave velocity (Rothman et al., 1974), with an accuracy of 10 mm. This was possible because only AE hypocenters satisfying the following conditions were located: (i) P-wave initial motions were detected by more than six

Figure 2.4 Photo of the loading system.

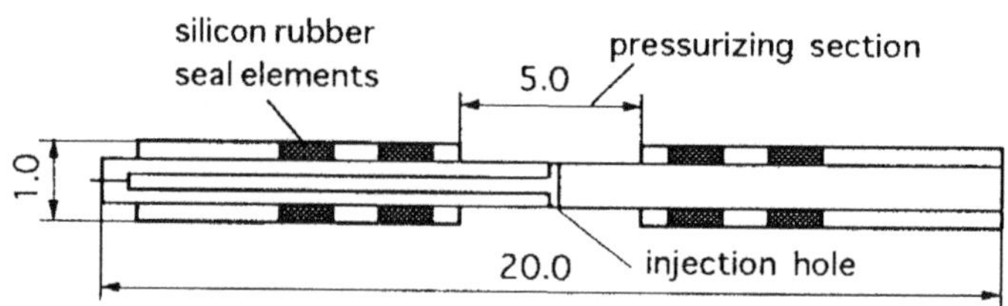

Figure 2.5 Mechanical-type double packer to inject water into the borehole (unit: cm) (Chen et al., 1998, Ishida et al., 2000a,b; Ishida, 2001).

sensors, and (ii) the standard deviation of differences between the observed and back-calculated arrival times was within 3 μs. As shown in Figure 2.7, 725, 472, 392, and 89 AE hypocenters were located during the HF of granitic rocks LL, L, M, and S, respectively.

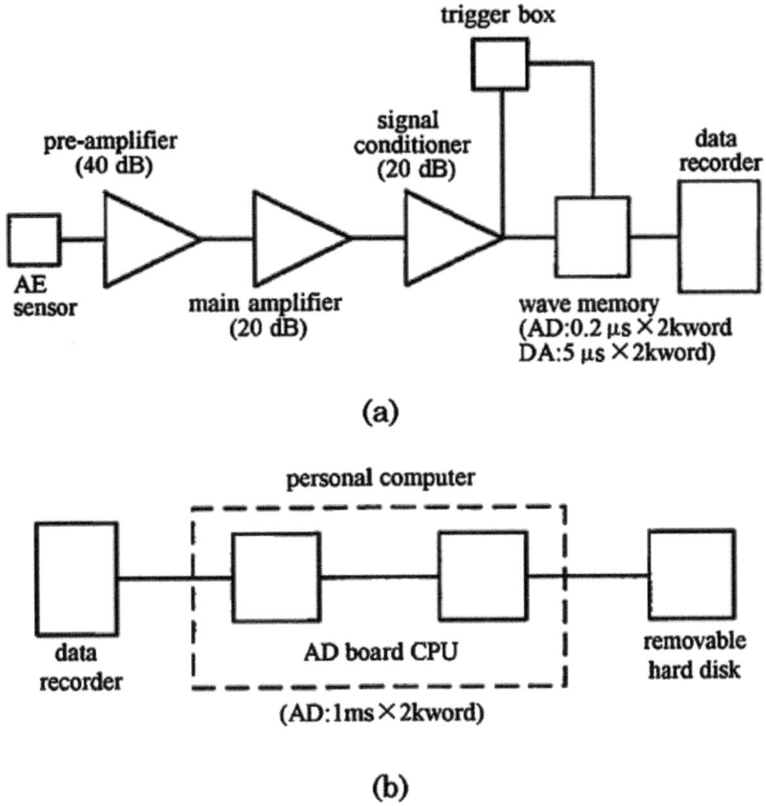

Figure 2.6 AE monitoring system: (a) recording system and (b) reproducing system (Chen et al., 1998).

Figure 2.8 shows changes in the *X*-, *Y*-, and *Z*-coordinates of the located AE hypocenters and the injected fluid pressure along elapsed time. The pressure records for all specimens indicate the following three stages:

i The injected pressure reached the peak pressure, i.e., the BD pressure, from 50 to 100 s after starting the injection.
ii The pressure was kept almost constant just after BD, from 100 to 250 s in the elapsed time. The constant pressure would correspond to the friction loss of water flow along the crack surfaces because the injection flow rate was kept constant after BD.
iii After being subjected to the constant pressure, the circuit of the injected pressure was opened, and the injected pressure returned to zero.

In addition, two AE bursts were recorded: the first burst was due to the initiation and extension of cracks at the point of the BD pressure at stage (i). The second burst was due to cracks closing during the pressure decrease at stage (iii). In the time interval between the two AE bursts, some AE events were located in the cases of specimens LL, L, and M, whereas no AE event was located in specimen S. When the numbers of located AE hypocenters were compared among the four specimens, it was found that the numbers decrease with the decrease of the grain size.

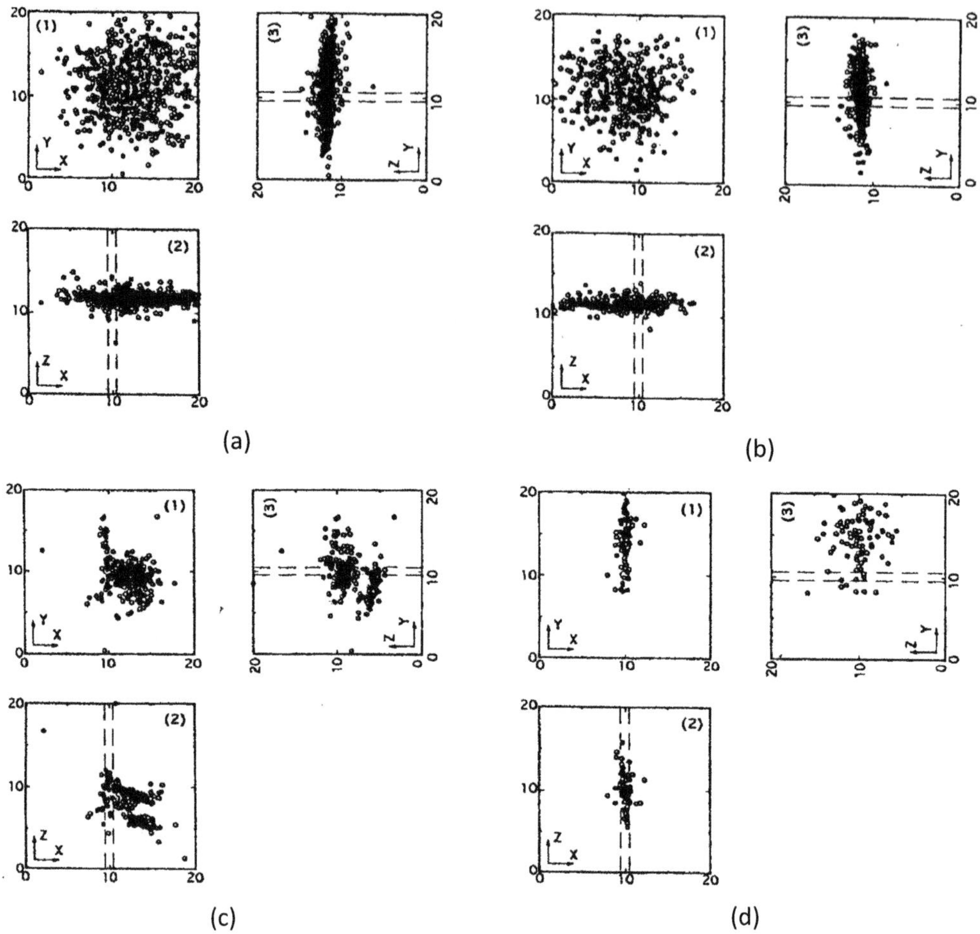

Figure 2.7 Located AE sources projected on the *XY* plane, *ZX*-plane, and *ZY*-plane (Chen et al., 1998, Ishida et al. 2000a,b, Ishida, 2001). (a) Specimen, LL. (b) Specimen, L. (c) Specimen, M. (d) Specimen, S.

Looking back at Figure 2.7, AE hypocenters located in specimens LL and L were widely spread in *X*- and *Y*-directions from the center to the specimen, while their distributions in the *Z*-direction were restricted to the proximity of the pressurized section.

Before discussing AE hypocenter distribution in specimen M, let's discuss those in specimen S. For specimen S, Figure 2.8 shows that AE hypocenters spread only in the *Y*- and *Z*-directions during the BD and the pressure release, while their distribution in the *X*-direction is restricted in the proximity of the borehole.

In contrast to AE hypocenter distributions in the other specimens, those in the specimen M are a little bit complicated. For specimen M, Figure 2.8 shows that, from 90 to 100 s in the elapsed time just after the BD, they spread from 5 to 15 cm in the *Y*-direction and from 5 to 12.5 cm in the *Z*-direction, while they made a cluster around 10 cm with some distributions up to around 15 cm from the center of the borehole in the *X*-direction. The hypocenter distribution indicates that a crack extended mainly along the *YZ* plane in the BD. On the other hand, after the BD, from 120 to 230 s, many AE hypocenters spread from 11 to 15 cm in the *X*-direction and from 4 to 11 cm in the *Y*-direction, while their distribution in the *Z*-coordinate was restricted between 4 and 8 cm from 120 to 140 s and restricted between 8 and 10 cm from 140 to 230 s. The hypocenter distribution indicates that two independent cracks extended along the *XY* plane along the coordinates $Z \fallingdotseq 6$ cm and $Z \fallingdotseq 9$ cm, respectively. The facts indicated that a crack along the *YZ* plane was induced just after the BD, and thereafter, two cracks along the *XY* plane along two different *Z*-coordinates were induced.

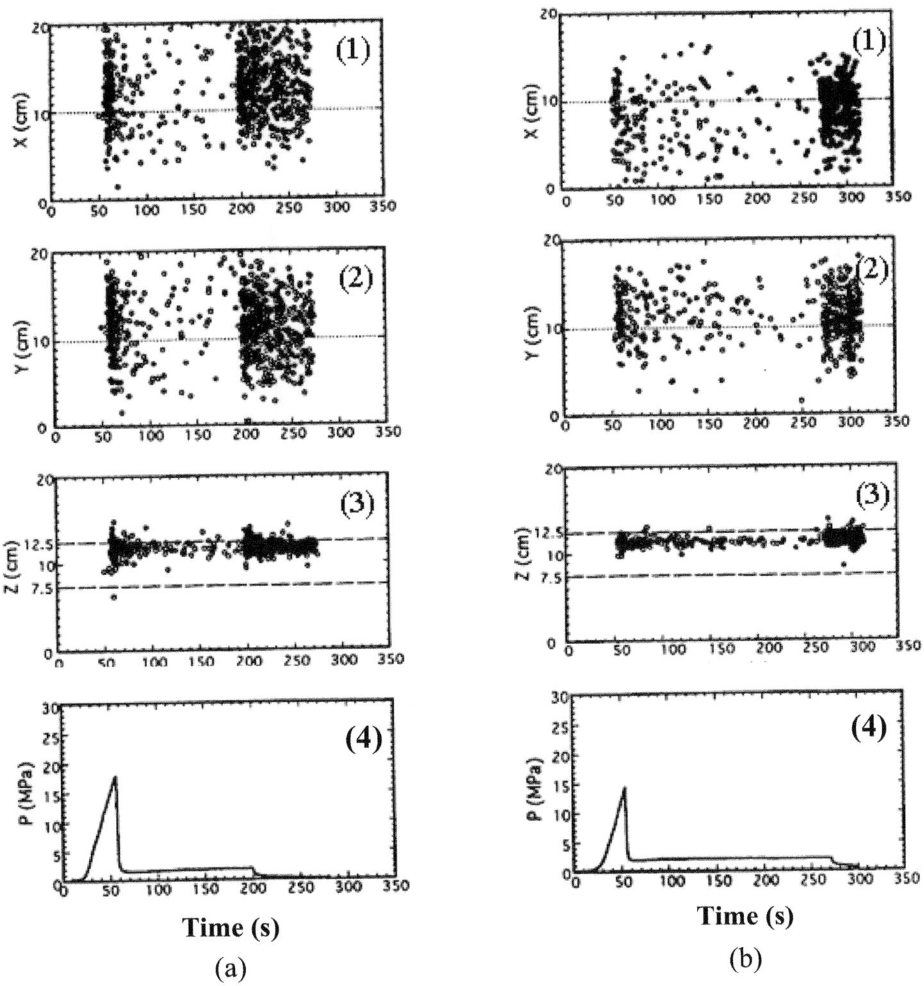

Figure 2.8 Changes of *X*-, *Y*-, and *Z*-coordinates of located AE hypocenters and the injected fluid pressure along elapsed time. The broken lines in plots (1) and (2) of Figure 2.8a–d indicate the *X*- and *Y*-coordinates of the center of the borehole. The sections between the two broken lines in plots (3) of the figures correspond to the sections in which the borehole walls were pressurized by the injected fluid (Chen et al., 1998; Ishida et al., 2000a). (a) Specimen, LL. (b) Specimen, L. (c) Specimen, M. (d) Specimen, S.

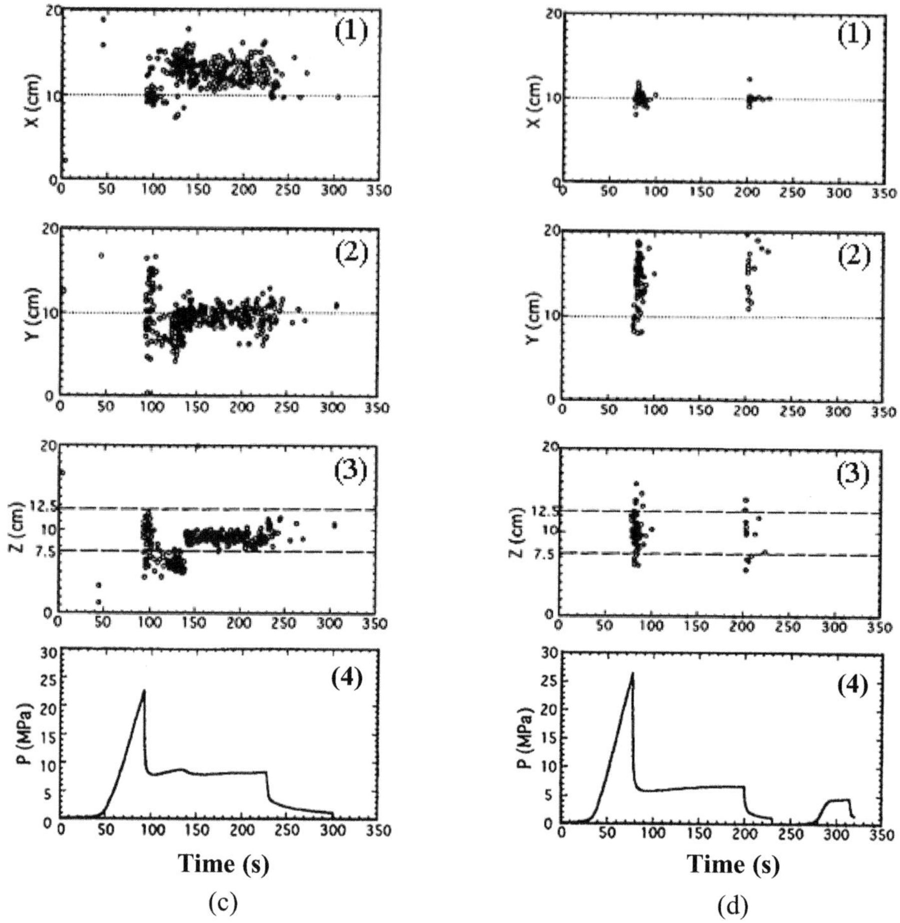

Figure 2.8 (Continued) Changes of X-, Y-, and Z-coordinates of located AE hypocenters and the injected fluid pressure along elapsed time. The broken lines in plots (1) and (2) of Figure 2.8a–d indicate the X- and Y-coordinates of the center of the borehole. The sections between the two broken lines in plots (3) of the figures correspond to the sections in which the borehole walls were pressurized by the injected fluid (Chen et al., 1998; Ishida et al., 2000a). (a) Specimen, LL. (b) Specimen, L. (c) Specimen, M. (d) Specimen, S.

2.2.2.2 *Induced crack geometries and magnitudes of BD pressure*

Cracks induced in the experiments were inspected by naked-eye observations on the surfaces of a cubic specimen and of a core taken around the borehole (i.e., overcoring), as shown in Figure 2.9. Figure 2.10 shows the geometries of induced cracks in the specimens. In Figure 2.10, the solid lines indicate the cracks observed on the surfaces of specimens with the naked eye, while broken lines indicate the cracks estimated from the located AE hypocenter distributions and observations on cores taken around the borehole.

1 In the specimens of LL and L, a horizontal crack was induced parallel to the *XY* plane along the rift plane. The crack reached two of the four lateral surfaces.

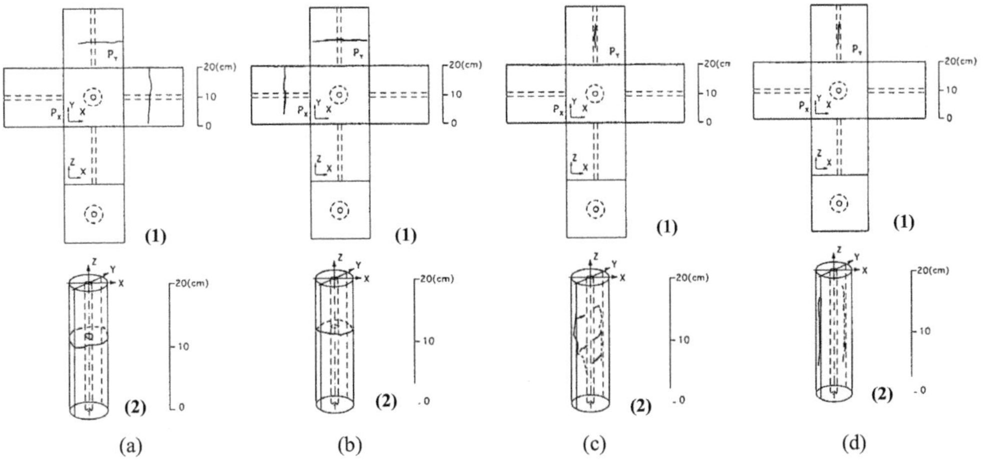

Figure 2.9 Naked-eye observations of cracks induced in the specimens. (1) Observations on the surfaces of a cubic specimen. (2) Observations on the surface of a core taken around the borehole (i.e., overcoring). (Chen et al., 1998). (a) Specimen, LL. (b) Specimen, L. (c) Specimen, M. (d) Specimen, S.

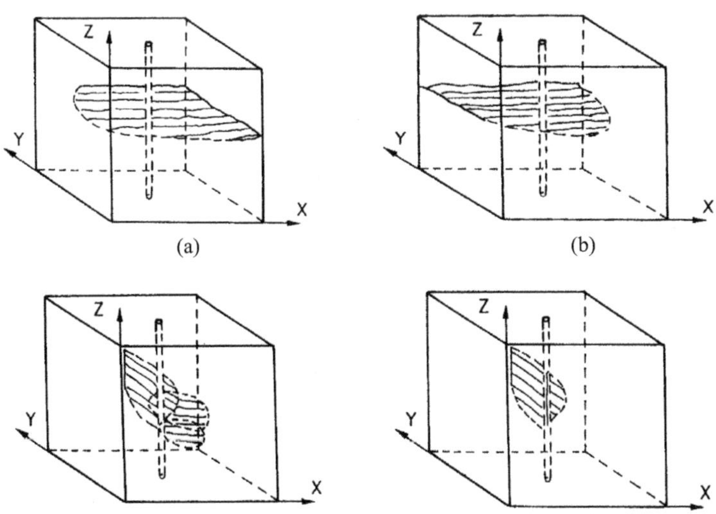

Figure 2.10 Geometries of induced cracks in the respective specimens (Chen et al., 1998; Ishida et al., 2000a, b, Ishida, 2001) (a) Specimen, LL. (b) Specimen, L. (c) Specimen, M. (d) Specimen, S.

2 In the specimen of M, three small cracks were induced from the borehole. One of the three was vertical and parallel to the *YZ* plane and reached one of the four lateral surfaces. The other two were horizontal and parallel to the *XY* plane, and they were small.
3 In the specimen of S, a vertical crack was induced that was parallel to the *YZ* plane and reached one lateral surface.

The distinctive feature of the crack geometries is that the horizontal cracks were induced and reached the lateral surfaces in specimens LL and L, while a vertical crack was induced in specimen S. On the other hand, in specimen M, not only one vertical crack similar to that of specimen S but also two horizontal small cracks similar to those of specimens LL and L were induced. This means that the cracks induced in specimen M show an intermediate feature between those of specimens LL and L and that of specimen S. Additional experiments under the same conditions confirmed that similar crack geometries were obtained depending on the grain sizes of the respective specimens.

Figure 2.11 shows the magnitudes of BD pressure along the order of grain size. The closed circles of the figure show the magnitudes of the additional experiments, which were conducted under the same conditions to confirm that the crack geometry was recreated in each of the specimens. The figure indicates that the magnitudes of the BD pressures increase with decreasing grain sizes, with the exception of specimen L. Table 2.2 shows tensile strengths obtained by the Brazilian test for the respective specimens. The tensile strengths are consistent with the tendency of the BD pressures, and they tend to increase with decreasing grain size.

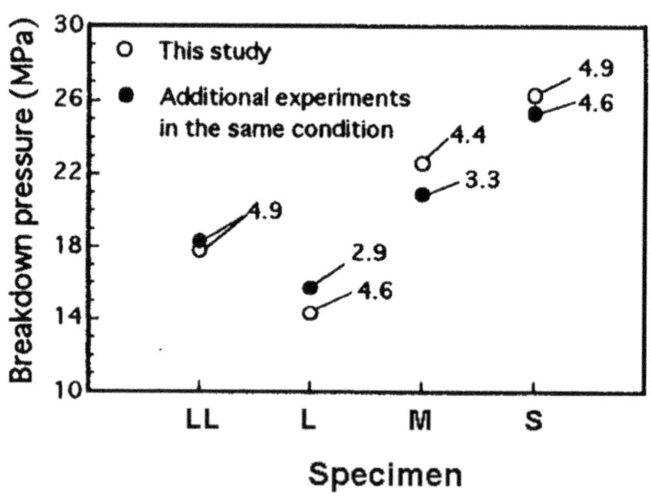

Figure 2.11 Magnitudes of BD pressure along the order of grain size. Numerals indicate flow rates (cm^3/min) (Chen et al., 1998; Ishida et al., 2000a).

Table 2.2 Tensile strengths (Chen et al., 1998, Ishida et al., 2000b; Ishida, 2001)

Specimen	*Tensile strength in the Brazilian test (MPa)*		
	T_X	T_Y	T_Z
LL	7.60 (0.15)	11.72 (2.74)	7.31 (0.44)
L	8.22 (0.39)	10.38 (0.36)	5.82 (0.54)
M	11.04 (0.27)	9.52 (0.49)	9.11 (0.53)
S	13.10 (0.41)	12.32 (0.79)	10.49 (0.65)

2.2.2.3 *Fault plane solutions of AE events recorded in the experiments*

In this subsection, fault plane solutions and ratios of polarities of P-wave initial motions of AE waveforms, which are explained in Section 1.4, are used to investigate the fracturing mechanism of HF.

In the experiments, only a few percent of the located AE events had polarities of P-wave initial motions that were clearly distinguished as "compressional" or "dilatational" at more than six sensors. Although the number of sensors is still too small to get a reliable fault plane solution, only for about five events in each specimen, their fracturing mechanisms were examined based on the polarities of P-wave initial motions projected onto a stereo net. Figure 2.12 shows a typical fault plane solution for each specimen; two solutions are shown only for specimen M. These AE events were all associated with the occurrence at the BD pressure.

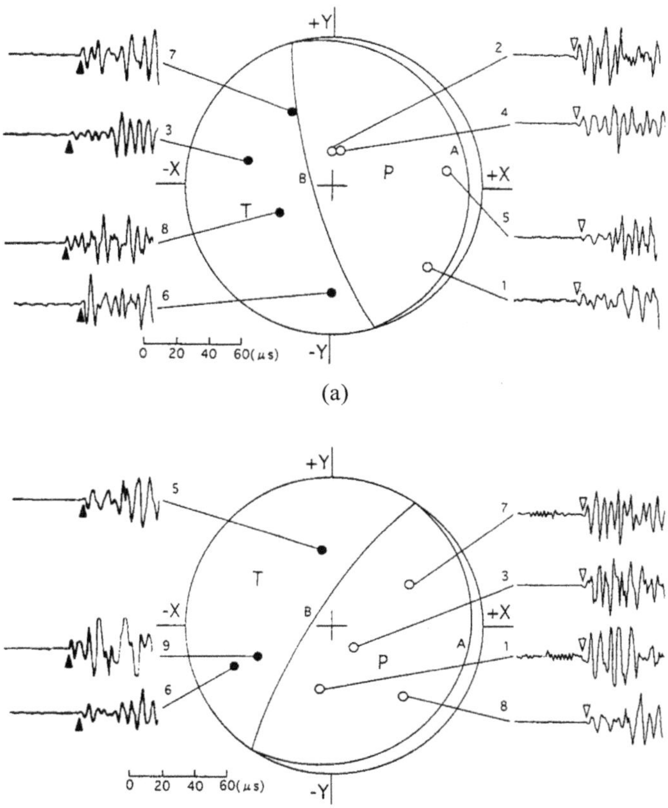

Figure 2.12 Fault plane solutions of AE events, projected onto a lower hemisphere of a Schmidt net. A solid circle indicates compression in P-wave initial motion, whereas an open circle indicates dilatation. The *X*-, *Y*-, and *Z*-coordinates in the captions indicate the location of an AE hypocenter (unit: cm), and *T* indicates an elapsed time when it is recorded (Unit: s) (Chen et al., 1998; Ishida et al., 2000b; Ishida, 2001). (a) Specimen, LL: (*X, Y, Z; T*) = (17.85, 5.09, 12.14; 60.22). (b) Specimen, L: (*X, Y, Z; T*) = (0.28, 11.74, 10.40; 58.61). (c) Specimen M, No.1 event: (*X, Y, Z; T*) = (13.11, 8.82, 5.87; 96.25). No.2 event: (*X, Y, Z; T*) = (9.89, 12.77, 9.76; 100.75). (d) Specimen, S: (*X, Y, Z; T*) = (9.88, 12.81, 9.64; 88.15).

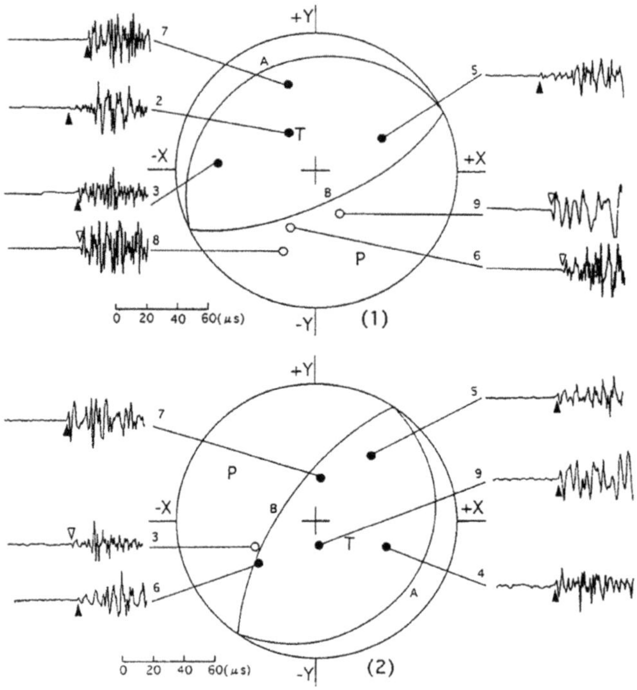

(c)

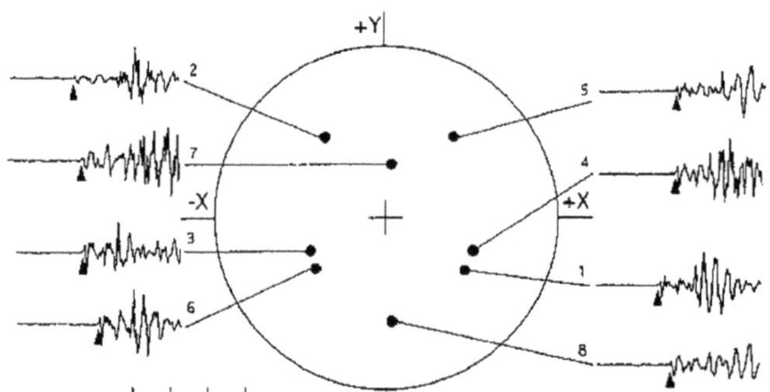

(d)

Figure 2.12 (Continued) Fault plane solutions of AE events, projected onto a lower hemisphere of a Schmidt net. A solid circle indicates compression in P-wave initial motion, whereas an open circle indicates dilatation. The *X*-, *Y*-, and *Z*-coordinates in the captions indicate the location of an AE hypocenter (unit: cm), and *T* indicates an elapsed time when it is recorded (Unit: s) (Chen et al., 1998; Ishida et al., 2000b; Ishida, 2001). (a) Specimen, LL: (*X, Y, Z; T*) = (17.85, 5.09, 12.14; 60.22). (b) Specimen, L: (*X, Y, Z; T*) = (0.28, 11.74, 10.40; 58.61). (c) Specimen M, No.1 event: (*X, Y, Z; T*) = (13.11, 8.82, 5.87; 96.25). No.2 event: (*X, Y, Z; T*) = (9.89, 12.77, 9.76; 100.75). (d) Specimen, S: (*X, Y, Z; T*) = (9.88, 12.81, 9.64; 88.15).

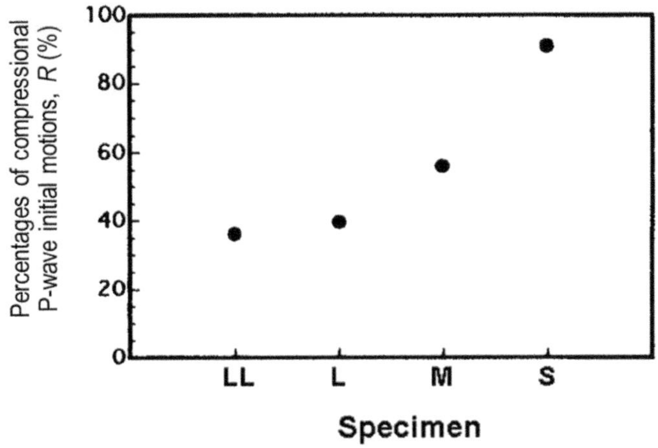

Figure 2.13 Percentages of compressional polarity of P-wave initial motions for the four granite specimens (Chen et al., 1998; Ishida et al., 2000b; Ishida, 2001).

In the specimens of specimen LL and L, as shown in Figure 2.12a and b, respectively, the fault plane solutions indicate a quadrant-type mechanism, thereby implying shear fracture. In both solutions, since the labeled "A" plane is close to the horizontal plane, it is almost parallel to a real macroscopic fractured plane induced in each specimen. Thus, the labeled "A" plane out of the two nodal planes is considered to correspond to the real fracture planes associated with the AE event in each specimen. Also, in specimen M, as shown in Figure 2.12c, both the fault plane solutions of the two AE events indicate a quadrant-type mechanism. However, the No. 2 event shows that five of the six sensors indicate polarity of compression, implying a tensile dominant mechanism. In specimen S, as shown in Figure 2.12d, all eight sensors indicate polarity of compression, implying tensile fracture.

Figure 2.13 shows percentages of compressional polarity, R, in P-wave initial motions for the four specimens. The data were corrected from the five events in each specimen; for each event in which the polarities of P-wave initial motions were clearly distinguished at more than six sensors, fracturing mechanisms were examined based on these data. This figure indicates that the percentage of compressional polarity, R, increases with decreasing grain size.

2.2.3 *Discussion*

The experimental results show that the crack geometries were quite different depending on the grain size. A horizontal crack along the rift plane was induced in specimens LL and L, having the larger grains, while a vertical crack was induced normal to the rift plane along the direction of the maximum confining pressure in specimen S, having the smallest grains. In specimen M, having an intermediate grain size, an intermediate crack geometry was induced. These results indicate that the effect of the rift plane increased with the increase in grain size.

One of the factors to control crack geometry might be the ratio of the Brazilian tensile strengths in the Z-direction to that in the X-direction, T_Z/T_X. Figure 2.14 shows the ratios with closed circles for the four specimens. Open circles in this figure indicate the ratio of P-wave velocity in the same directions. These ratios were calculated from the numbers shown in Tables 2.1 and 2.2. The ratios of T_Z/T_X seem to correspond to the ratios of Vz/Vx. If the ratios of T_Z/T_X control the crack geometry, the ratios should decrease with the grain sizes of the specimens.

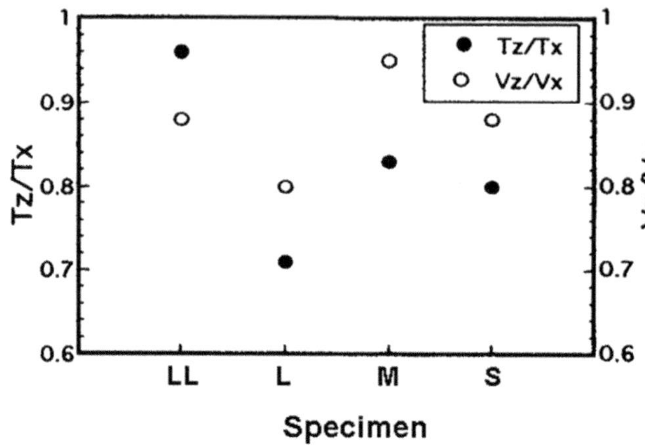

Figure 2.14 Ratios of tensile strengths in the Z-direction to those in the X-direction are shown by closed circles. Open circles also show ratios of P-wave velocities in the same directions (Chen et al., 1998; Ishida et al., 2000b).

However, the tendency cannot be found in this figure. Thus, other properties related to tensile strength should control the crack geometry.

Fracturing in HF most likely initiates from a defect or a microcrack in a pressurized section. It is expected that the larger the grain size becomes, the larger the size of a defect along the boundary of the grains and in the grains themselves becomes, and the smaller the tensile strength becomes. This expectation is consistent with the fact that the magnitudes of BD pressure decreased with the increase of grain size. Therefore, it could be considered that in the specimens LL and L, a crack initiated from a defect and extended by connecting defects oriented along the rift plane. On the other hand, in specimen S, due to the small influence of the defects, a crack initiated in the direction of the maximum tensile stress around the hole following the solution of the elastic theory indicated by Kirsch (1898) and extended in the direction.

This consideration is also supported by the fact that shear mechanism AE events are dominant in specimens LL and L, in contrast with the dominance of tensile mechanism AE events in specimen S.

The number of located AE hypocenters also increased with the increase in grain size. This was most likely due to differences in the roughness of induced crack surfaces because the crack surface of a larger grain size specimen was much rougher than that of a smaller one, and rougher surfaces likely cause more AE events by rubbing each other. This tendency agreed with the results of the uniaxial experiments conducted by Eberhardt et al. (1999), which showed that the number of detected AE events decreases markedly with the decrease of the grain size.

The fault plane solutions of AE implied shear-type fracturing in specimens LL and L, while they implied tensile-type fracturing in specimen S. In addition, in specimen M of the intermediate grain size, they showed the intermediate fracture mechanism between the larger-grained specimens LL and L and the smaller-grained specimen S. The results indicate the tendency that the dominant fracturing mechanism of HF becomes tensile rather than shear with decreasing grain size, although the number of sensors is not enough to confirm the tendency.

2.2.4 Concluding remarks

In this section, to investigate the influence of grain size on induced crack geometry and fracturing mechanism of HF, the experimental results using four 20 cm-cubic granitic rock specimens of different grain sizes are shown. The obtained results are summarized as follows.

i A horizontal crack along the rift plane was induced in specimens LL and L having the larger grains, while a vertical crack was induced normal to the rift plane along the direction of the maximum confining pressure in specimen S having the smallest grains. In the intermediate grain size specimen M, cracks having intermediate crack geometry were induced.
ii Magnitudes of BD pressure decreased with the increase of grain size. Because fracturing in HF most likely initiates from a defect or a microcrack in a pressurized section, it is expected that the larger the grain size becomes, the larger the size of a defect along the boundary of the grains and in the grains themselves becomes, and the smaller its tensile strength becomes.
iii The number of located AE hypocenters also increased with the increase of grain size. This was most likely due to differences in the roughness of induced crack surfaces because the crack surface of a larger grain size specimen was much rougher than that of a smaller one, and rougher surfaces likely cause more AE events by rubbing each other.
iv The fault plane solutions of AE events indicated the tendency that the dominant fracturing mechanism of HF becomes tensile rather than shear with the decrease of grain size, although the number of sensors is not enough to confirm the tendency.

2.3 HF experiment in impermeable acrylic resin

The acrylic resin, which is the same as polymethyl methacrylate (PMMA), is transparent and impermeable, and it could also be considered an extremely fine-grain material. Sasaki et al. (1988) conducted an HF experiment using it and elucidated that all recorded AE events indicated tensile fracturing mechanisms. This result is consistent with the tendency that the dominant fracturing mechanism of HF becomes tensile rather than shear with decreasing grain size. In this section, the HF experiment in acrylic resin conducted by Sasaki et al. (1988) is introduced.

2.3.1 Specimen, experimental setup, and AE monitoring method

2.3.1.1 Specimen and experimental setup

In the experiment, a cubic ($20 \times 20 \times 20\,cm^3$) of impermeable acrylic resin was used as a specimen. A 1 cm diameter hole was drilled along the Z-direction in the chosen Cartesian coordinate system shown in Figure 2.15. P-wave velocities measured along the X-, Y-, and Z-directions were almost the same from 2.75 to 2.76 km/s, and thus, this suggested that the specimen is isotropic and homogeneous.

Two horizontal confining pressures, 2.45 MPa (25 kgf/cm^2) in the X-direction and 4.41 MPa (45 kgf/cm^2) in the Y-direction, were applied to the specimen through flat jacks inserted between a loading frame and the specimen. The loading system to inject water and apply the confining pressures is the same as previously shown in Figures 2.3 and 2.4. A mechanical-type double packer similar to that shown previously in Figure 2.5, but having a length of sealed-off section adjustable from 10 to 60 mm, was used to inject water in the borehole.

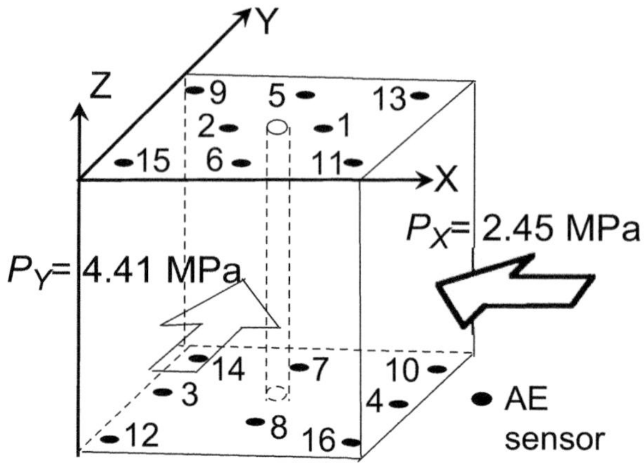

Figure 2.15 Acrylic resin specimen, loading condition, setting positions of AE sensors, and coordinate system.

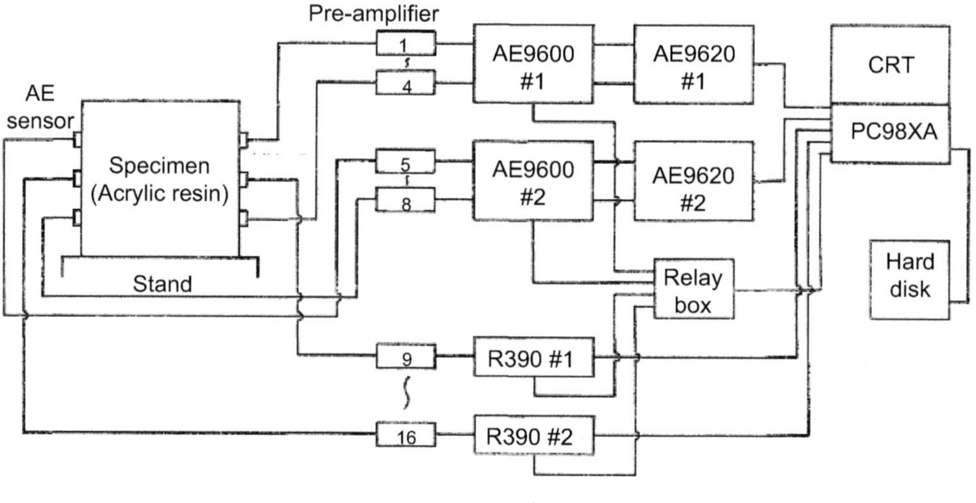

Figure 2.16 Block diagram of the AE recording system (Sasaki et al., 1988),

The oil pressure induced by a hydraulic pump was converted into the injection water pressure through the booster pump. The water pressure was measured by a transducer set on the injection pipe close to the packer. The injection flow rate was kept at a constant rate of 0.85 mL/min.

2.3.1.2 *AE monitoring method*

To monitor AE events, 16 AE sensors were set, 8 on the upper and 8 on the lower surface, respectively, as shown in Figure 2.15. The sensor is the same as that used in the experiments shown in the previous section, 2.2; that is, it has a cylindrical shape with a 1.8 cm diameter and 1.85 cm length, and a resonance frequency of 500 kHz (Type AE904E, NF Corporation).

AE waveforms were monitored using a system a little bit more advanced than that used in the previous section, as shown in Figure 2.16.

The signals detected at all 16 AE sensors were amplified 40 dB through pre-amplifiers. After that, 8 signals out of the 16 signals were amplified and filtered in frequency through two local processors (Type AE9600, NF Corporation). One of the local processors monitored triggering times for four channels by setting a threshold amplitude, and only when the maximum difference of the triggering times of the four channels is within a setting time, waveforms of the four channels were temporarily stored in the wave memory (Type AE9620, NF Corporation). The wave memory has a bit length of 10 bits for amplitude resolution, 2 Mbytes for memory capacity, 0.5 μs for a sampling time, and 1 ms for the minimum dead time between AE waveforms. The wave memory has a storage capacity of 1 k-word waveform data per channel for an AE event, and thus, it can store 256 AE events without stopping (Sato and Katsuyama, 1985). The continuous AE monitoring was once stopped just after the waveform data of 256 AE events were stored, and the data were transferred through a personal computer (Type PC98XA, NEC Corporation) to a hard disk. AE signals detected at the remaining eight channels were led to the other two types of wave memory (Type R390, Sony Corporation), and the only waveform data discriminated as useful by an external triggering signal from the two local processors (Type AE9600) were transferred to the hard disk. However, due to the long dead time of around 8 s necessary for the data transfer, the number of AE events whose waveform data could be recorded during an experiment was lower than 10% of that recorded through the other wave memories (Type AE9620). Due to this limitation, AE events were located mainly using the data recorded through the wave memories (Type AE9620), whereas the fracturing (focal) mechanisms were analyzed using the data added with those recorded through the other type two wave memories (Type R390).

In this experiment, besides the digital AE monitoring system explained above, an analog system using a tape recorder (Type RTP802A, Kyowa Electronic Instruments CO., LTD.) was used to continuously record the injection pressure and the AE signals for comparison of the relation between them.

2.3.2 *Results*

2.3.2.1 *Pressure record and AE activity*

Figure 2.17 shows the change in injected water pressure along the elapsed time of the HF experiment for the acrylic resin. In the experiment, the pressure suddenly decreased after showing the BD pressure at 36.5 MPa (372 kg/cm^2) because the fracture was induced in the pressuring section and extended beyond the seal section of the packer. Figure 2.18 shows AE activity at the BD. This figure elucidated that no AE activity was found before the BD, and after only a few large AE events were recorded for 10 or 20 m-s immediately after the BD, AE activity had completely ceased.

Sasaki et al. (1988) conducted similar HF experiments for granite to compare the results to those for the acrylic resin. They showed that, unlike in the acrylic resin, AE activity in granite started at 70%–80% of the BD pressure, and after showing high activity at the BD, it continued for one or two minutes after the BD. In addition, the amplitudes of AE events in acrylic resin were much larger than those in granite. The observations suggest that a larger crack extended by one fracture in the acrylic resin, whereas small cracks extended step by step in granite.

2.3.2.2 *Crack extension and fracturing mechanism*

In the experiment, by adjusting the interval of the pressurizing section of the packer along the hole, the HF crack could be extended stepwise a total of 12 times. Because the acrylic resin was

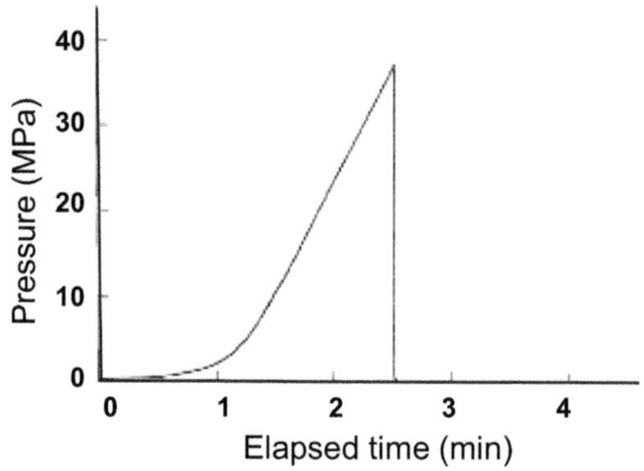

Figure 2.17 Change of injected water pressure along elapsed time of the HF experiment for acrylic resin (Sasaki et al., 1988).

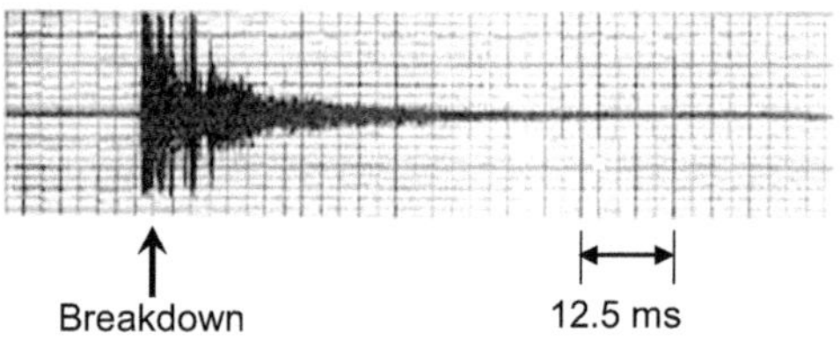

Figure 2.18 AE activity at the BD (Sasaki et al., 1988).

transparent, we could identify a region of each crack extension with intermittent small BDs. Crack traces delineating the region of each crack extension showed a conchoidal structure. Ten AE hypocenters out of 12 intermittent small BDs were located with the usual non-linear least-squares method using P-wave arrival times.

Figure 2.19a and b shows located AE hypocenters with crack traces observed through the transparent specimen with the naked eye. The numbers in the figures correspond to the respective intermittent small BDs shown in Figure 2.19c. Only 10 AE hypocenters out of 12 BDs were located. The locations of AE hypocenters moved farther stepwise from the pressurizing section, following the order of crack extensions with the respective intermittent BDs.

The crack extended in the X-direction normal to the direction of the maximum confining compressive stress, which is inconsistent with the elastic theory. This crack extension was probably due to residual stress after annealing of the resin specimen and others.

2.3.2.3 Fracturing mechanism deduced from polarity of P-wave initial motion

Figure 2.20 shows waveforms of the first AE event shown in Figure 2.17. The respective box shows a waveform recorded at sensors No. 1 through No. 16; however, only Nos. 5 and 7 do not show waveforms due to trouble with transmitting signals of the channels. A closed triangle on each waveform indicates an arrival time of the P-wave initial motion. A calibration test was conducted, in which a small steel ball was dropped onto the upper surface of a thick steel plate having sensors glued onto its lower surface. All recorded waveforms at the test showed

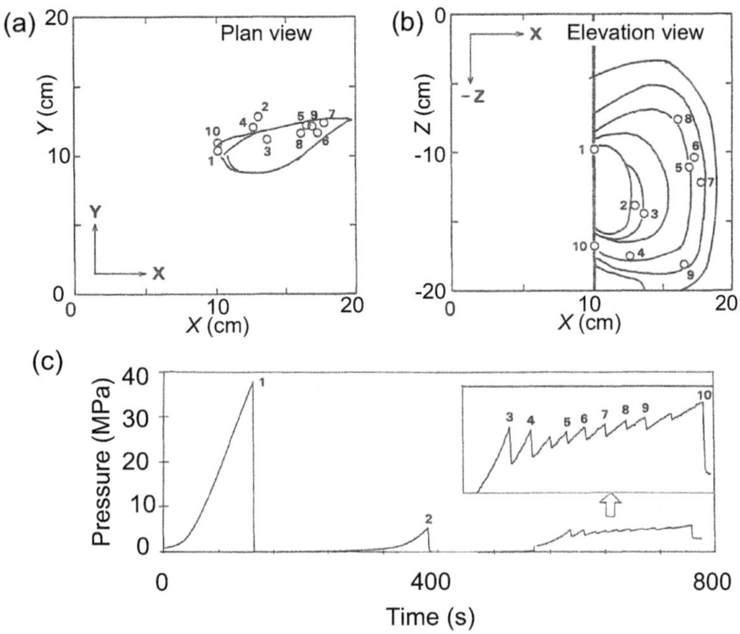

Figure 2.19 Crack shapes and located AE hypocenters with BDs in the HF experiment for the acrylic resin. (a–b) Plan view and elevation view of crack shapes observed with the naked eye (bold lines) and located AE hypocenters (open circles). (c) Pressure record of injected water. Each number of BDs corresponds that of the located AE hypocenter (Sasaki et al., 1988).

downward traces. Thus, it was found that compression (C) corresponded to a downward trace and dilation to an upward trace. Figure 2.20 indicates that all the P-wave initial motions show downward traces, that is, compression, although those of sensors Nos. 6 and 10 are slightly ambiguous.

Figure 2.21 shows the polarity distribution of P-wave initial motions projected onto a lower hemisphere of a focal sphere. The waveform recorded at each sensor is also shown with a sensor number. Because the polarities show compression at all sensors, it is difficult to obtain a solution of a quadrant type, suggesting a shear fracture, as shown in Figure 2.12a and b, and No. 1 of c shown in the previous section, 2.2. The other nine AE events also showed a similar polarity distribution, where the P-wave initial motions at all sensors are compression. The polarity distribution belongs to the pattern shown in Figure 1.3b, suggesting the tensile displacement shown in Figure 1.3a. Thus, we can conclude that the crack extensions by the intermittent BDs in the acrylic resin specimen were all induced by tensile fracture.

2.3.3 *Discussion and concluding remarks*

In the previous section, 2.2, we showed the results of HF experiments conducted in four 20 cm-cubic granitic rock specimens having different grain sizes. The fault plane solutions of AE events indicated that the dominant fracturing mechanism of HF tends to change from shear to tensile with decreasing grain size.

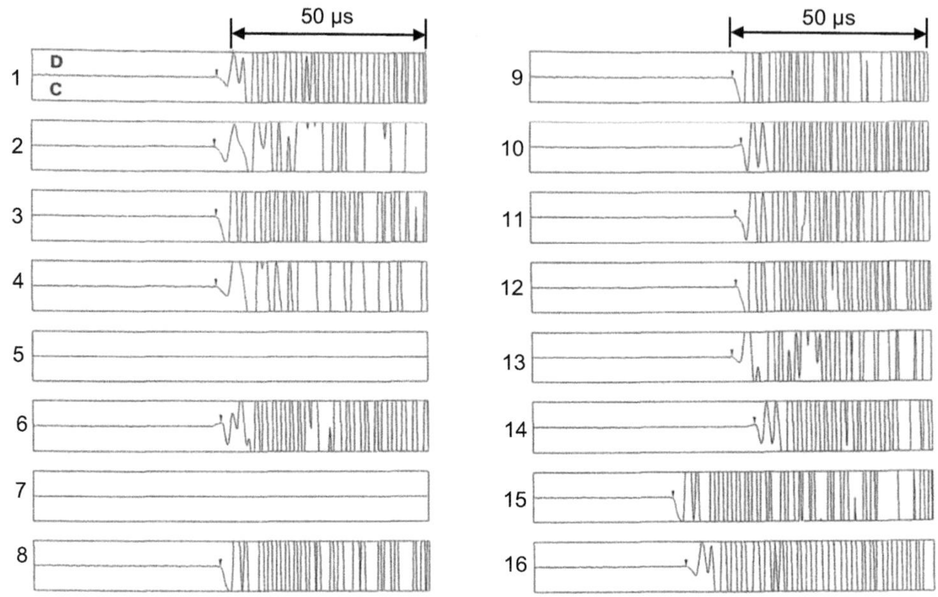

Figure 2.20 Typical example of AE waveforms recorded at 16 sensors during the HF experiment of acrylic resin. The AE waveforms were recorded at the first BD labeled with No. 1 in Figure 2.17. The number at the left side of each box in this figure indicates a sensor number. Solid triangles indicate arrival times of P-wave first motions. C: compression. D: dilatation (Sasaki et al., 1988).

In this section, we showed the results of HF experiments conducted in the impermeable acrylic resin, which could be considered as an extremely fine-grain material. In the experiment, HF cracks could be extended stepwise with intermittent small BDs, and a region of each crack extension was observed with the naked eye because the acrylic resin specimen was transparent. The located AE hypocenters moved farther stepwise from the pressurizing section, following the order of the crack extensions observed with the respective intermittent BDs. On the fracturing mechanism, the polarities of P-wave initial motions of the AE events show compression at all sensors, strongly suggesting that the crack extensions by the intermittent BDs were all induced by tensile fracture. Thus, through the experiments, the tendency that the dominant fracturing mechanism of HF changes from shear to tensile with decreasing grain size in granites is confirmed to be consistent also in the acrylic resin, which could be considered as an extremely fine-grain material.

2.4 Fracturing mechanism of HF crack extension along and normal to rift plane of granite

Recently, Naoi et al. (2018) conducted HF experiments to make a crack extend along and normal to the rift plane of granite samples using an acrylic resin (methyl methacrylate) having a fluorescent agent as fracturing fluid. They observed regions dyed with the fluorescent agent in comparison to the distributions of located AE events. Following the experiments, Yamamoto

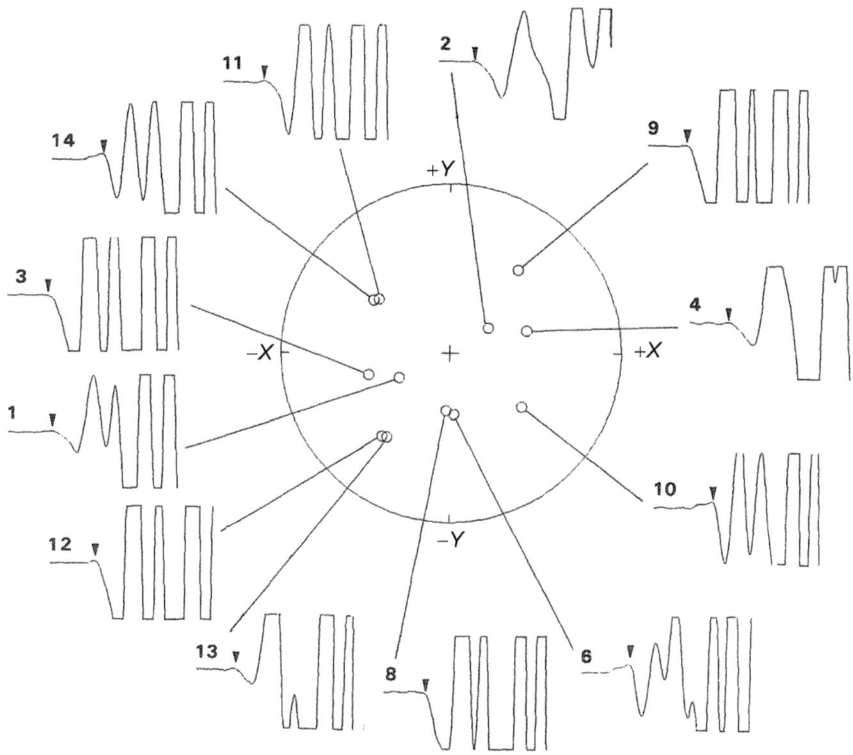

Figure 2.21 Polarity of P-wave initial motion projected onto the lower hemisphere of a focal sphere (equal area projection of stereonet) and each waveform with a sensor number (Sasaki et al., 1988).

et al. (2019) applied a moment tensor analysis to the AE data obtained in the experiments and compared the fracturing mechanism when a crack extends along the rift plane to that when normal to the rift plane.

In this section, I will mainly introduce the results of Yamamoto et al. (2019), adding some information from Naoi et al. (2018), because the results help us to understand the fracturing mechanism depending on rock properties. Please note that the fracturing fluid used in the experiments was not water but an acrylic resin. However, because the viscosity, 0.80 mPa·s, is almost the same as that of water, 1.0 mPa·s, we will discuss the results under the assumption that the fracturing mechanism by the acrylic resin is the same as that of water.

2.4.1 *Specimen, experimental setup, and AE monitoring method*

2.4.1.1 *Specimen and experimental setup*

In the experiment, seven rectangular blocks ($8.5 \times 8.5 \times 17\,cm^3$) of Kurokami-jima granite, which is Cretaceous granite produced in Kurokami-jima Island in the Seto Inland Sea in Southwestern Japan, were used. A 1 cm diameter hole was drilled at the center of each block along the X-direction in the chosen Cartesian coordinate system, as shown in Figure 2.22. A packer with a

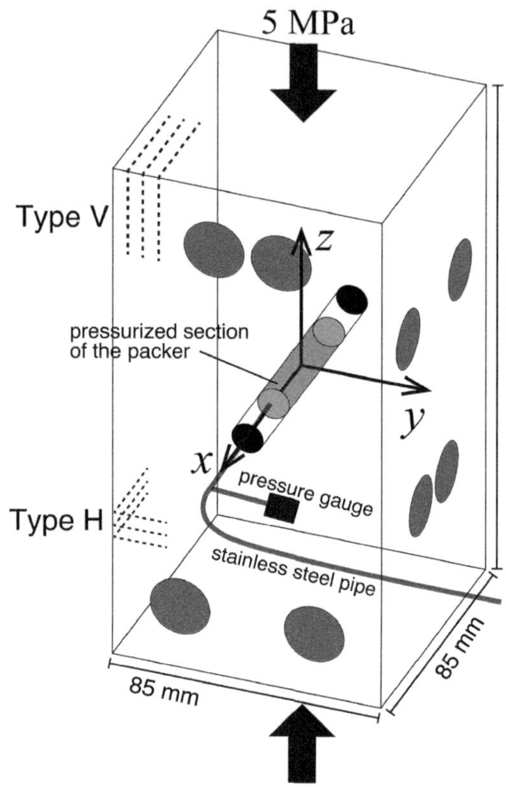

Figure 2.22 Positions of AE sensors, loading direction, and the coordinate system. The origin of the coordinate system corresponds to the center of the specimen. Gray circles represent the positions of the AE sensors, which are shown only on the front side for clarity (Yamamoto et al., 2019).

30 mm pressurizing section sealed with O-rings at each end was inserted into the hole to inject a fracturing fluid. As the fracturing fluid, an acrylic resin (methyl methacrylate) of viscosity 0.80 mPa·s containing a fluorescent agent was used.

The experiments were conducted under the uniaxial loading condition by using samples in which the rift plane was either perpendicular (Type H specimen) or parallel (Type V specimen) to the expected direction of fracture propagation (i.e., direction of the loading axis). Here, the rift plane is one of the three inherent orthogonal planes on granite, and it may become the weakest plane in the fracturing due to many microcracks existing along the plane, as explained in the last part of Section 2.1. Based on the difference of P-wave velocities, V_p, measured along the x-, y-, and z-directions shown in Table 2.3, Naoi et al. (2018) estimated that the rift plane is situated in the plane whose normal direction corresponds to that of the lowest P-wave velocity. Thus, they classified the blocks into two types (Figure 2.22): Type V (KJG05, 07, 10), in which the rift plane is parallel to the x–z plane (i.e., vertical rift plane in the experimental setting), and Type H (KJG06, 08, 09, 11), in which the rift plane is parallel to the x–y plane (horizontal rift plane).

The P-wave velocities were measured with S-wave velocities, Vs, measured before drilling the injection hole. No significant velocity change was observed under loading for an experiment after drilling the hole.

Table 2.3 P-wave velocities of specimens measured along the *x*-, *y*-, and *z*-directions (Naoi et al., 2018)

Specimen No.	*Type*	*Vx (km/s)*	*Vy (km/s)*	*Vz (km/s)*
KJG05	V	5.80	5.34	5.57
KJG06	H	5.09	5.09	4.40
KJG07	V	5.08	4.55	5.00
KJG08	H	5.67	5.47	5.11
KJG09	H	5.52	5.54	5.08
KJG10	V	5.50	5.18	5.54
KJG11	H	5.67	5.64	5.03

HF experiments were conducted under a uniaxial loading of 5 MPa along the *z*-axis. The acrylic resin was provided to the packer from a cylindrical piston, which was pushed by the pressure of water injected from a syringe pump at a constant flow rate of 2 mL/min. The injected fluid pressure was measured by a pressure gauge set on the pipeline near the packer (Figure 2.22), and the data were recorded every 10 μs.

2.4.1.2 *AE monitoring method*

As shown in Figure 2.22, 16 AE sensors (R15; Physical Acoustic Corp., the resonance frequency, 150 kHz) were attached to each specimen using thermo-fusible wax (Electron Wax; Furuuchi Chemical Corp.). During the experiments, signals were continuously recorded at a sampling time of 0.1 μs using a 14-bit analog-to-digital converter (PXIe-5170R; National Instruments Corp.) after amplifying by 52 dB through a 0.02–3.00 MHz bandpass filter. The maximum range of the analog-to-digital converter was set to 2 V in peak-to-peak voltage.

From these data, an AE catalog containing the hypocenter coordinates was created. AE events were identified using a method based on the ratios of the short-term and long-term averages (Allen, 1978). For each cut-out waveform, the P-wave arrival time was automatically detected using the algorithm of Takanami and Kitagawa (1988), and their hypocenters were located using a least-squares algorithm in which the P-wave anisotropy was considered (Rothman et al., 1974). In the iterative hypocenter-determination procedure, arrival times that had large residuals were removed using a bi-weight function. From the results, "well-located events" satisfying the following conditions were selected: (1) eight or more arrival times remained after the final iteration, and (2) the maximum standard error was less than 40 mm among the three principal axes of an error ellipsoid calculated from the variance–covariance matrix. Then, the events were relocated using the double-difference algorithm (Waldhauser and Ellsworth, 2000). Details of the catalog development were reported by Naoi et al. (2018).

2.4.2 Results

2.4.2.1 *Pressure record and AE hypocenter distribution*

Figure 2.23 shows fluid pressure, AE activity, and the *z*-coordinates of AE hypocenters with time in experiment KJG07 (Type V). The total time for the experiment was 1,054 s from the start of fluid injection to the end of AE data recording, although the figure is focused on the 250 s interval around the BD. The BD occurred at $t = 455$ s. The AE activity started before the BD,

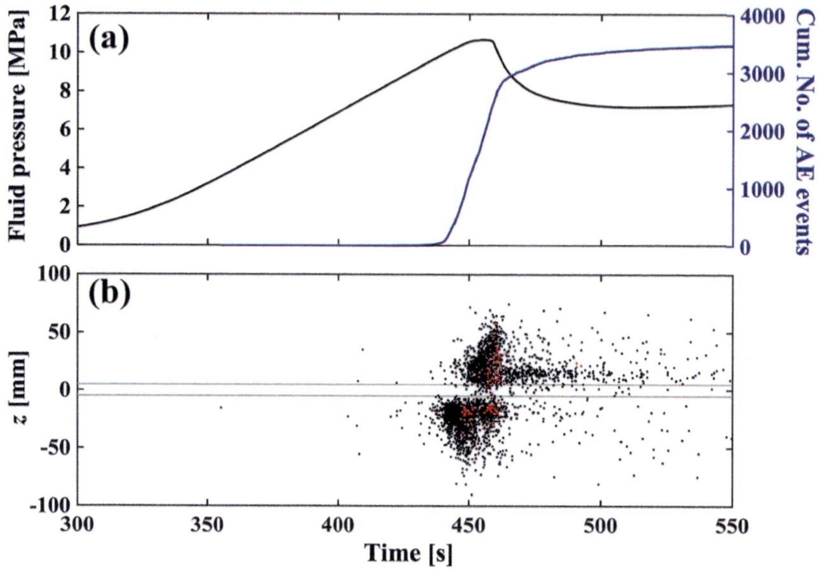

Figure 2.23 Fluid pressure, AE activity, and z-coordinates of AE hypocenters with time in experiment KJG07 (Type V). (a) Black line indicates fluid pressure, whereas the blue line indicates the cumulative numbers of well-located AE events. (b) The z-coordinates of well-located AE hypocenters by the double-difference method. Red dots show AE events whose moment tensors were determined, which satisfy the condition, VR ≥ 50% (Yamamoto et al., 2019).

and the hypocenters expanded outwards with time from the injection hole. Figure 2.24 shows the AE hypocenter distribution obtained in this experiment. The hypocenters exhibit a bi-wing distribution along the *z*-direction from the pressurizing section (Figure 2.24c). This is consistent with the fluid penetration region observed after the HF experiment, and the features described above occurred in all seven granite samples (Naoi et al., 2018).

2.4.2.2 Moment tensor inversion for AE events

Seismic moment tensors for the AE events were estimated from the polarities of the P-wave first motions and their amplitudes. The directivity of an AE sensor is generally unknown, and its sensitivity depends on the specific coupling condition between the specimen and the sensor. This makes it difficult to invert the moment tensor because it requires wave amplitude information. To overcome these problems, the directivity and the relative sensitivity of individual AE sensors were estimated using a procedure, which is described in detail in Yamamoto et al. (2019).

Figure 2.25 shows source-type plots of the obtained moment tensor solutions for experiments KJG05-11. The AE events whose P-wave polarities could be read at six or more channels were analyzed. Although various source-type plots have been suggested in previous studies, the "simplified decomposition method" (Vavryčuk, 2015) was adopted here.

In this analysis, the amplitudes were used after the correction of the sensor directivity and coupling conditions. For the obtained solutions, the variance reduction (VR) was estimated as

$$\mathrm{VR} = \left(1 - \frac{\sum \left| A - A^{\mathrm{th}} \right|^2}{\sum \left| A \right|^2} \right) \times 100\ (\%) \tag{2.1}$$

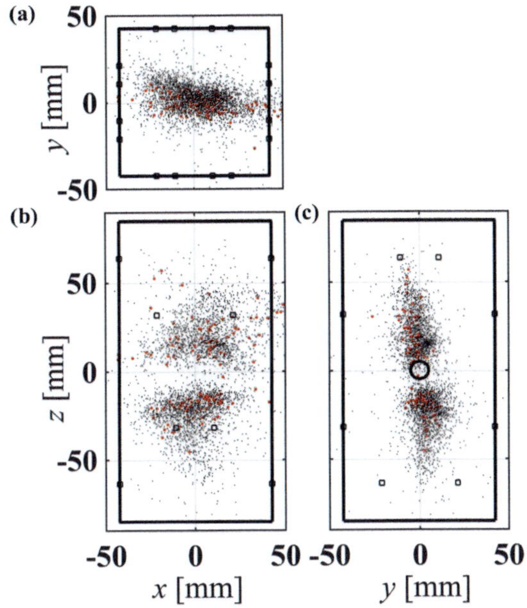

Figure 2.24 Distribution of well-located AE hypocenters obtained by the double-difference method in experiment KJG07 (Type V). Projections onto (a) *x–y*, (b) *x–z*, and (c) *y–z* planes. The bold-black rectangle corresponds to the size of the sample. Small rectangles show the sensor positions. Red dots show 142 AE events whose moment tensors were determined, which satisfy the condition, VR ≥ 50% (Yamamoto et al., 2019).

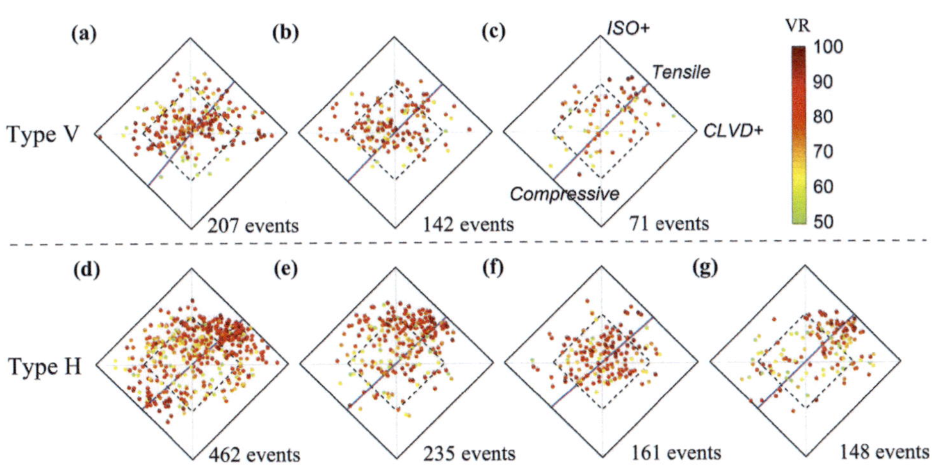

Figure 2.25 Source-type plot for the moment tensors obtained for each experiment. Solutions in the square area surrounded by the dashed lines correspond to events having DC components of 50% or more (i.e., $|c| \le 0.5$). Colors represent the VR of each solution. Ideal shear–tensile events (Vavryčuk, 2001, 2011b) correspond to the solutions on the blue line, which are determined by the ratio of *Vp* to *Vs* (Vavryčuk, 2015). Results for (a) KJG05 (Type V), (b) KJG07 (Type V), (c) KJG10 (Type V), (d) KJG06 (Type H), (e) KJG08 (Type H), (f) KJG09 (Type H), and (g) KJG11 (Type H) (Yamamoto et al., 2019).

where A is the amplitude of the P-wave after the correction of coupling and incident angle effects, and A^{th} is the theoretical amplitude. Thus, the solution with a larger percentage of VR is more reliable. In Figure 2.25, only the solutions having VR ≥ 50% are shown.

In these plots, solutions near the origin correspond to shear events that have large DC (double-couple) components, and the DC percentage declines with distance from the origin. Shear–tensile events (Vavryčuk, 2001, 2011), in which both tensile and shear displacements occur on a single fault plane, correspond to solutions along a line determined by the ratio of V_p to V_s (Vavryčuk, 2015). The line estimated from the average V_p and V_s among the X-, Y-, and Z-directions is shown in blue in Figure 2.25. On this blue line, the part in the first quadrant corresponds to tensile events (crack opening), whereas the part in the third quadrant corresponds to compressive events (crack closure). Although shear, tensile, and compressive AE events occurred in both Type V and Type H samples (see Figure 2.25), the shear type was dominant for Type V (KJG05, 07, 10; Figure 2.25a and c), whereas the tensile-type was dominant for Type H (KJG06, 08, 11; Figure 2.25d, e, g). Many compressive events were also observed for Type H, especially for KJG06 (Figure 2.25d).

2.4.2.3 Crack orientations

For the obtained moment tensors, the crack orientations were evaluated. Before that, tensile dominant AE events were selected using the following consistency coefficient (Vavryčuk, 2011);

$$c = \text{sign}\left(\frac{\text{ISO}}{\text{CLVD}}\right)\left|1-\frac{\text{DC}}{100}\right| \tag{2.2}$$

where ISO, CLVD, and DC are the percentages of each component of a moment tensor after the decomposition. The value of c takes between −1 and 1, and a moment tensor solution is consistent with the shear–tensile model if $c \geq 0$ (Vavryčuk, 2011). The value of c approaches 0 when the shear component becomes large, and approaches 1 when the tensile or compressive component becomes large. In the following analysis, only AE events in which $c \geq 0$ were considered.

For the selected events, the fault normal direction n was calculated as

$$\boldsymbol{n} = \sqrt{\frac{M_1 - M_2}{M_1 - M_3}}\boldsymbol{e}_1 + \sqrt{\frac{M_3 - M_2}{M_3 - M_1}}\boldsymbol{e}_3 \tag{2.3}$$

and the dislocation direction v as

$$\boldsymbol{v} = \sqrt{\frac{M_1 - M_2}{M_1 - M_3}}\boldsymbol{e}_1 - \sqrt{\frac{M_3 - M_2}{M_3 - M_1}}\boldsymbol{e}_3 \tag{2.4}$$

according to Vavryčuk (2011), where M_1, M_2, and M_3 are the maximum, intermediate, and minimum eigenvalues of a moment tensor, respectively. Eigenvectors $\boldsymbol{e}_1$ and $\boldsymbol{e}_3$ correspond to M_1 and M_3, respectively. $\boldsymbol{n}$ and $\boldsymbol{v}$ are complementary solutions, such as the nodal planes of a DC component, and cannot be determined uniquely; that is, two candidates of a source model exist for a pair of $\boldsymbol{n}$ and $\boldsymbol{v}$.

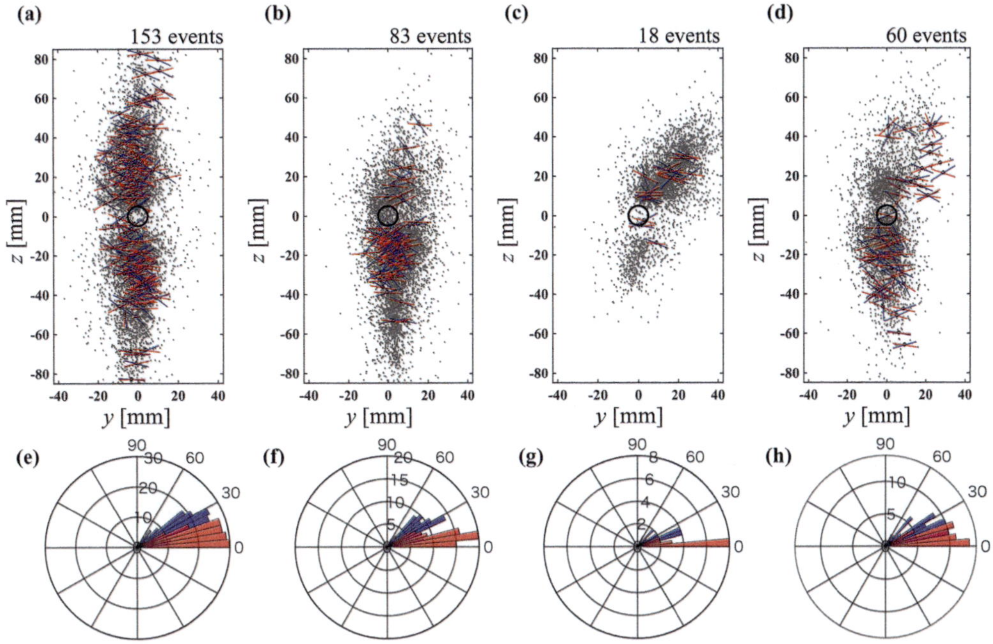

Figure 2.26 Fault normal direction ***n*** and dislocation direction ***v***, and polar coordinate histogram of their dip angles for the experiments of Type H specimens, in the order of KJG06, KJG08, KJG09, and KJG11. (a–d) *y–z* projections of ***n*** and ***v*** directions. (e–h) Polar histogram. These are plotted only for the moment tensors that have $c > 0.5$ and are situated in the first quadrant of the source-type plot (i.e., tensile dominant events). For each pair of ***n*** and ***v***, the vector with a higher dip angle is shown in blue and the other in red (red bins overlay blue ones). Hypocenters relocated by the double-difference method, including events whose moment tensor was not determined, are overlaid (gray dots) (Yamamoto et al., 2019).

For the results of the experiments for Type H specimens (KJG06, 08, 09, 11), Figure 2.26a–d shows the *y–z* projections of the ***n*** and ***v*** directions of tensile dominant events, which are defined as solutions exhibiting $c > 0.5$ and situated within the first quadrant of the source-type plot. Polar coordinate histograms of the dip angles of the events are shown in Figure 2.26e–h. Most instances of ***n*** and ***v*** for the tensile events are subparallel to the horizontal (*x–y*) plane. This indicates that most of the tensile dominant events had a fracture plane along the loading axis (*z*-axis).

Yamamoto et al. (2019) showed the same diagrams also for the compressive-dominant events ($c > 0.5$ and situated in the third quadrant) and shear-dominant events (i.e., $0 \leq c \leq 0.5$). Through examination and discussion, they concluded that tensile events along the loading axis were dominant for Type H specimens. In addition, as the reason that only the case of KJG09 did not show tensile dominance despite being a Type H specimen as shown in Figure 2.25f, they indicated that the crack extension was deduced to be inclined by around 38° from the AE distribution in Figure 2.26c and this may have caused a large number of shear events rather than tensile due to development of shear stress on planes inclined from the principal compressive stress.

2.4.3 *Concluding remarks*

In the section, the author introduced HF experiments under uniaxial loading conditions in granite blocks where the rift plane was orthogonal (Type H) or parallel (Type V) to the fracture direction expected from the stress condition around the injection hole. In the experiments, AE events induced by the fracturing were monitored, and their moment tensors were estimated by carefully correcting sensor directivities and coupling conditions between respective sensors and the specimen. For Type H, tensile events along the loading axis were dominant, whereas shear events were dominant for Type V. This difference implies that the dominant fracturing mechanisms of AE events induced by HF are significantly affected by the relation between the direction of the fracture propagation and that of pre-existing microcracks.

2.5 Discussion on the fracturing mechanism

In this chapter, to discuss the effects of rock properties on HF, the three types of experiments were introduced. At first, the experiments for granite specimens having four different grain sizes were introduced in Section 2.2. The second, the HF experiment in the impermeable acrylic resin block, which could be considered as an extremely fine-grain material, was presented in Section 2.3. The third, the experiments to compare the fracture propagation between along and normal to the rift plane by setting granite specimens taken from the same block under uniaxial vertical loading conditions where the rift plane was parallel (Type V) or orthogonal (Type H) to the fracture direction expected from the stress condition around the injection hole were introduced in Section 2.4.

The results elucidated that fracture propagation normal to the rift plane induced tensile dominant AE events in the specimen, S, having the smallest grain size in Section 2.2 and those of Type H (i.e. KJG06, KJG08, and KJG 11 excluding KJG9) in Section 2.4, although there was the difference in cutting direction of the rift plane; in the former, the fracture propagation along the hole cut the rift plane, whereas in the latter the propagation in a radial direction from the hole cut the rift plane. In addition, HF in the acrylic resin also induced AE events showing tensile fracturing mechanisms.

On the other hand, fracture propagation along the rift plane induced shear-dominant AE events in the specimens, L and LL, having the larger grain size in Section 2.2 and those of Type V (i.e., KJG05, KJG07, and KJG10) in Section 2.4.

In this section, we will discuss the reasons why the difference between tensile and shear in the fracturing mechanism is induced by the difference in the fracture propagation direction for the rift plane.

2.5.1 *Mechanism of tensile fracturing*

As shown in Section 1.3, the tangential tensile stress around the borehole wall becomes larger with the increase of the injected fluid pressure. When it reaches a tensile strength of rock, the tensile fracture is initiated on the wall of the vertical borehole, and the crack extends to the horizontal maximum stress direction. The theory proposed by Hubbert and Willis (1957), following the traditional elastic theory, states that HF cracks are created by tensile fracturing.

This theory can explain the fracturing mechanism only for the initiation of the crack on the borehole wall. Once the HF crack extends, stress around the borehole is disturbed by the crack, and it becomes difficult to calculate the stress condition around the crack using the traditional elastic theory. However, after the crack extends, the injection fluid penetrates into the crack, and the fluid pressure applies to the crack surfaces as shown in Figure 1.2. If the rock around the borehole is impermeable and has no microcrack, the fluid pressure applied to the crack surfaces wedges open and extends the crack by inducing tensile fracturing at the crack tip. The crack extension causes the displacement normal to the crack extending direction, as shown in Figure 1.4a in Section 1.4, and results in the P-wave radiation pattern as shown in Figure 1.4b.

This most likely corresponds to the cases in which tensile fracturing dominated, such as the experiment of specimen S, having the smallest grain size in Section 2.2, that of impermeable acrylic resin in Section 2.3, and those of crack extension normal to the rift plane (i.e. Type H: KJG06, KJG08, and KJG 11 excluding KJG9) in Section 2.4.

2.5.2 *Mechanism of shear fracturing*

To discuss the mechanism of shear fracturing in HF, the results of a three-point bending test help us to understand the cause of shear-dominant fracturing. Kao et al. (2011) conducted a three-point-bend fracture test on a Charcoal granite specimen measuring 217 mm long, 73 mm high, and 32 mm thick in the x–y–z-directions with a 4 mm notch, by controlling crack mouth opening displacement of the notch at a constant rate of 5×10^{-5} mm/s, as shown in Figure 2.27. They monitored AE events during the experiment with eight AE sensors (Physical Acoustics model S9225) attached to the front and back surfaces of the specimen, four on each surface, surrounding a region adjacent to the notch. AE events detected by the sensors having a frequency response of 0.1–1.2 MHz were recorded through transient recorders (LeCroy Corporation model 6848) having 8-bit resolution and a sampling time of 50 ns.

Kao et al. (2011) recorded a total of 1711 AE events and located 638 events with an error of less than 2.5 mm. The crack paths observed on the surfaces of the specimen agree with the projected (the z-coordinate is not included) AE locations, as shown in Figure 2.28.

To find microcrack orientations, they obtained the moment tensors for 393 AE events out of the located 638 events after careful calibration of sensor sensitivity, including the coupling

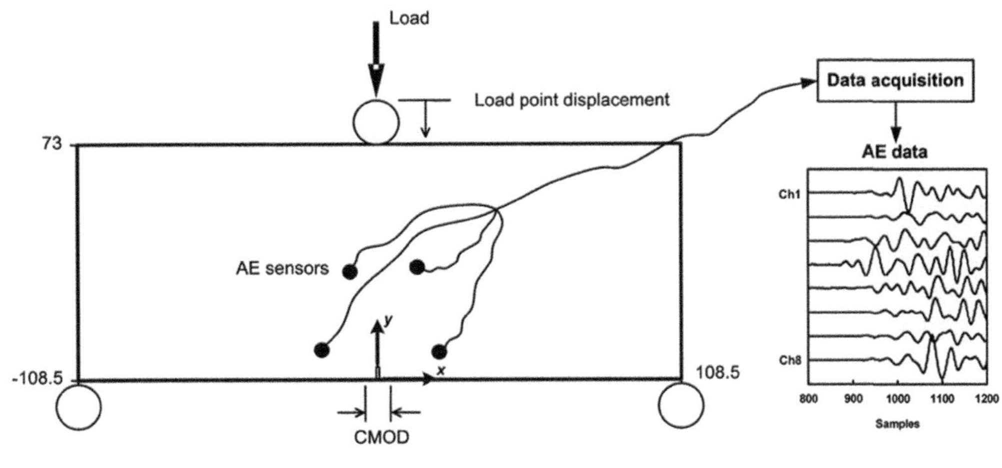

Figure 2.27 Configuration of the three-point-bend fracture test and an example of obtained AE data (Kao et al., 2011).

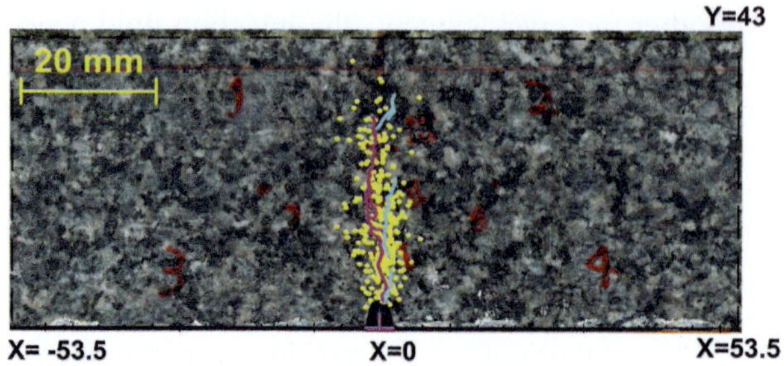

Figure 2.28 Global fracture observed after the test, mapped onto the *x–y* plane with z = 0 (red) and z = 32 mm (blue). Yellow dots represent AE locations (Kao et al., 2011).

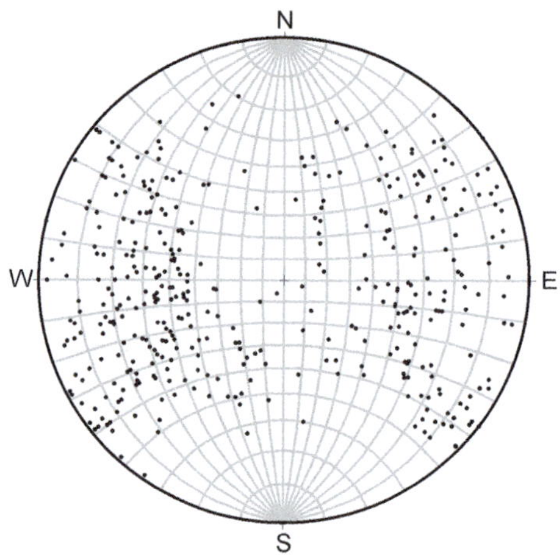

Figure 2.29 Stereonet representation of microcrack normals, ***n***; East and South align with x- and z-axes (Kao et al., 2011).

effect between each AE sensor and the specimen surface, using a signal of glass capillary break under the conditions of the actual experiment. The moment tensors have an average relative error of less than 10%, as determined by the normalized difference at each sensor location between the measured displacement and the calculated displacement. From the moment tensor, they calculated a displacement vector ***b*,** a microcrack normal ***n***, and an angle between the vectors ***b*** and ***n*** defined as 2α. The angle α can take any angle from 0 to 90°. When the angle α equals zero, ***b*** is parallel to ***n***, and the case corresponds to a pure opening mechanism. The case that $\alpha=90°$ corresponds to a pure closing mechanism, whereas the case that $\alpha=45°$ corresponds to a pure shear mechanism. Because the two vectors are not uniquely determined and are interchangeable due to the symmetry of the moment tensor, it is necessary to introduce other information to identify the orientation of the two vectors. In this test, a tensile macroscopic crack extended in the *y*-direction perpendicular to the *x*-direction, and therefore, the microcrack normal ***n*** should be aligned with the *x*-axis. Thus, from the two vectors, a vector more aligned to the *x*-axis was taken as the microcrack normal ***n***. Figure 2.29 shows the microcrack's normal ***n*** directions plotted on a lower hemisphere stereo net, where East and South align with the *x*- and *z*-axes, i.e., viewed from a positive *y*-direction. The figure elucidates that the microcrack normals were subhorizontal close to the *X*-direction, but widely scattered. From the results, Kao et al. (2011) suggested that microcrack orientations are distorted from the extending direction of a macroscopic crack by the influence of inhomogeneities such as different minerals, grain boundaries, and crystal defects. Kao et al. (2011) analyzed the results further and found that the crack mode angles, α, for all characterized events were within 45°±15°, and the average value of α was 45°. This is interesting because this means that all the events were locally shear-dominant. A similar mechanism was also observed in true-triaxial tests (Carvalho and Labuz, 2002; Pettitt and King, 2004).

This mechanism most likely corresponds to the cases that shear fracturing dominated in the HF experiments, when a macroscopic fracture extended in a direction along the rift plane in the experiments of the specimens, L and LL, having the larger grain size in Section 2.2 and those of Type V (i.e. KJG05, KJG07, and KJG10) in Section 2.4. This mechanism also can explain the fact that shear-dominant AE events were induced with the majority of tensile dominant events, even when a macroscopic fracture extended in a direction normal to the rift plane, in the experiments of specimen S, having the smallest grain size in Section 2.2 and those of Type H (i.e. KJG06, KJG08, and KJG 11) in Section 2.4. The mechanism implies that shear-dominant small fracture associated with an extension of a macroscopic tensile fracture is a natural consequence of tortuosity, and the local fracturing mechanism, that is the mechanism of the AE event, does not necessarily reveal the nature of the macro failure mode.

On the other hand, in volcanic earthquake swarms, significant parts of seismic events show a shear mechanism, although magma intrusions or eruptions characterize many events. To explain the observation, Hill (1977) proposed a conceptual model that magma intrudes into the weak planes lying along the direction of the maximum horizontal compressive stress, σ_H, among many weak planes prevailing in a volcanic region, as shown in Figure 2.30. Using a model like this, Hill (1977) stated that the magma intrusion forming a dike would accompany some tensile fracturing; in contrast, shear fracture would form conjugate faults, connecting the tips of dikes, as shown with symbols *A* and *B* in Figure 2.30.

The granite specimens used in the HF experiments shown in this chapter should also have many small inherent microcracks and defects lying along the rift plane. Among these, in the experiments of the specimens, L and LL, in Section 2.2, and those of Type V specimens (i.e., KJG05, KJG07, and KJG10) in Section 2.4, cracks extended along the rift plane, accompanied

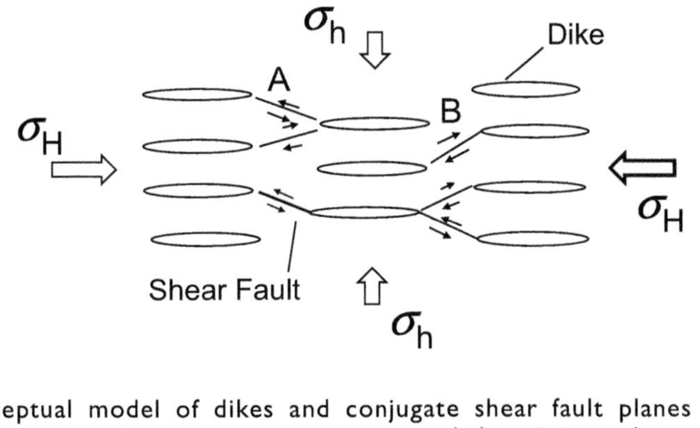

Figure 2.30 Conceptual model of dikes and conjugate shear fault planes under the maximum horizontal compressive stress, σ_H, and the minimum horizontal compressive stress, σ_h. The original model (Figure 1 in Hill, 1977) of this was proposed to explain the fracture mode of volcanic earthquake swarms.

by shear-dominant fracturing. Most of the shear-dominant fracturing likely occurs in the parts connecting tips of microcracks and defects lying along the rift plane, in a similar way as indicated by symbols A and B in Figure 2.30.

2.6 Conclusions

In this chapter, to investigate the effects of rock properties on HF, at first, the HF experiments for granite specimens having four different grain sizes were introduced. Under the same bi-axial confining stress condition, a macroscopic HF crack extended along the rift plane, cutting across the injection hole in the specimens having the larger grains, LL and L. In contrast, the crack extended along the injection hole in a plane normal to the rift plane in the specimen having the smallest grains, S. AE events with the crack extension along the rift plane showed a shear-dominant mechanism. In contrast, those normal to the rift plane showed tensile dominance. In the intermediate grain size specimens, M, cracks having intermediate crack geometry were induced according to the grain sizes, and both shear-dominant and tensile dominant AE events were recorded.

Following the experiments, in Section 2.3, the HF experiment in the impermeable acrylic resin block, which could be considered as an extremely fine-grain material, was introduced. In the experiment, HF cracks extended stepwise with intermittent small BDs, and every stepwise crack extension induced AE event showing a complete tensile fracturing mechanism without exception.

In Section 2.4, the HF experiments to compare the difference between crack extensions along the rift plane and those normal to the rift plane were introduced. In the experiments, granite specimens taken from the same block were set under uniaxial vertical loading conditions where the rift plane was parallel (Type V) or orthogonal (Type H) to the direction of crack extension expected from the stress condition around the injection hole. AE events with the crack extension along the rift plane (Type V) showed a shear-dominant mechanism, whereas those normal to the rift plane (Type H) showed a tensile dominant. This tendency is consistent with the first experiments for the specimens having four different grain sizes.

The results of the three types of experiments suggest that crack extension along the rift plane induces shear-dominant fracture, whereas the crack extension normal to the rift plane or in the intact material having extremely fine grains induces tensile dominant fracture. In Section 2.5, the discussion was made on the reason why such a difference in fracturing mechanism between tensile and shear is induced.

The concluding remarks on the fracturing mechanism are the following:

1 The stress condition on a wall of an injection hole derived from the elastic theory indicates that the initiation of HF crack is induced by tensile fracture under the assumption that rock is isotropic, homogeneous, and impermeable. Thus, at least, the HF crack initiation is most likely induced by tensile fracturing as proposed by Hubbert and Willis (1957).
2 After an HF crack extends, although the stress condition around the hole cannot be estimated by the elastic theory, the injection fluid penetrates the crack and the fluid pressure applies on the crack surfaces as well as on the injection hole wall. If the rock around the hole is impermeable and has no microcrack, the fluid pressure applied to the crack surfaces wedges open and extends the crack by inducing tensile fracturing at the crack tip.

   However, the actual rock mass around an injection hole has many microcracks and defects, particularly in granite among various types of rock.

   Thus, the following concluding remarks for shear fracturing are important.

3 The results of the three-point-bend fracture test by Kao et al. (2011) suggested that microcrack orientations are distorted from the extending direction of a macroscopic crack by the influence of inhomogeneities such as different minerals, grain boundaries, and crystal defects.
4 In the experimental results by Kao et al. (2011), almost all the AE events distorted from the extending direction of a macroscopic crack showed a shear-dominant mechanism. From this fact, they suggest that a local shear-dominant fracture associated with the extension of a macroscopic tensile fracture is a natural consequence of tortuosity, and the local fracturing mechanism does not necessarily reveal the nature of the macro failure mode.
5 Thus, the macroscopic crack extension of HF along a direction where many pre-existing cracks align like the rift plane easily induces shear-dominant AE events. Even in the case that a macroscopic crack extends in a direction normal to the rift plane, shear-dominant AE events can also be induced with most tensile dominant events.
6 The other mechanism proposed by Hill (1977), explaining shear seismic events in volcanic earthquake swarms, suggests that shear fractures connecting the tips of weak planes are induced when fluid intrudes into the weak planes lying along the direction of the maximum compressive stress.
7 This mechanism also enhances shear-dominant fracturing when HF macroscopic cracks extend along the rift plane because many shear fractures are induced by connecting the tips of many pre-existing microcracks that align along the rift plane and are likely prevalent in a volcanic region.

References

Allen, R.V. (1978) Automatic earthquake recognition and timing from single traces. *Bull. Seism. Soc. Am.*, 68, 1521–1532.

Baisch, S., Rothert, E., Stang, H., Vörös, R., Koch, C. & McMahon, A. (2015) Continued geothermal reservoir stimulation experiments in the Cooper Basin (Australia). *Bull. Seism. Soc. Am.*, 105(1), 198–209. https://doi: 10.1785/0120140208

Baisch, S., Vörös, R., Weidler, R. & Wyborn, D. (2009) Investigation of fault mechanisms during geothermal reservoir stimulation experiments in the Cooper Basin, Australia. *Bull. Seism. Soc. Am.*, 99(1), 148–158. https://doi.org/10.1785/0120080055

Baisch, S., Weidler, R., Vörös, R., Wyborn, D. & DeGraaf, L. (2006) Induced seismicity during the stimulation of a geothermal HFR reservoir in the Cooper Basin, Australia. *Bull. Seism. Soc. Am.*, 96(6), 2242–2256. https://doi.org/10.1785/0120050255

Carvalho, F. & Labuz, J. F. (2002) Moment tensors of acoustic emissions in shear faulting under plane-strain compression. *Tectonophysics*, 356, 199–211.

Chen, Q., Ishida, T., Sasaki, S., Matsunaga, I. & Mizuta, Y. (1998) Influence of grain size of granite on hydraulic fracturing mechanism, *J. JSCE (Jpn. Soc. Civ. Eng.)*, 589(III-42), 179–194 (in Japanese with English abstracts)

Dale, T. N. (1923) The commercial granites of New England, I, U.S. *Geol. Surv. Bull.*, 738, 1–97.

Eberhardt, E., Stimpson, B. & Stead, D. (1999) Effect of grain size on the initiation and propagation thresholds of stress-induced brittle fractures. *Rock Mech. Rock Eng.*, 32(2), 81–99.

Evans, K. F., Moriya, H., Niitsuma, H., Jones, R., H., Phillips, W. S., Genter, A., Sausse, J., Jung, R. & Baria, R. (2005) Microseismicity and permeability enhancement of hydrogeologic structures during massive fluid injections into granite at 3 km depth at the Soultz HDR site. *Geophys. J. Int.*, 160, 388–412.

Hampton, J., Gutierrez, M., Matzar, L. Hu, D. & Frash, L. (2018) Acoustic emission characterization of microcracking in laboratory-scale hydraulic fracturing tests. *J. Rock Mech. Geotec. Eng.*, 10, 805–817. https://doi.org/10.1016/j.jrmge.2018.03.007

Hill, D. P. (1977) A model for earthquake swarms, *J. Geophy. Res.*, 82, 1347–1352.
Hubbert, M., K. & Willis, D., G. (1957) Mechanics of hydraulic fracturing. *Trans. Am. Inst. Min. Metall. Pet. Eng.*, 210, 153–168.
Hummel, N. & Shapiro, S. A. (2013) Nonlinear diffusion-based interpretation of induced microseismicity: a Barnett Shale hydraulic fracturing case study. *Geophysics,* 78(5), B211–B226. https://doi.org/10.1190/GEO2012-0242.1
Ishida, T. (2001) Acoustic emission monitoring of hydraulic fracturing in laboratory and field, *Constr Build Mater.*, 15, 283–295.
Ishida, T., Labuz, J. F., Manthei, G., Meredith, P. G., Nasseri, M. H. B., Shin, K., Yokoyama, T. & Zang, A. (2017) ISRM (International Society for Rock Mechanics) suggested method for laboratory acoustic emission monitoring, *Rock Mech. Rock Eng.*, 50(3), 665–674. https://doi.org/10.1007/s00603-016-1165-z
Ishida, T., Mizuta, Y., Matsunaga, I., Sasaki, S. & Chen, Q. (2000a) Effect of grain size in granitic rock on crack extension in hydraulic fracturing. In: *Proceedings of 4th North American Rock Mechanics Symposium*, Seattle, pp. 1105–1111.
Ishida, T., Sasaki, S., Matsunaga, I., Chen, Q. & Mizuta, Y. (2000b) Effect of grain size in granitic rocks on hydraulic fracturing mechanism. In: *Trends in Rock Mechanics (Proceedings of Sessions of Geo-Denver 2000), Geotechnical Special Publication No.102,* ASCE Press, Reston, VA, USA pp.128–139.
Kaieda, H., Fujimitsu, Y., Yamamoto, T., Mizunaga, H., Ushijima, K. & Sasaki, S. (1995) AE and mise-a-la-masse measurements during a 22 days water circulation test at Ogati HDR site, Japan. In: *Proceedings of World Geothermal Congress*, Florence, pp. 2695–2700.
Kao, C.S., Carvalho, F. C. S. & Labuz, J. F. (2011) Micromechanisms of fracture from acoustic emission. *Int. J. Rock Mech. Min. Sci.*, 48, 666–673.
Kirsch, C. (1898) Die Theorie der Elastizität und die Bedürfnisse der Festigkeitslehre. *Z. VDI*, 42, 797–807. (in German).
Lockner, D. & Byerlee, J. D. (1977) Hydrofracture in Weber sandstone at high confining pressure and differential stress. *J. Geophys. Res.*, 82, 2018–2026.
Majer, E. L. & Doe, T. W. (1986) Studying hydrofractures by high frequency seismic monitoring. *Int. J. Rock Mech. Min. Sci. & Geomech. Abstr.*, 23, 185–199.
Matsunaga, I., Kobayashi, H., Sasaki, S. & Ishida, T. (1993) Studying hydraulic fracturing mechanism by laboratory experiments with acoustic emission monitoring, *Int. J. Rock Mech. Min. Sci. & Geomech. Abstr.*, 30, 909–912.
Maxwell, S. C. & Cipolla, C. (2011) What does microseismicity tell us about hydraulic fracturing? In: *SPE Annual Technical Conference and Exhibition*, Denver, Colorado, USA, SPE 146932.
Naoi, M., Chen, Y., Nishihara, K., Yamamoto, K., Yano, S., Watanabe, S., Morishige, Y., Kawakata, H., Akai, T., Kurosawa, I. & Ishida, T. (2018) Monitoring hydraulically-induced fractures in the laboratory using acoustic emissions and the fluorescent method. *Int. J. Rock Mech. Min. Sci.*, 104, 53–63. https://doi.org/10.1016/j.ijrmms.2018.02.015
Pettitt, W. S. & King, M. S. (2004) Acoustic emission and velocities associated with the formation of sets of parallel fractures in sandstones. *Int. J. Rock Mech. Min. Sci.,* 41(suppl 1), 151–156.
Rothman, R. L., Greenfield, R. J. & Hardy, Jr. H. R. (1974) Errors in hypocenter location due to velocity anisotropy, *Bull. Seism. Soc. Am.,* 64, 1933–1996.
Rutledge, J., T., Phillips, W., S. & Mayerhofer, M., J. (2004) Faulting induced by forced fluid injection and fluid flow forced by faulting: an interpretation of hydraulic-fracture microseismicity, Carthage Cotton Valley gas field, Texas. *Bull. Seism. Soc. Am.,* 94(5), 1817–1830.
Sano, S., Kudo, Y. Mizuta, Y. (1992) Experimental determination of elastic constants of Oshima granite, Barre granite, and Chelmsford granite. *J. Geophys. Res.*, 97(B3), 3367–3379.
Sasaki, S. (1997) Microseismic activity induced during hydraulic fracturing experiment at the Hijiori hot dry rock geothermal energy site, Yamagata, Japan. In: *Proceedings of 4th International Symposium on Rockburst and Seismicity in Mines*, Krakow, pp. 403–407.
Sasaki, S. (1998) Characteristics of microseismic events induced during hydraulic fracturing experiments at the Hijiori hot dry rock geothermal energy site, Yamagata Japan. *Tectonophysics*, 289, 171–188.

Sasaki, S., Matsunaga, I., Kobayashi, H. & Ishida, T. (1988) *Mechanism of Fracturing Inferred from Acoustic Emission.* CRIEPI (Central Research Institute of Electric Power Industry), Abiko, Chiba, Japan, U88034 (in Japanese with English abstracts)

Sato, Y. & Katsuyama, K. (1985) Development of a system for the source location of AE events. *Mine Safety*, 31(4), 169–177 (in Japanese)

Stanchits, S., Surdi, A., Gathogo, P., Edelman, E. & Suarez-Rivera, R. (2014) Onset of hydraulic fracture initiation monitored by acoustic emission and volumetric deformation measurements. *Rock Mech. Rock Eng.*, 47, 1521–1532. https://doi.org/10.1007/s00603-014-0584-y

Takanami, T. & Kitagawa, G. (1988) A new efficient procedure for the estimation of onset times of seismic waves. *J. Phys. Earth.*, 36, 267–290.

Talebi, S. & Cornet, F. H. (1987) Analysis of the microseismicity induced by a fluid injection in a granitic rock mass. *Geophys. Res. Lett.*, 14, 227–230.

Vavryčuk, V. (2001) Inversion for parameters of tensile earthquakes. *J. Geophys. Res.*, 106, 16339–16355.

Vavryčuk, V. (2011) Tensile earthquakes: theory, modeling and inversion. *J. Geophys. Res.*, 116. https://doi.org/10.1029/2011JB008770

Vavryčuk, V. (2015) Moment tensor decomposition revisited. *J. Seismol.*, 19, 231–252.

Waldhauser, F. & Ellsworth, L.W. (2000) A double difference earthquake location algorithm: method and application to the Northern Hayward fault, California. *Bull. Seism. Soc. Am.*, 90, 1353–1368.

Yamamoto, K., Naoi, M., Chen, Y., Nishihara, K., Yano, S., Kawakata, H., Akai, T., Kurosawa, I. & Ishida, T. (2019) Moment tensor analysis of acoustic emissions induced by laboratory-based hydraulic fracturing in granite. *Geophys. J. Int.*, 216, 1507–1516. https://doi.org/10.1093/gji/ggy493

Chapter 3

Fracturing mechanism depending on viscosity of fracturing fluid

3.1 Introduction

The previous chapter discussed how the properties of rock texture affect the features of crack extension and the hydraulic fracturing (HF) mechanism, based on the results of laboratory experiments. On the other hand, the fracturing fluid's viscosity also influences the crack extension characteristics and the fracturing mechanism.

Field experience in petroleum engineering indicates that HF using high-viscosity fluid increases fracturing area and production (Howard and Fast, 1970). Thus, the viscosity of the fracturing fluid has often been used to control the growth of HF. The primary purposes of the fracturing fluid are as follows: (1) to wedge open and extend fractures hydraulically, and (2) to transport and distribute the proppant along the fracture. The first fluids used in HF operations were oil-based fluids, which were prepared by thickening gasoline with napalm, an aluminum fatty acid salt (Howard and Fast, 1970). Since then, in addition to oil-based fluids, water-based fluids, alcohol-based fluids, foams, and emulsion fluids have been developed and used. Recently, fracturing fluids have become more complex, containing various additives (Gidley et al., 1989).

On the other hand, as shown in the introduction of the previous chapter (Section 2.1), there is a conflict between tensile fractures suggested by theory (e.g., Hubbert and Willis, 1957; Haimson, 1978; Mizuta et al., 1987) and shear failure deduced from recorded waveforms of microseismicity (e.g., Baria and Green, 1986; Sasaki, 1995; Talebi and Cornet, 1987), and this conflict has not yet been resolved. Maxwell and Cipolla (2011) stated that the discrepancy between the theory and observed microseismicity is one of the paradoxes of HF.

The features of crack extension and the fracturing mechanism, depending on the viscosity of the fracturing fluid, were investigated in laboratory experiments by Matsunaga et al. (1993) and Sasaki (1995). They recorded AE (acoustic emission) waveforms using 16 sensors to locate the sources and obtain the fault plane solutions. From these solutions, they examined fracturing mechanisms in granite, marble, andesite, and acrylic resin with injections of water and viscous oil. Although they presented some interesting results, there was a lack of consistency due to the numerous variables involved. Chen et al. (1996) and Ishida et al. (1997, 2004) expanded on their studies by focusing on the effect of fracturing fluid itself and its viscosity. The experimental results are described in the next section of this chapter, "3.2 Effect of injected fracturing fluid and its viscosity on HF", which mainly follows the studies of Chen et al. (1996) and Ishida et al. (1996, 2004).

On the other hand, Zoback et al. (1977) demonstrated in their laboratory experiment that viscous fluid injection can generate a new crack in the direction of maximum compressive stress, independent of the pre-existing joint. These results provide valuable and important insights into actual operations of HF in fields with various pre-existing joints. However, the mechanism

DOI: 10.1201/9781003282631-3

of generating a new crack in a rock mass with a pre-existing joint has not been fully clarified. Building on the experiments by Zoback et al. (1977), Nakayama et al. (2002) and Ishida et al. (2003) conducted experiments to monitor AE associated with new crack generation in a granite specimen with a pre-existing joint. The experimental results are described in the third section of this chapter, "3.3 New crack generation by viscous fluid injection in a specimen with a pre-existing crack", following Nakayama et al. (2002) and Ishida et al. (2003).

3.2 Effect of injected fracturing fluid and its viscosity on HF

3.2.1 *Introduction*

One of the characteristic features of HF is that injected fluid invades the crack induced by the fracturing process, continuously applying pressure to the crack tip. As such, the viscosity of the fracturing fluid should affect both crack extension and the overall mechanism of HF. To investigate the effect of fluid viscosity, Ishida et al. (1996, 2004) conducted HF experiments on cubic granite specimens, using transmission oil for a car with a viscosity of 80 mPa·s and tap water with a viscosity of 1 mPa·s (1 mPa·s $= 1.0\times10^{-3}$ Pa·s $= 1.0\times10^{-2}$ dyn·s/cm^2 = 1 cP), comparing their results. In addition to these experiments, Chen et al. (1996) and Ishida et al. (1997, 2004) also investigated the effect of the fracturing fluid itself, specifically the pressure it applies to the crack tip, by conducting experiments to induce cracks through the pressurization of a urethane sleeve. This method allowed for the realization of dry HF without the use of fracturing fluid, under the same laboratory settings, The experimental results are described and discussed in this section, following the works of Chen et al. (1996) and Ishida et al. (1996, 1997, 2004).

3.2.2 *Specimen, experimental setup, and AE monitoring method*

3.2.2.1 *Specimen and experimental setup*

Small cubical specimens (19 cm × 19 cm × 19 cm) of Kurokami-Jima granite having a 2 cm diameter central hole were used as shown in Figure 3.1.

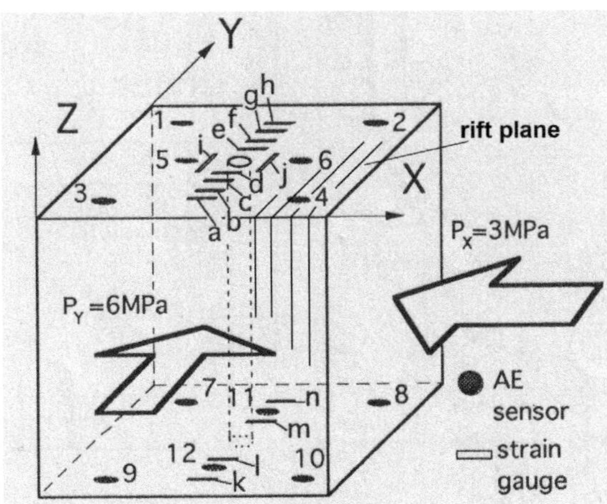

Figure 3.1 Specimen, loading condition, positions of AE sensors and strain gauges, and coordinate system. (Slightly modified Ishida et al., 1996.)

Three experiments were conducted at a constant injection rate using (i) water injection, (ii) viscous oil injection, and (iii) pressurization via a urethane sleeve. The granite specimen's inherent rift plane (see the last paragraph of Section 2.1 for more information on "rift plane") was oriented to correspond to the *YZ*-plane in the chosen Cartesian coordinate system. P-wave velocities measured along the *X*-, *Y*-, and *Z*-directions of the specimen for the water injection were 4.93, 5.14, and 5.20 km/s, those for the viscous oil injection were 4.93, 5.21, and 5.28 km/s, and those for the pressurization via a urethane sleeve were 4.98, 5.16, and 5.27 km/s, respectively. The slight differences among the specimens were probably due to the inherent specimen heterogeneity.

Twelve AE sensors (Type R-15, Physical Acoustic Corporation) were placed on the top and bottom surfaces of the specimen as shown in Figure 3.1. The sensor has a cylindrical shape measuring 1.74 cm in diameter and 1.63 cm in length.

As shown in Figure 3.1, strain gauges were also glued on two surfaces of a specimen to detect induced cracks. On the upper surface, eight strain gauges, denoted "a" through "h", were glued at 1.5 cm intervals from the hole along the *Y*-direction, because cracks were expected to extend in this direction. In the *X*-direction, two strain gauges denoted "i" and "j" were glued at 1.5 cm apart from the hole wall. On the lower surface, four strain gauges denoted "k" through "n" were glued at 3 cm intervals from the hole in the *Y*-direction. All strain gauges (Tokyo Sokki Co. Ltd., Type FLA-10-11) were 10 mm long and placed in a direction to measure tangential strains around the hole. If water or oil seeped into a gauge during injection, they would make the gauge electrically unstable and result in trouble with measurement; hence, all strain gauges are coated with impermeable material.

In all experiments, two horizontal confining pressures, 3 MPa in the *X*-direction and 6 MPa in the *Y*-direction, were applied to the specimen using flat jacks, as shown in Figure 3.2. To set AE

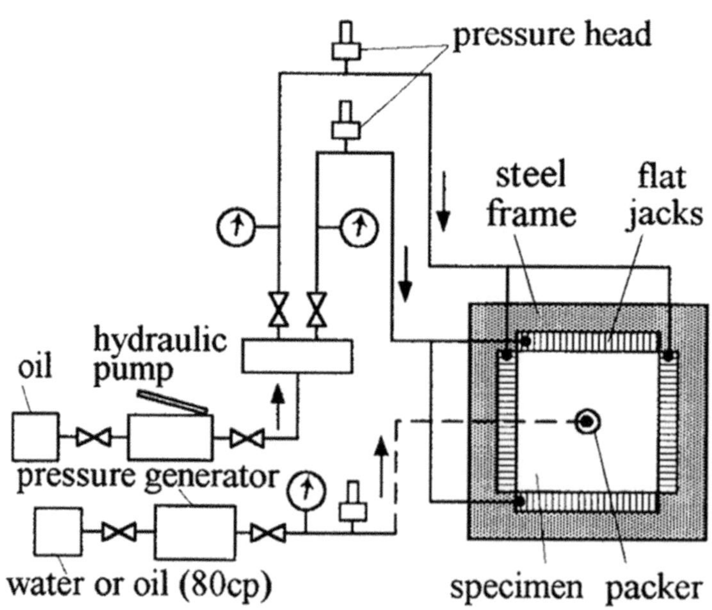

Figure 3.2 Loading system (Chen et al., 1996).

sensors, no confining pressure was applied in the Z-direction. Under such conditions, the linear elastic theory leads to tangential compressive stress along the wall of the hole, 15 MPa in the Y-direction, and 3 MPa along the X-direction (Kirsch, 1898). Since the rift plane is also oriented parallel to the YZ-plane, pressurizing the borehole should result in a planar fracture in that same plane.

A flat jack that we selected to apply confining pressure has the form of two deformable steel rectangular plates that are welded around the periphery. A feeder tube inserted in the edges of the plates allows pressurization with oil. Figure 3.3 shows the flat jack and the loading frame. The flat jacks can apply uniform loading to the specimen surfaces by fitting over the specimen surfaces with smooth

Figure 3.3 Flat jack and loading frame.

bending of the surfaces of the flat jack plates, even if the two specimen surfaces facing each other are not exactly parallel. When we compare the loading distribution between the plates and the flat jack loading using "Prescale", which is the film to measure pressure and pressure distribution provided by FUJIFILM Corporation (https://www. fujifilm.com/pl/en/business/inspection/measurement-film/prescale), we confirmed that the plate loading usually concentrates on one side of the specimen surface whereas the flat jack achieves uniform loading over the specimen surfaces. This is the reason we selected the flat jacks for the loading of confining pressure. Figure 3.4 shows a view of a specimen set in the loading frame with four flat jacks inserted between the specimen and the loading frame.

Figure 3.4 View of a specimen set in the loading frame with the four flat jacks between them (Ishida, 1999).

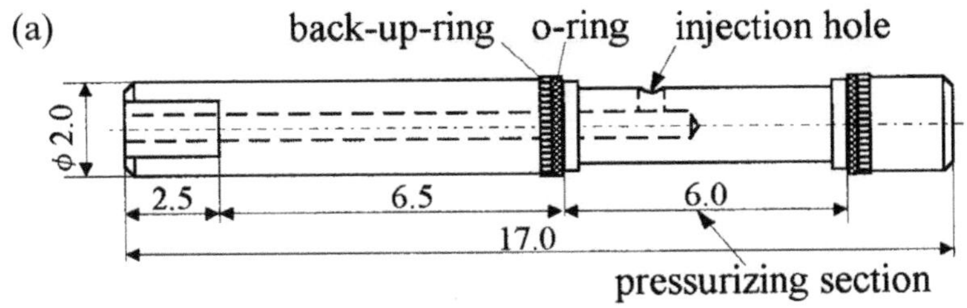

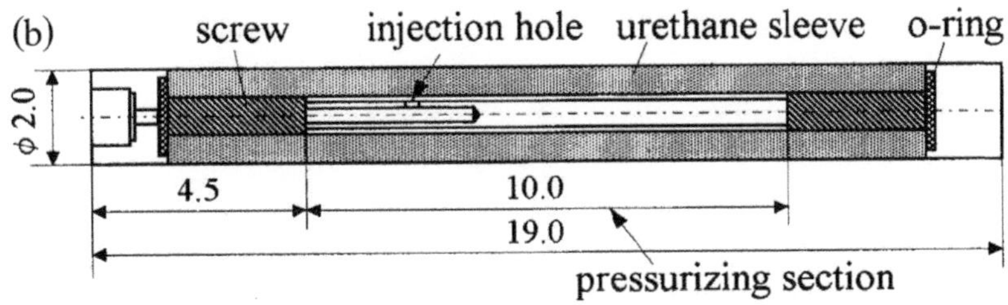

Figure 3.5 Packer and urethane sleeve inserted into the boreholes (unit: cm) (Chen et al., 1996). (a) Packet for water and oil injection. (b) Urethane sleeve.

(a)

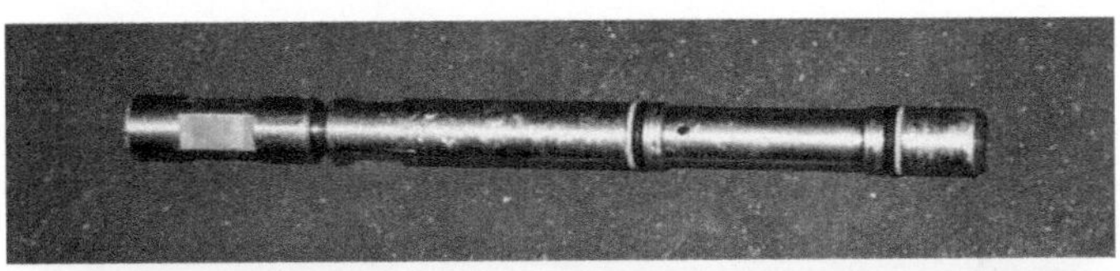

(b)

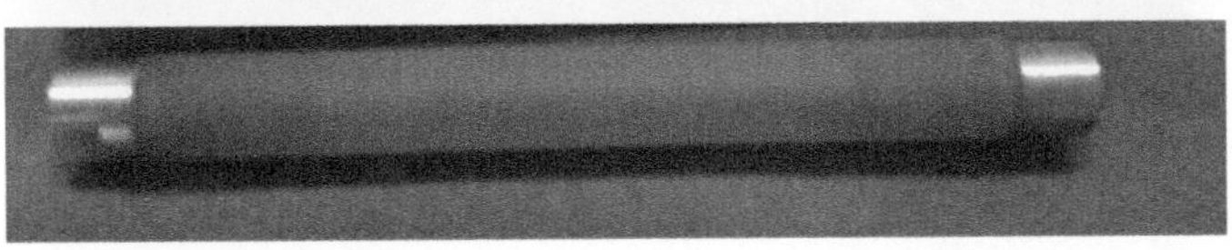

Figure 3.6 Photo of the packer and the urethane sleeve. (a) Packer for water and oil injection. (Ishida, 1999). (b) Urethane sleeve.

As shown in Figure 3.5a, a mechanical-type double packer was used to inject water and oil at a flow rate of approximately 10 mL/min within a 6.0 cm sealed-off section. The water has a viscosity of 1 mPa·s, while the oil has a viscosity of 80 mPa·s (transmission oil for a car, Mahha Super Hypoid Multi, GL-5 80W-90, Fuji Kosan Co. Ltd.). In the third test, a 10 cm long urethane sleeve, as shown in Figure 3.5b, was inserted into the hole and inflated with oil at about the same rate, prohibiting direct contact of the injected fluid with the borehole rock surface. Figure 3.6a shows the packer for water and oil injection, and Figure 3.6b shows the urethane sleeve.

The fluid injection was achieved by a hand-operated screw pump, while a hand pump was used to inflate the urethane sleeve. The borehole pressure and the flat jack pressures were monitored by a Bourdon gauge, and they were recorded at four-second intervals. Figure 3.7 shows a view of pressurizing injected fluid with the screw pump.

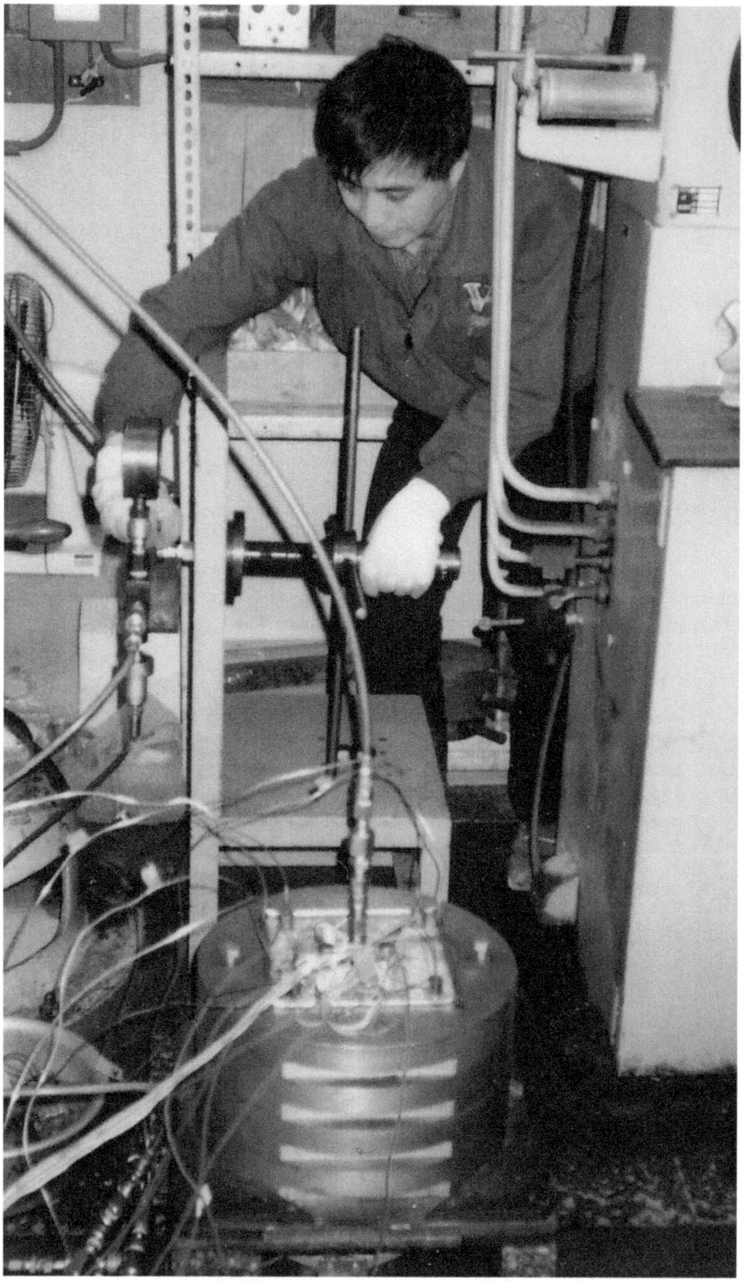

Figure 3.7 View of pressurizing injected fluid with a hand-operated screw pump (Ishida, 1999).

3.2.2.2 AE and strain monitoring method

Two independent systems were employed to monitor the AE events, one for recording and one for reproducing the signals, as shown in Figure 3.8. For the recording part, the AE signals detected at the sensor having a resonance frequency of 150 kHz were amplified by 70 dB through

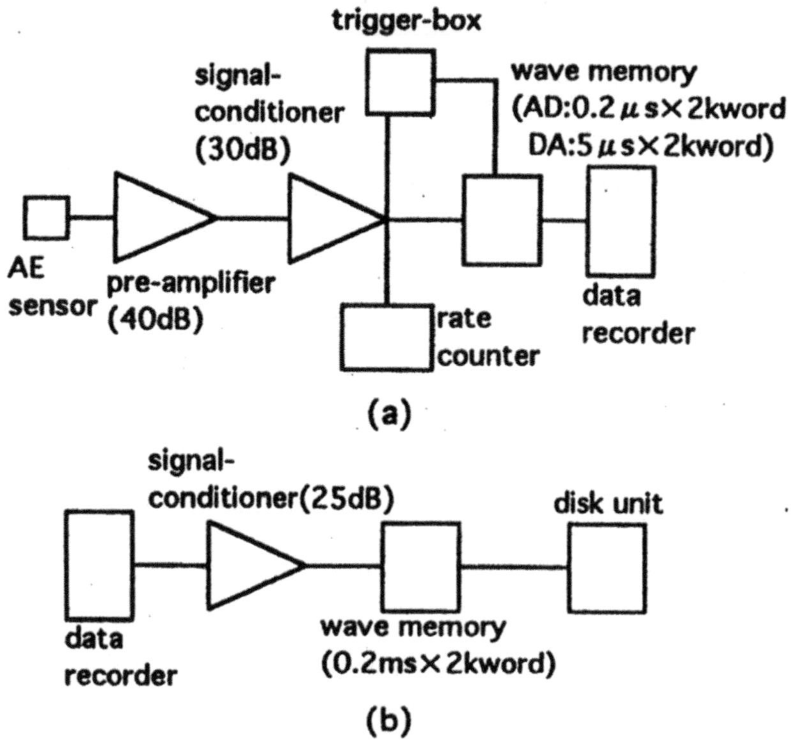

Figure 3.8 AE monitoring system. (a) Recording system. (b) Reproducing system (Chen et al., 1996).

a pre-amplifier and a signal conditioner. When a signal from one of the three sensors (1, 6, and 10 in Figure 3.1) exceeded a threshold level, a wave memory was triggered with a recording length of 2 kwords (kilo-words) and 8 bit resolution in amplitude. After the signal was digitized using a sampling interval of 0.2 μs (micro-second), an analog signal was produced again using a sampling interval of 5 μs and recorded on tape. This procedure was used to avoid dead time associated with the direct storage of digital data and to capture as many AE events as possible. The time axis of the signal waveforms was thus enlarged 25 times, and a 150 kHz sensor resonance frequency was converted to 6 kHz so that it could be recorded on the tape, as the frequency band available for the recorder was limited only from 0 Hz (direct current) to 20 kHz at its highest tape speed of 76 cm/s. After the experiments, the signals were reproduced from the tape, using the system shown in Figure 3.8b, at a speed of 1.2 cm/s, whereas that of the recording system was 76 cm/s. The time axis of the waveform was thereby further enlarged 64 (=76/1.2) times, in addition to the 25 times enlargement during the course of the A/D and D/A procedures performed in the wave memory. Therefore, the reproduced signals were 1,600 (=25 × 64) times slower than the real waveforms. The reproduced signals were digitized using a sampling interval of 0.2 ms and stored on floppy disks through a GP-IB cable. The signal conditioner and the memory have low-frequency characteristics and differ from those used in the recording system.

The strains measured at the strain gauges glued on the specimen's top and bottom surface were recorded at four-second intervals in a personal computer through the automatic digital strain meter.

3.2.3 *Results*

3.2.3.1 Pressure and AE count rate

Figure 3.9 shows pressure changes with time in the borehole and flat jack, and the associated AE count rates for the three experiments. The thick (bold) curves indicate the injected fluid pressures, *P*, while thin and dotted lines indicate the confining pressures of the flat jacks in two perpendicular directions, respectively; the bar diagrams show AE count rates taken every 30 s. It should be pointed out that the scales of the lateral and the left longitudinal axes are not the same for the three figures.

In Figure 3.9a, the pressure drops were recorded along the pressure curves as indicated by the open arrows. Pressurization often had to be stopped to supply additional fracturing fluid, because the hand-operated screw pump had only a small reservoir. In addition, pressurization had to be stopped every 330 s to replace a tape with a new one in the analog tape recorder for AE monitoring. All these interruptions caused large pressure drops, while the observed smaller jags were probably due to inherent variations in the manual control. The fluid injection was interrupted immediately after BD (breakdown), indicated by closed arrows. BD pressure is defined as the peak pressure immediately before the large sudden drop due to the leak of fracturing fluid into an induced HF crack. The rate of pressure decay immediately after BD was closely observed. After the pressure decrease had leveled off, the pressurization cycle was started again and repeated three times.

Although the open and closed arrows are not shown in Figure 3.9b, similar jags and pressure drops were observed here as well. Note that the lateral axis in the case of oil injection is three times larger due to the fact that the oil pressure needed much more time to decay after BDs.

In Figure 3.9a–b, the flat jack pressures, P_X, show some sudden increases corresponding to the BDs. These pressure increases, ranging from 0.3 to 0.5 MPa in the case of oil injection, are much larger than those found for water injection. With inducing cracks, the invasion of injected

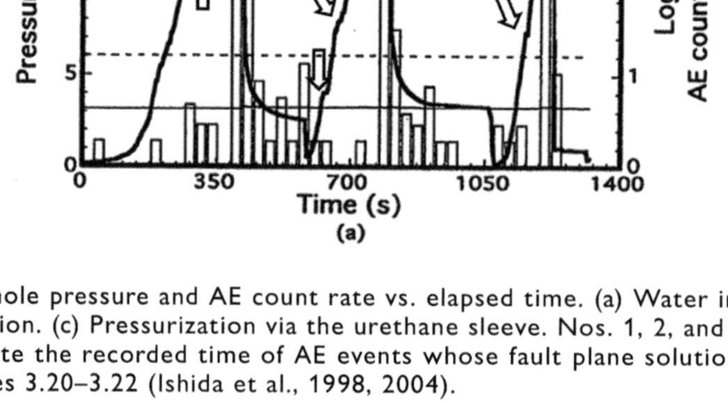

(a)

Figure 3.9 Borehole pressure and AE count rate vs. elapsed time. (a) Water injection. (b) Oil injection. (c) Pressurization via the urethane sleeve. Nos. 1, 2, and 3 in each figure indicate the recorded time of AE events whose fault plane solutions are shown in Figures 3.20–3.22 (Ishida et al., 1998, 2004).

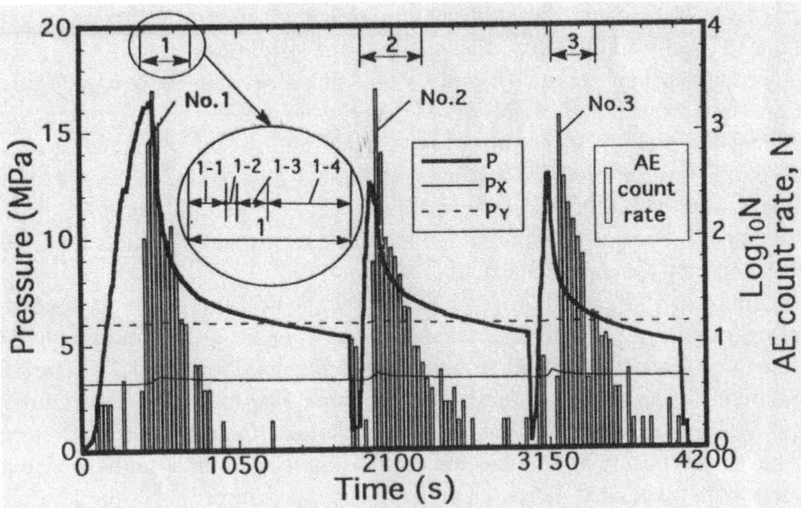

(b)

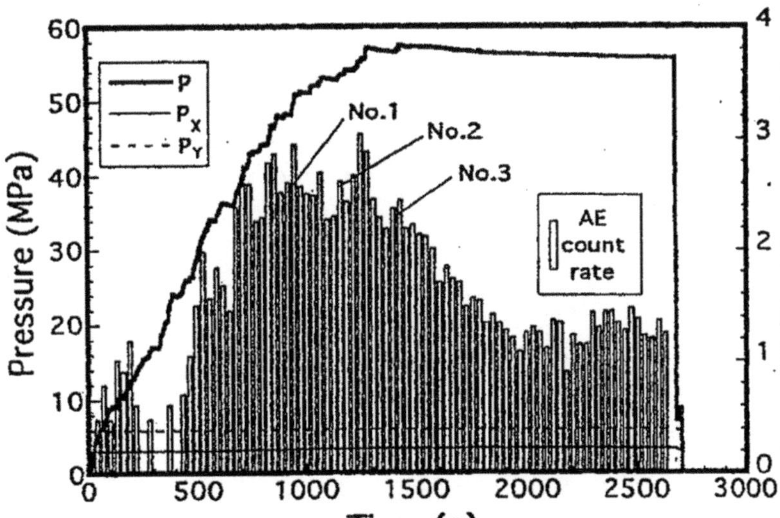

(c)

Figure 3.9 (Continued) Borehole pressure and AE count rate vs. elapsed time. (a) Water injection. (b) Oil injection. (c) Pressurization via the urethane sleeve. Nos. 1, 2, and 3 in each figure indicate the recorded time of AE events whose fault plane solutions are shown in Figures 3.20–3.22 (Ishida et al., 1998, 2004).

viscous oil most likely widened these induced cracks in the *Y*-direction more than in the case with water. This would expand the specimen in the *X*-direction and increase the flat jack pressure. As described later, this fact was corroborated by crack observations by the naked eye on the lateral surfaces as well as on the cores subsequently taken by co-axially drilling around the borehole (i.e., overcoring), and, in addition, strain changes measured by strain gauges glued along the induced cracks on the top surfaces of both the specimens.

On the other hand, when pressurizing via a urethane sleeve, as shown in Figure 3.9c, any remarkable pressure drops like those observed in the water and the oil injection never occurred. Some jags along the pressure curve were most likely due to stepwise crack extensions. The pressure curve indicates oil pressure applied inside the urethane sleeve. The magnitude of the pressure actually applied on the borehole wall is expected to be only about 88 % of this value due to the urethane sleeve participating in the load-bearing capacity (Ishida et al., 1997), based on the formulation by Crouch and Starfield (1983).

The bar diagrams in all figures (a–c) of Figure 3.9 indicate AE count rates per 30-s intervals, using a semilogarithmic scale. The rates were counted when a signal, amplified by 70 dB, from one among the three sensors (1, 6, and 10) exceeded a threshold level of 1.2 V. The bar diagrams for water and oil injection show substantial AE corresponding to the respective three BDs, suggesting intensive crack extension within these short periods. On the other hand, we cannot find such an AE burst when using a urethane sleeve. This indicates that a urethane sleeve extended cracks stepwise with associated increases in the injected pressure.

3.2.3.2 *AE sources and observed visible cracks*

AE hypocenters were located considering the anisotropy of the P-wave velocity (Rothman et al., 1974), with the accuracy expected to fall within 10 mm. This was possible because only AE events satisfying the following conditions were considered: (i) the P-wave arrival times could be read at six or more sensors, and (ii) the standard deviation of differences between the observed and back-calculated arrival times was within 3 μs. As shown in Figures 3.10–3.12, 139, 218, and 262 AE hypocenters were located in the courses of water injection, oil injection, and

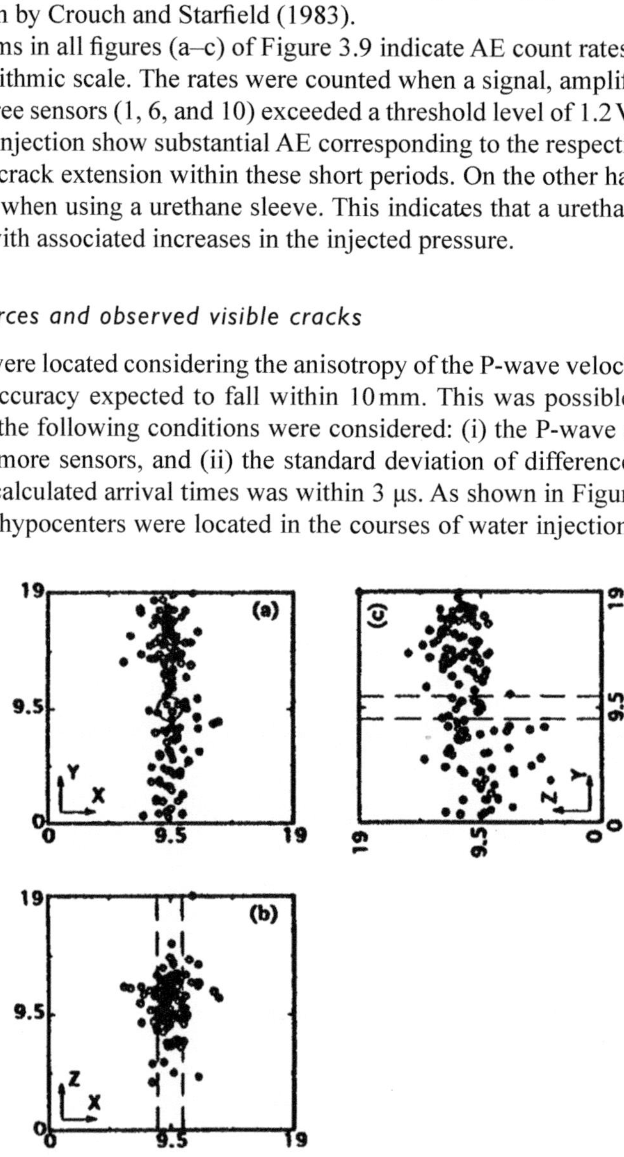

Figure 3.10 Location of AE hypocenters in water injection. Figures (a–c) show projections on *XY*, *XZ*, and *YZ* planes, respectively. Broken lines indicate the position of the borehole (Chen et al., 1996).

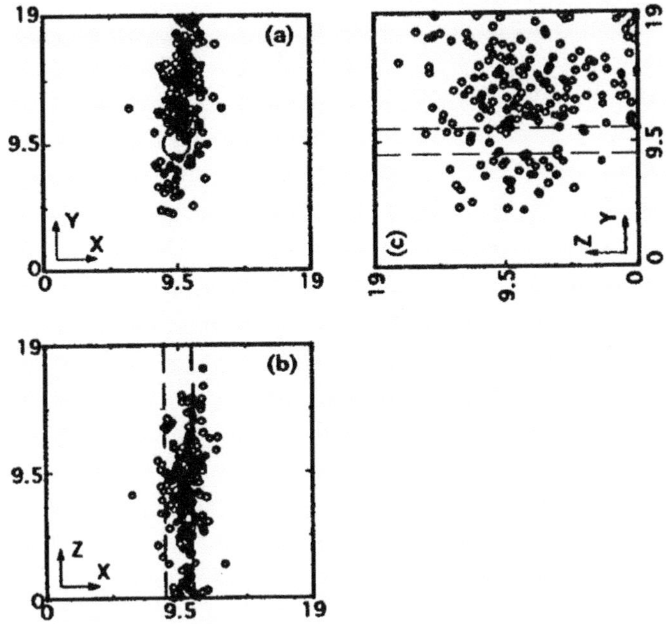

Figure 3.11 Location of AE hypocenters in oil injection. Projections on (a) *X-Y* plane, (b) *X-Z* plane, and (c) *Y-Z* plane (Ishida et al., 1996).

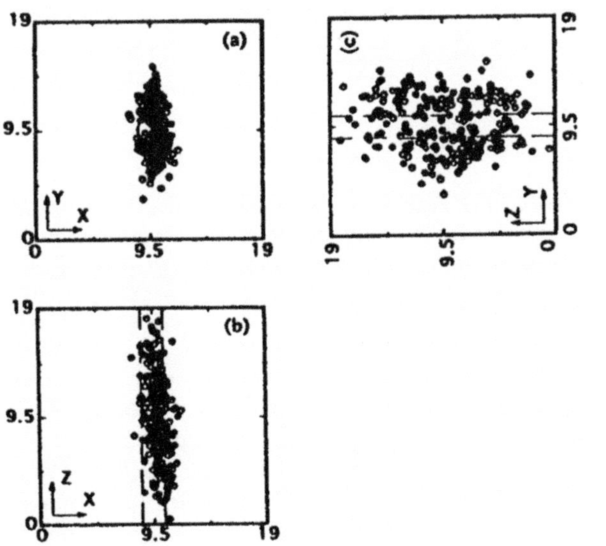

Figure 3.12 Location of hypocenters in pressurization via the urethane sleeve. Projections on (a) *X-Y* plane, (b) *X-Z* plane, and (c) *Y-Z* plane (Chen et al., 1996).

pressurization via the urethane sleeve, respectively. In all experiments, cracks extended along the *YZ*-plane, corresponding to the granite specimen's rift plane and also corresponding to the direction of maximum confining compressive stress.

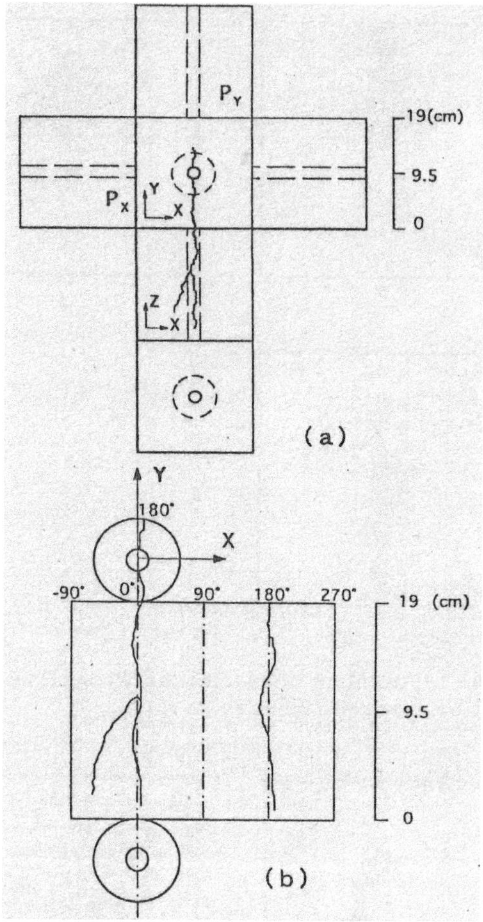

Figure 3.13 Sketches of visible cracks in water injection shown on unfolded planes. Upper figure (a) shows cracks observed on the surfaces of the specimens. Broken-line circles indicate the positions of cores drilled for inspection of cracks in the specimens. The lower figure (b) shows cracks observed on the surfaces of these extracted cores. (Chen et al., 1996).

As seen in the *XY*-plane projections of Figures 3.10a, 3.11a, and 3.12a, the AE hypocenters due to water injection spread from the hole toward two opposite surfaces; in the case of oil injection, they spread to only one surface. In contrast to these, the use of a urethane sleeve concentrated the AE hypocenters to the vicinity of the hole, although the maximum applied borehole pressure was around three times larger than those of the water and the oil injection.

Figures 3.13–3.15 show visible cracks observed on the surfaces of the specimens and the surfaces of cores subsequently taken by co-axially drilling around the hole to inspect cracks extending inside the specimens. As can be seen, all visible cracks correspond to the distributions of the located AE hypocenters. Cracks induced by oil injection were clear to the naked eye due to their large aperture; in contrast, cracks by water injection and via pressurization of the urethane sleeve could only be detected after careful and close observation due to their small aperture.

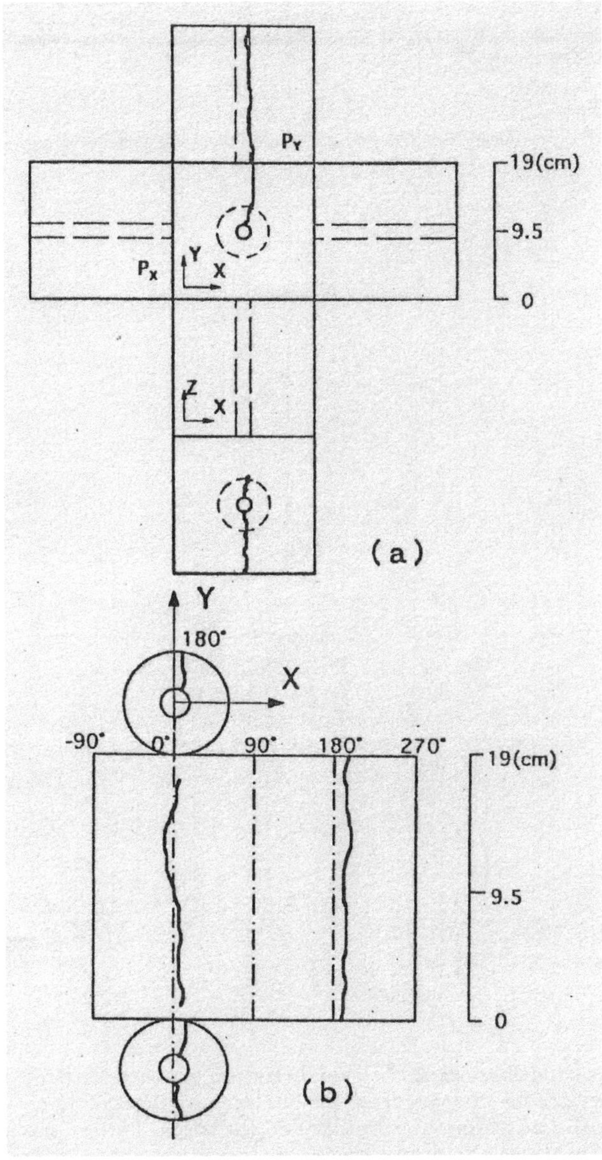

Figure 3.14 Sketches of visible cracks in oil injection Upper figure (a) shows cracks observed on the surfaces of the specimens. Broken-line circles indicate the positions of cores drilled for inspection of cracks in the specimens. The lower figure (b) shows cracks observed on the surfaces of these extracted cores (Ishida et al., 1996).

3.2.3.3 *Strain change*

Since cracks were induced from the hole in the *Y*-direction as expected, the four gauges denoted "e", "f", "g", and "h" in Figure 3.1 were properly placed to straddle the crack induced visible on the top surface in the water and oil injection. Figure 3.16 shows the temporal strain changes

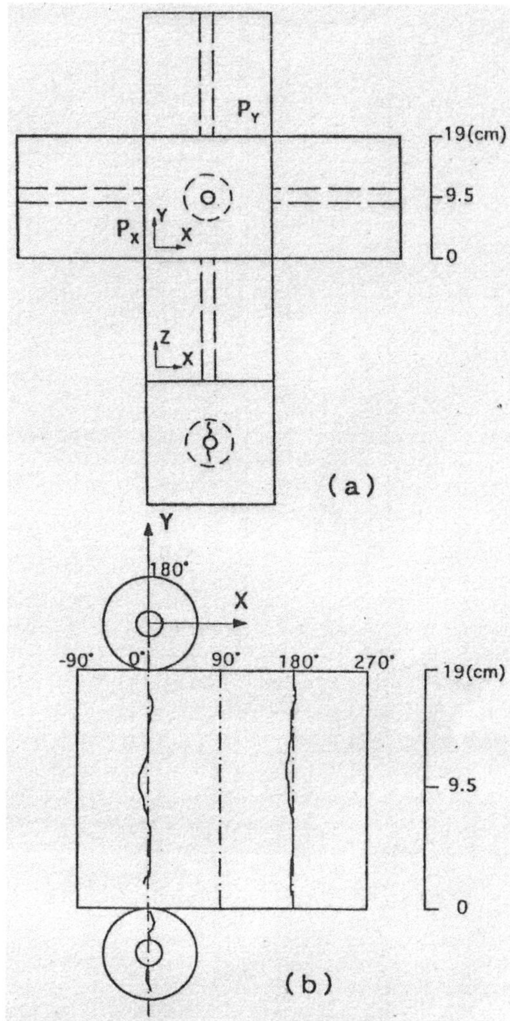

Figure 3.15 Sketches of visible cracks in pressurization via the urethane sleeve. Upper figure (a) shows cracks observed on the surfaces of the specimens. Broken-line circles indicate the positions of cores drilled for inspection of cracks in the specimens. The lower figure (b) shows cracks observed on the surfaces of these extracted cores (Chen et al., 1996).

measured with the gauges, and the strains were always kept in tensile (Ishida et al., 1996). Here, the stain changes in pressurization via the urethane sleeve are not shown because notable strain changes were not recorded, corresponding to the fact that no visible crack was observed on the top surface, as shown in Figure 3.15. In the water and oil injection, three strain peaks were recorded corresponding to BDs, that is, three drops of injected pressures, accompanying the crack generation or extension.

Figure 3.16a shows the strain change with time in the case of the water injection. This indicates that the closer the gauge position to the hole, the greater the strain magnitude. This is

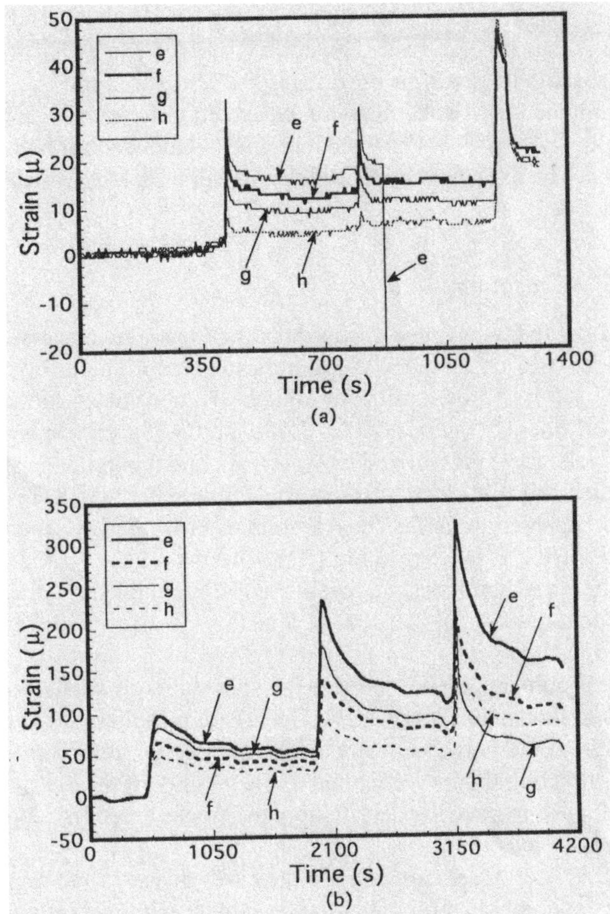

Figure 3.16 Strain changes measured with the gauges glued to straddle the crack on the surface. (a) Water injection. (b) Oil injection. (Refer to the gauge positions in Figure 3.1 and the visible cracks observed on the surfaces in Figures 3.13 and 3.14) (Ishida et al., 1996).

reasonable because the closer to the hole, the larger the crack aperture becomes. Only gauge "e" scaled out in a compressive direction after showing the second peak. Probably, water seeped through the induced cracks and wet the strain gauge, resulting in a drop in its electric resistance. In Figure 3.16b for the oil injection, the longitudinal axis is seven times larger and the lateral axis is three times larger than those for water injection. The order of the strain magnitudes almost follows the case with the water injection, although the strain gauge, "g", switched places with "f" from the first peak to the third peak, while "g" changed places with "h" after the third peak.

Comparing the magnitudes of the three peaks, the second peak is larger than the first peak, and the third peak is larger than the second peak. This indicates that the crack aperture increased step by step with the three pressure drops. Comparing the peak magnitudes between the water and oil injection, those for oil are three to seven times larger than those for water.

Although the strain changes measured with gauges, "a", "b", "c", and "d", which were glued on the opposite side for the hole, are not shown, they showed somewhat complicated behavior and did not show the same tendency as the gauges, "e", "f", "g", and "h". In oil injection, the difference can be explained by the fact that cracks extend only to the side of the latter gauges, as described later. However, with water injection, cracks extended on both sides alternatively; the difference could not be explained by the crack extension feature, which was different in the case of oil injection.

3.2.3.4 *Change of AE locations with time*

Figures 3.17–3.19 show the *X*-, *Y*-, and *Z*-coordinates of the sources versus elapsed time. The broken lines in Figures 3.17a, 3.18a, and 3.19a indicate *X*-coordinates of the central borehole, and those in Figures 3.17b, 3.18b, and 3.19b indicate *Y*-coordinates of that. The sections between the two broken lines in Figures 3.17c, 3.18c, and 3.19c correspond to the sections in which the borehole walls were pressurized by water, oil, and the urethane sleeve, respectively.

In the case of water and oil injection, shown in Figures 3.17–3.18, respectively, bursts of AE were found, accompanied by BDs. The AE sources moved away from the borehole in the *Y*-direction with any of the respective three BDs. The movement in the *X*-direction was restricted to the vicinity of the borehole, and that in the *Z*-direction was restricted to the proximity of the pressurized section; although for the oil injection, it was a little larger in the negative *Z*-direction, indicating a crack extension along the *YZ*-plane.

Pressurization via a urethane sleeve led to AE hypocenters distributed along the time axes without AE bursts, as shown in Figure 3.19. The AE hypocenters moved in the *Y*-direction to more than ± 5 cm from the borehole, while in the *X*-direction they were restricted to only $\pm$ 1.5 cm, and in the *Z*-direction they were limited to the vicinity of the pressurized section. Without contradiction to the AE hypocenter distribution, the cracks extended along the *YZ*-plane, like the cases of water and oil injection.

The following differences are remarkable. In the cases of water and oil injection, AE hypocenters spread from the borehole throughout the specimen, within short periods corresponding to the respective BDs. In contrast, in the case of pressurization via a urethane sleeve, when the fracturing fluid is not allowed to invade the fracture, AE hypocenters spread stepwise but remained in the vicinity of the hole without AE burst or BD.

3.2.3.5 *Difference between water and viscous oil in an alternative change of crack extending direction*

In the case of water injection, the AE hypocenters moved away from the borehole, in the negative *Y*-direction with the first and second BD, and switched to the positive *Y*-direction with the third BD, as shown in Figure 3.17b. In contrast to this, when viscous oil was injected, the crack extending direction never changed and was always in the positive *Y*-direction, as shown in Figure 3.18b.

The changes of a crack extending direction in water injection are more clearly shown in Figure 3.20, projecting locations of AE hypocenters during the respective three BDs, BD1, BD2, and BD3, on the *XY*-plane (Ishida et al., 1998). The crack extended in the minus *Y*-direction during BD1 and BD2, whereas in the plus *Y*-direction during BD3, which corresponds to that shown in Figure 3.17b. Figure 3.21 shows AE sources during BD1 into three smaller time steps, BD1-1, BD1–2, and BD1–3, along the elapsed time, to investigate the changes of the

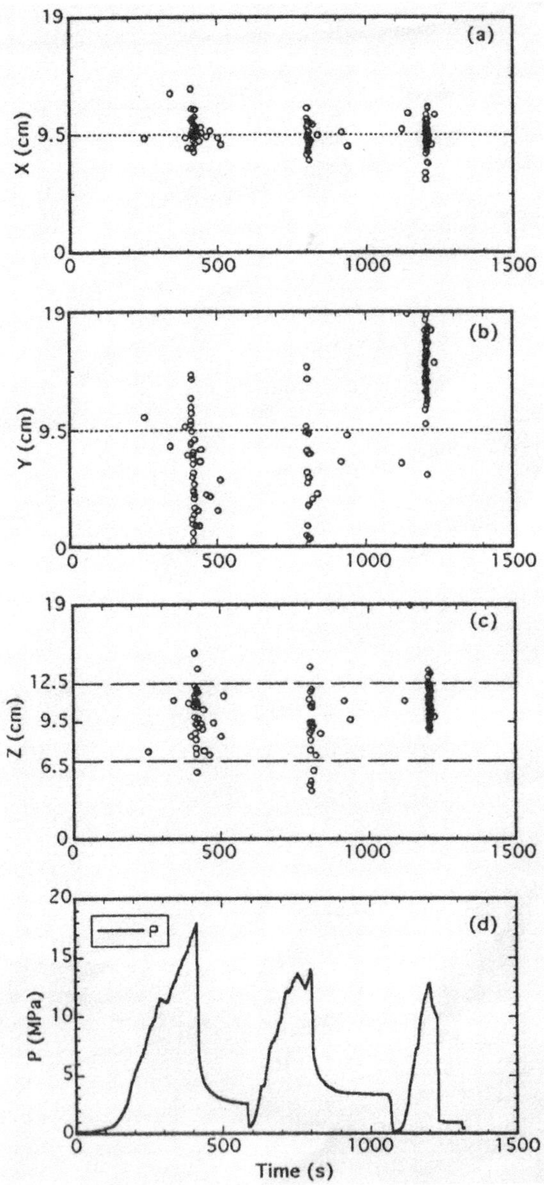

Figure 3.17 Time-space distributions of AE hypocenters in water injection. Three figures (a–c), respectively, show movements of AE hypocenters in *X*-, *Y*-, and *Z*-directions versus elapsed time. Figure (d) shows pressure vs. elapsed time, which is the same as Figure 3.9a, for facilitating comparisons with the movements of the AE hypocenters (Chen et al., 1996).

crack extending direction within the period of BD1. This reveals that, even within the period of BD1, AE hypocenters distributed in the plus *Y*-direction during BD1-1, while they spread in the minus *Y*-direction during BD1–2 and BD1–3. The alternative change in the crack extending

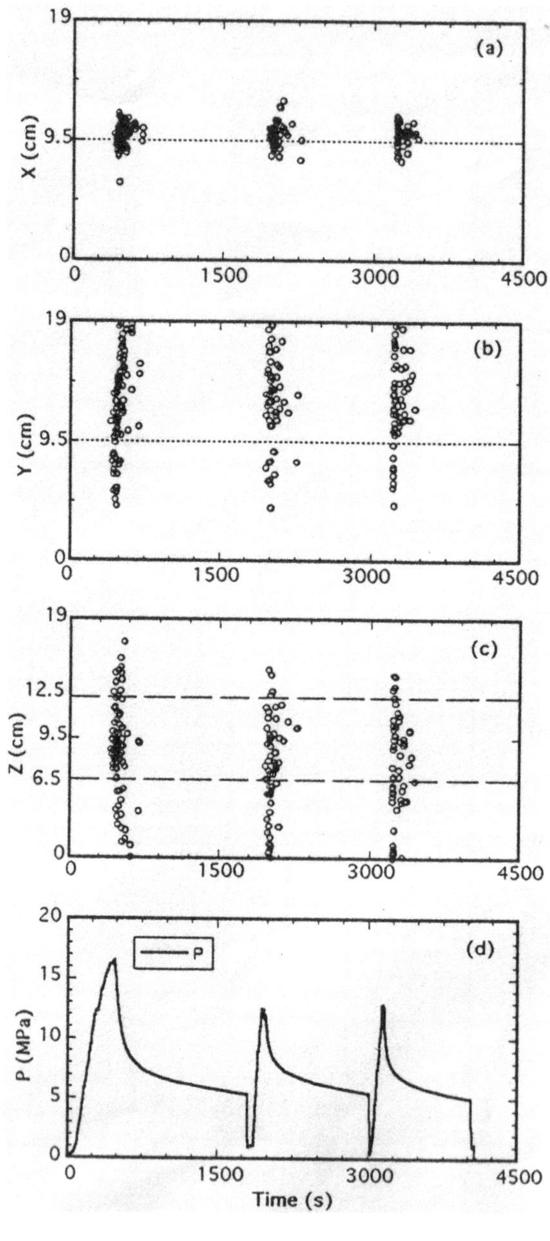

Figure 3.18 Time-space distributions of AE hypocenters in oil injection. Three figures (a–c), respectively, show movements of AE hypocenters in *X*-, *Y*-, and *Z*-directions versus elapsed time. Figure (d) shows pressure vs. elapsed time, which is the same as Figure 3.9b, for facilitating comparisons with the movements of the AE hypocenters (Ishida et al., 1998).

direction was also observed within a period of BD2, although it was not observed within that of BD3. The fact indicates that, with water injection, the crack extending direction tends to change easily with the respective BDs and within a period of each BD.

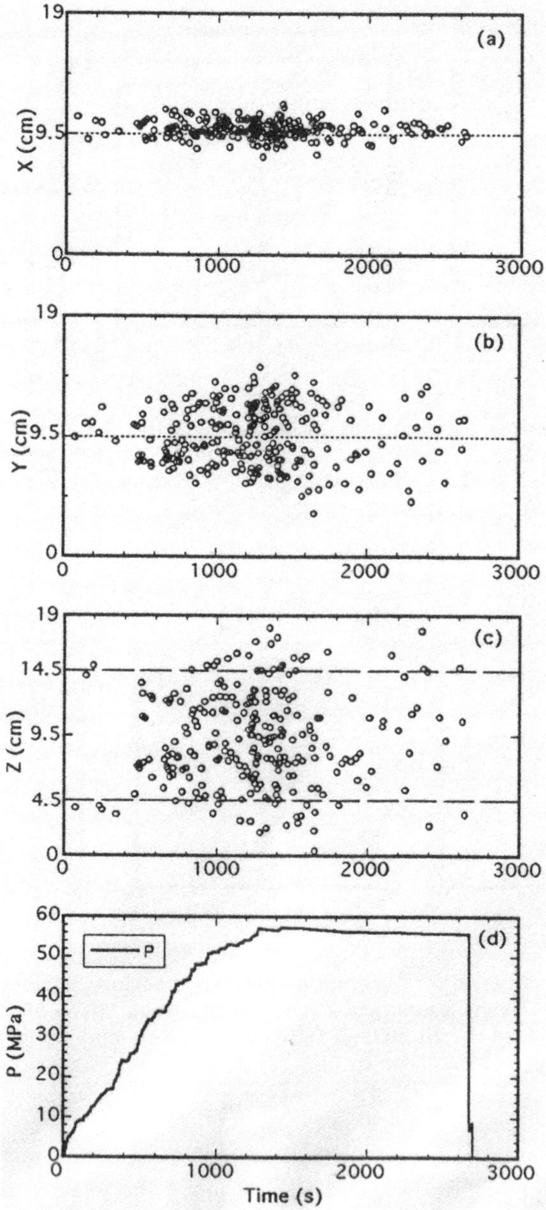

Figure 3.19 Time-space distributions of AE hypocenters in pressurization via the urethane sleeve. Three figures (a–c), respectively, show movements of AE hypocenters in *X*-, *Y*-, and *Z*-directions versus elapsed time. Figure (d) shows pressure vs. elapsed time, which is the same as Figure 3.9c, for facilitating comparisons with the movements of the AE hypocenters (Chen et al., 1996).

Similar to those for water injection, Figure 3.22 shows AE source locations projected on the *XY*-plane during the respective three BDs, BD1, BD2, and BD3, in oil injection (Ishida et al., 1998). The crack extending direction never changed and was always in the positive

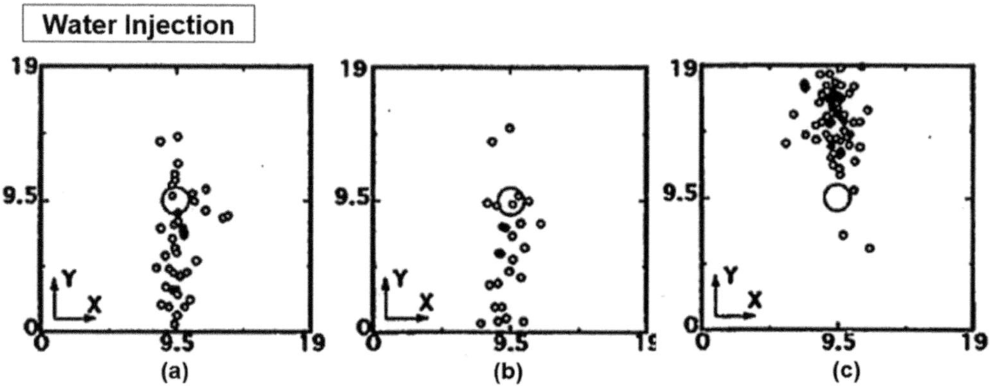

Figure 3.20 AE hypocenter distributions in water injection projected on the *XY*-plane with respective BDs. (Time spans are shown in the upper part of Figure 3.9a); (a) BD1 (254.87–425.65 s), (b) BD2 (801.82–808.26 s), (c) BD3 (1122.1–1234.5 s) (Ishida et al., 1996).

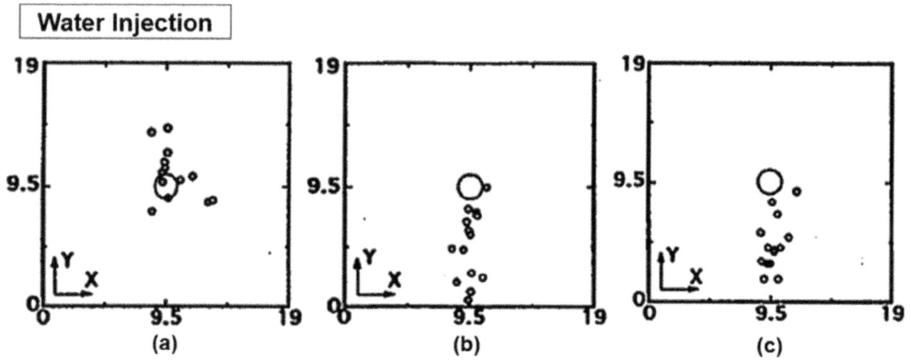

Figure 3.21 AE hypocenter distributions in water injection projected on the *XY*-plane within BD1 (time spans are shown in the upper part of Figure 3.9a); (a) BD1-1 (254.87-416.24 s), (b) BD1-2 (430.45-426.63 s), and (c) BD1-3 (417.42-425.97 s) (Ishida et al., 1996).

Y-direction for all three BDs. Figure 3.23 shows AE sources during BD1 into four smaller time steps, BD1-1, BD1–2, BD1–3, and BD1–4 along the elapsed time. From the figure, we can see the following features: AE hypocenters distributed only around the borehole in BD1-1, whereas they spread to the plus *Y*-direction largely than to the minus in BD1–2; in BD1–3, they distributed throughout the span between the borehole and the plus side surface, and in BD1–4, they mostly distributed near the plus side surface. The same tendency was also observed within periods of BD2 and BD3. The facts indicate that, in contrast to water injection, in oil injection, the crack extending direction neither changes with the respective BDs nor within a period of each BD.

The difference between water and oil injection in the alternative change of a crack extending direction is most likely attributed to the difference in viscosity between water and oil.

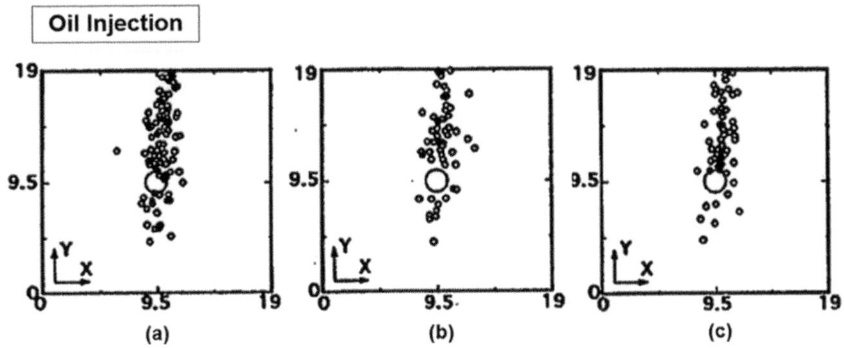

Figure 3.22 AE hypocenter distributions in oil injection projected on the *XY*-plane with respective BDs. (Time spans are shown in the upper part of Figure 3.9b); (a) BD1 (416.26–708.03 s), (b) BD2 (1961.7–2278.4 s), and (c) BD3 (3223.5–3442.1 s) (Ishida et al., 1996).

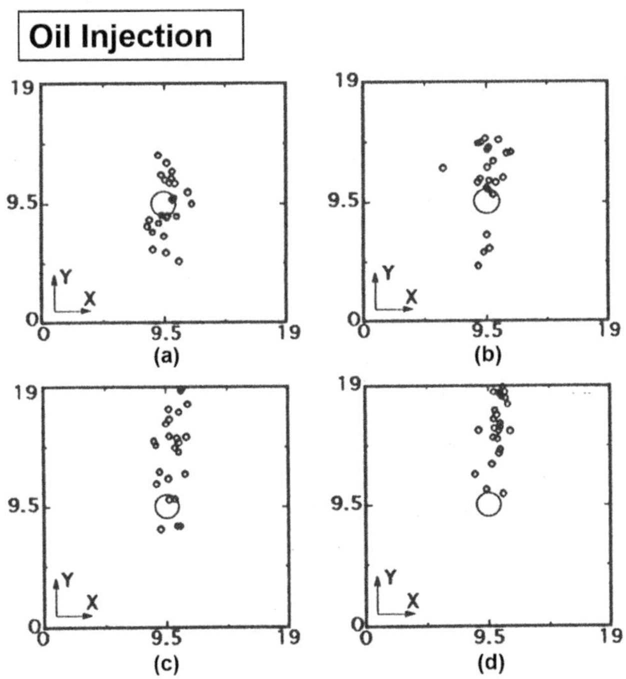

Figure 3.23 AE hypocenter distributions in oil injection projected on the *XY*-plane within BD-1 (time spans are shown in the upper part of Figure 3.9b); (a) BD1-1 (416.26–471.00 s), (b) BD1–2 (471.66–482.91 s), (c) BD1–3 (483.37–528.63 s), and (d) BD1–4 (529.03–708.03 s) (Ishida et al., 1996).

The granite specimen consists of mineral grains like felspar, quartz, mica, and others and has various microcracks and weak planes between the grains and within the grains themselves. On the other hand, the crack observation with the naked eye, the strain gauge measurements, and pressure changes in the flat jacks at BDs all indicated that the oil injection generated cracks with a wider aperture than the water injection. Thus, it is considered that viscous oil fracturing tends to generate thick and planar cracks with few branches, ignoring various microcracks and weak planes between the grains and within the grains themselves, while water tends to generate thin and wavelike cracks with many secondary branches affected by them. Considering that crack extension in HF likely starts at a defect having low strength in rock, it seems reasonable that a viscous oil injection tended not to change the crack extending direction; in contrast, a water injection tended to change it easily.

3.2.3.6 *Fault plane solutions of AE events*

Only approximately 5% of the located AE events had polarities of P-wave initial motions that were clearly distinguished between compression and dilatation at six or more sensors. Thus, only for seven events in each experiment, we examined their fracturing mechanism through fault plane solution based on the polarities of P-wave initial motions projected onto a stereo net. Figures 3.24–3.26 show three typical examples of them for each experiment. The recorded time of each AE event is shown in Figure 3.9 with the labels 1, 2, and 3, indicating a respective bar in which the AE event was counted. Figure 3.9a–b lets us know that the AE events labeled as Nos. 1, 2, and 3 in water and oil injections were all associated with the BDs. Amplitudes of waveforms were normalized to the maximum amplitude of each waveform to show their first motion polarities as clearly as possible.

For the case of water injection, shown in Figure 3.24, all the fault plane solutions indicated a quadrant-type mechanism, thereby implying shear fracturing. To check the validity of the solutions, let us examine the directions of nodal planes and the directions of the P- and T-axes shown in Figure 3.24a. The letter positions, “P” and “T”, indicate the directions of the P- and T-axes, which correspond the directions of the maximum compression and the minimum compression associated with the fracture inducing the AE event, respectively. In this solution, a vertical plane delineated as a line segment labeled “A” and a horizontal plane labeled “B” could be chosen as the two nodal planes. One out of them corresponds to the “real” fracture plane causing the AE event. Although the macroscopic crack extended along a vertical *YZ*-plane, it is actually composed of many small jagged planes. A similar macroscopic crack extension composed of many small jagged planes is also observed in a three-point-bending test (Kao et al., 2011) as discussed in Section 2.5.2, “Mechanism of shear fracturing”, in the previous chapter. Consequently, one could assume that the AE event was caused on the vertical plane A, to form an angle of 38 degrees to the *Y*-direction. In this experiment, a pressure of 6 MPa was applied in the *Y*-direction, 3 MPa in the *X*-direction, and no pressure in the *Z*-direction. If the size of the specimen is large enough in comparison with the borehole diameter and a rift plane did not affect the crack extension, a crack initiating along a vertical *YZ*-plane in the vicinity of the borehole wall should re-orient itself at a distance from the borehole and extend along a horizontal *XY*-plane normal to the *Z*-direction of the minimum principal stress in which no confining pressure was applied. Under such a crack extension condition, the P-axis of an AE event at a distance from the borehole should be in the *Y*-direction, whereas the T-axis, in the *Z*-direction. However, because the specimen size was limited and the rift plane was seated in the vertical *YZ*-plane, a macroscopic crack extended along the *YZ*-plane to the side surfaces of the specimen without re-orientation. Similar discussions can be carried out for the fault plane solutions of Figure 3.24b–c.

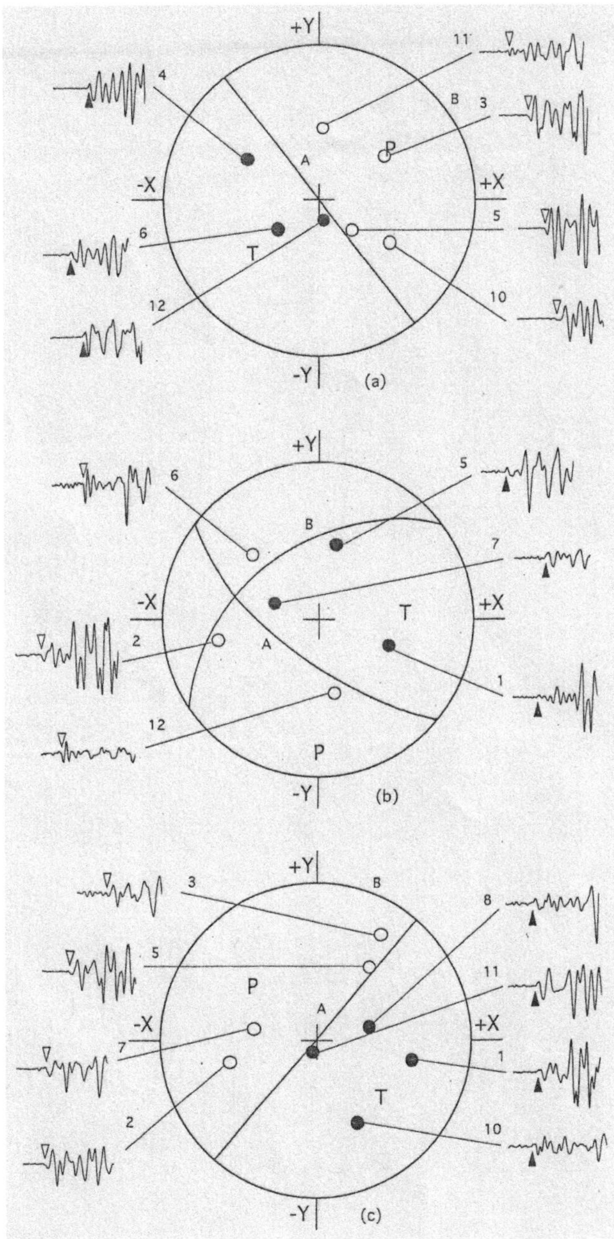

Figure 3.24 Fault plane solutions of AE events in water injection. (a) No. 1 event, (*X*, *Y*, *Z*; *T*) = (10.33, 4.27, 10.99; 424.92). (b) No. 2 event, (*X*, *Y*, *Z*; *T*) = (9.16, 15.58, 11.44; 1207.92). (c) No. 3 event, (*X*, *Y*, *Z*; *T*) = (10.00, 15.46, 12.43; 1209.04). The *X*-, *Y*-, and *Z*-coordinates indicate the location of an AE source (unit: cm), and *T* indicates the occurrence time along the elapsed time (unit: s). The fault plane solutions are projected onto a lower hemisphere of a Schmidt net. The solid circles indicate compression in P-wave initial motion, whereas the open circles indicate dilatation (Chen et al., 1996).

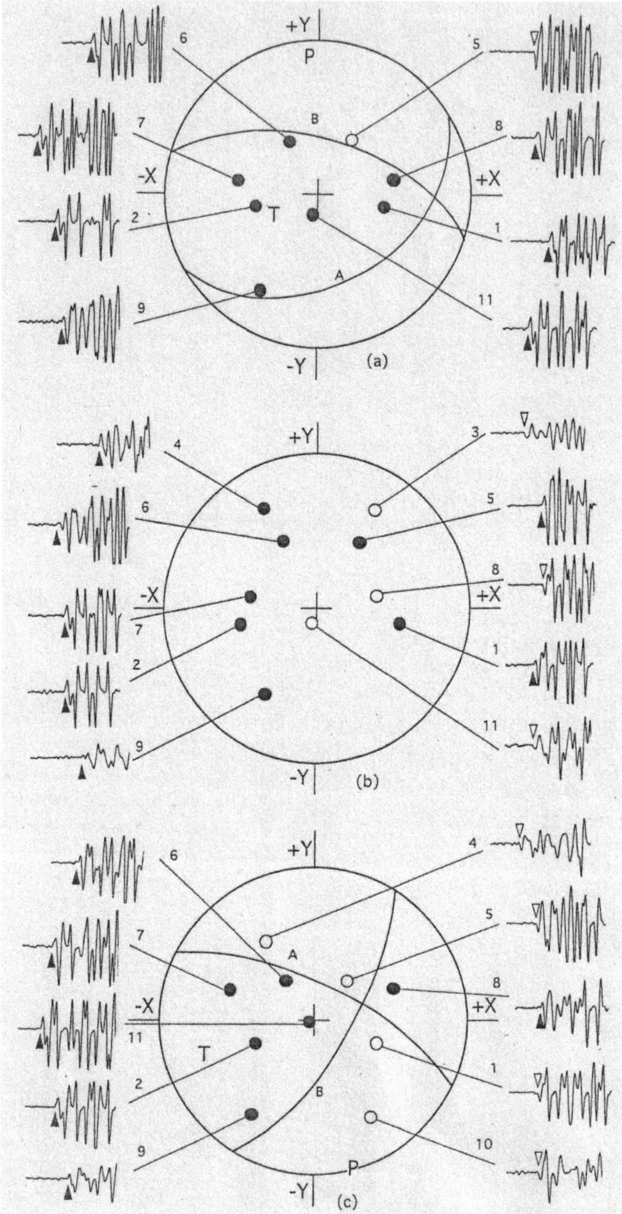

Figure 3.25 Fault plane solutions of AE events in oil injection. (a) No. 1 event, (*X*, *Y*, *Z*; *T*) = (9.75, 12.22, 11.68; 890.37). (b) No. 2 event, (*X*, *Y*, *Z*; *T*) = (9.53, 12.23, 14.39; 1480.03). (c) No. 3 event, (*X*, *Y*, *Z*; *T*) = (9.73, 7.40, 4.51; 1169.50) (Ishida et al., 1996).

Although pressurization via a urethane sleeve induced cracks only in the vicinity of the borehole, these cracks extended along a vertical *YZ*-plane without bending, similar to the case of water injection. Thus, using a discussion similar to the case of water injection, the fault plane solutions in Figure 3.26 are also validated, thereby implying the dominance of shear fracturing.

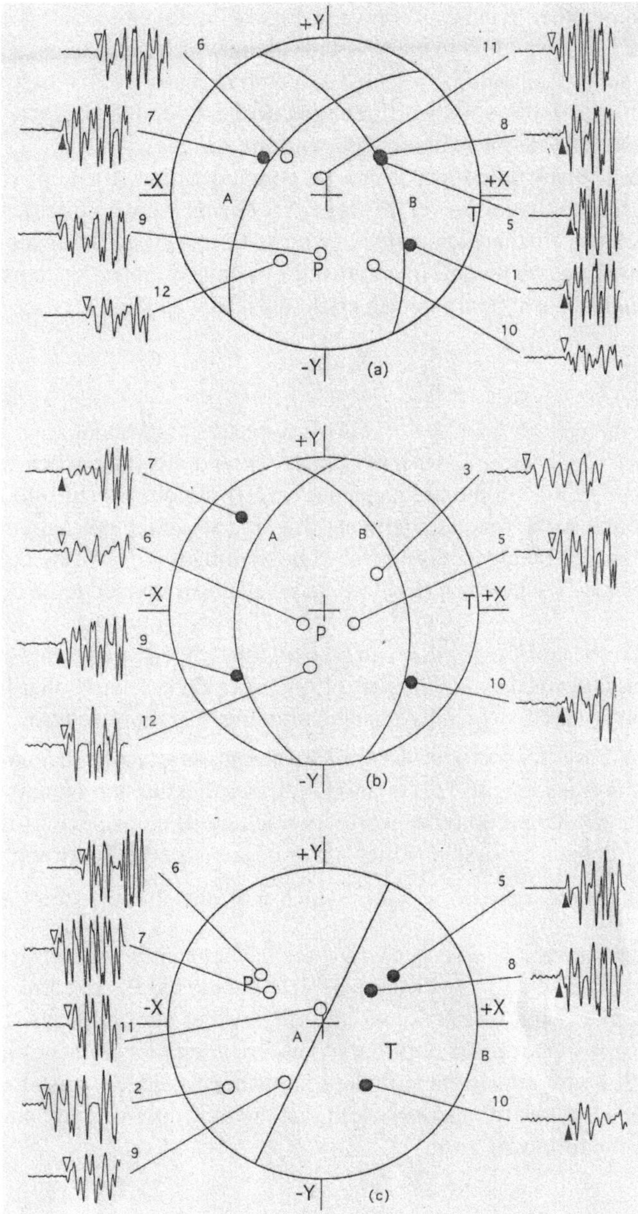

Figure 3.26 Fault plane solutions of AE events in pressurization via the urethane sleeve. (a) No. 1 event, (*X*, *Y*, *Z*; *T*) = (9.75, 12.22, 11.68; 890.37). (b) No. 2 event, (*X*, *Y*, *Z*; *T*) = (9.73, 7.40, 4.51; 1169.50). (c) No. 3 event, (*X*, *Y*, *Z*; *T*) = (9.53, 12.23, 14.39; 1480.03) (Chen et al., 1996).

In contrast, for the case of oil injection, as shown in Figure 3.25, many sensors detected compression in P-wave initial motions rather than dilatation, thereby implying the dominance of tensile fracturing. For some events, as shown in Figure 3.25b, the dominance of compression in P-wave initial motions makes it difficult to obtain two nodal planes based on a shear-type mechanism. For some other events, as shown in Figure 3.25a, although two nodal planes were obtained, neither of them was close to the *YZ*-plane along which the macroscopic crack extended; as a result, such a solution should be rejected as being unreasonable. Finally, Figure 3.25c implies shear mechanisms with consistent directions of the *P*- and *T*-axes, considering the nodal plane "B" as the real fracture plane associated with the AE event.

Based on the fault plane solutions, it can be concluded that, for both water injection and pressurization via the urethane sleeve, all large AE events whose polarities of P-wave initial motions could be clearly distinguished at six or more sensors were associated with fractures in which the shear mode was dominant. In contrast, in viscous oil injection, tensile-type fracturing is dominant, although shear-type fracturing could exist as well.

3.2.4 Discussion

By comparing the three experiments, the following results are obtained: (I) cracks extend rapidly by injection of water and oil, whereas cracks extend stepwise when fluid penetration is prohibited in pressurization via the urethane sleeve; (II) viscous oil tends to generate thick and planar cracks, whereas water tends to generate thin and wavelike cracks with many secondary branches; (III) shear-type mechanisms appear to be dominant in water injection and pressurization via a urethane sleeve, whereas a tensile-type mechanism seemed to be dominant in viscous oil injection.

The first result on the crack extension can be explained by comparing the stress intensity factors at a crack tip, as described by Ishida et al. (1997) after Zoback and Pollard (1978). The left of Figure 3.27 depicts the case in which hydrostatic pressure, σ, is applied throughout an ellipsoidal crack surface. In this case, $K_I = \sigma\sqrt{\pi\ell}$, where K_I is the stress intensity factor of mode I at a crack tip and ℓ is half of the crack length. This relation indicates that the stress intensity factor at a crack tip increases with crack length; in other words, a crack never stops once it starts to extend. The right of Figure 3.27 depicts the case in which a pair of point loads, F, are applied at the center of an ellipsoidal crack. In this case, $K_I = \frac{F}{\sqrt{\pi\ell}}$, which indicates that the stress intensity factor at a crack tip decreases with crack length; that is, a crack never extends without an additional increase in the applied loads. Needless to say, the case depicted in the left figure corresponds to fracturing in which fracturing fluid, such as water or viscous oil, invades into an induced crack, whereas that in the right figure corresponds to fracturing via pressurization of a urethane sleeve.

The second result is also consistent with the established traditional fluid-loss equation. Carter (1957) assumed for a fracture of uniform width that the fluid-loss velocity normal to the fracturing faces, v_l, takes the following form:

$$v_l = \frac{K_l}{\sqrt{t-\tau}} \tag{3.1}$$

where K_l is the overall fluid-loss coefficient, t is the current time, and τ is the time when filtration starts. The overall fluid-loss coefficient, K_l, includes the following three effects: (i) viscosity and relative-permeability effects of fracturing fluid; (ii) reservoir-fluid viscosity-compressibility effects; and (iii) wall-building effects. All these effects may act simultaneously and determine

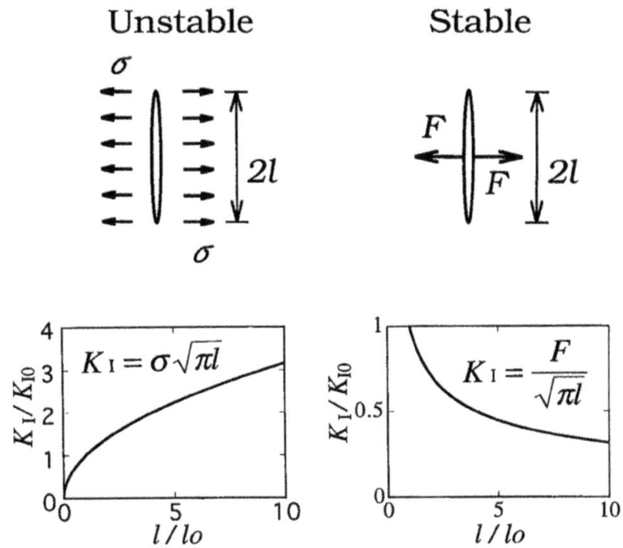

Figure 3.27 Stable and unstable crack growth from a viewpoint of stress intensity factor of mode I at a crack tip (Ishida et al., 1997, after Zoback and Pollard, 1978).

the actual coefficient, K_l. On the other hand, when the fracturing fluid viscosity is individually considered, Howard and Fast (1957) related the coefficient, K_l, to the viscosity, μ, in the following relation:

$$K_l \propto \frac{1}{\sqrt{\mu}} \tag{3.2}$$

This relation indicates that high-viscosity fracturing fluid results in low fluid-loss velocity.

In the case of oil injection, the fluid-loss velocity, v_l, in other words, the leak-off from the generated fracture plane, should be low, because the viscosity, μ, was high. As a result, the pressure in the induced fracture easily increased with the viscous oil injection, resulting in a thick and planar crack extension. On the other hand, in the case of water injection, because the leak-off was high due to the low viscosity of the fracture fluid, thin and wavelike cracks having many secondary branches were generated easily with the leak-off of the slick water. Consequently, the observation of cracks is consistent with the discussion derived from Carter's equation.

Regarding the third result on the fracturing mechanism, the observed difference between water and oil injection can be explained by Hill's model for volcanic earthquake swarms, shown in Figure 2.30 in the previous chapter. Although many volcanic seismic events characterized by magma intrusions or eruptions show a dominance of compression in P-wave initial motions on seismographs of many stations, implying a tensile mechanism, other events show a usual quadrant-type distribution, implying a shear mechanism. Based on this model, Hill (1977) proposed that magma intrudes into one of the weak planes lying along the direction of the maximum compressive stress, among many weak planes prevailing in a volcanic region. The magma intrusion forming a dike would induce some tensile fracturing, whereas shear fractures would form conjugate faults, connecting the tips of dikes, as indicated by symbols A and B in Figure 2.30. It is important to recall that the viscosity of magma is much higher than that of water (Dingwell, 1995). The granite specimens used in the present experiments should also have many small inherent defects lying along the direction of maximum compressive stress associated with their rift plane. Thus, tensile fracturing observed during viscous oil injection would correspond to phenomena similar to a dike formation. On the other hand, shear fractures dominant during water injection would correspond to the fracture connecting the tips of dikes.

In contrast to these, the fracturing induced by pressurization via a urethane sleeve might correspond to usual seismic events, because these were caused under compressive stresses without fluid intrusion. Therefore, exhibiting a normal quadrant-type distribution of P-wave initial motions, implying shear fracturing, may be reasonable.

Recalling the experience at the Cornwall geothermal site, viscous fluid injection resulted in much less seismic activity (J. -C. Roegiers, Oral communication, 1997); no microseismic event was recorded at the beginning of the viscous fluid injection without an event located within 30 m of the injection well (Baria and Green, 1986). These observations agree with the results described in this section: viscous fluids tend to induce tensile fractures, and microseismic events from tensile fractures are usually undetectable (Boler and Spetzler, 1986; Hayashi et al., 1988). The tendency is also demonstrated by a recent numerical simulation using the distinct element method (Shimizu et al., 2011), as shown in Chapter 4 of this book. In addition, the viscous gel used for the fracturing fluid itself might make it difficult for the vibration to arrive at seismometers by absorbing most of the energy emitted from fracturing (J. -C. Roegiers, Oral communication, 1997).

3.2.5 *Concluding remarks*

We conducted HF experiments on cubic granite specimens to examine the effect of fracturing fluid itself and fluid viscosity, comparing conventional water injection, viscous oil injection, and pressurization via a urethane sleeve. Based on the results of these experiments, the following can be concluded.

1 When using water and oil, cracks extended rapidly with BDs and substantial AE bursts. On the other hand, pressurization via a urethane sleeve without fracturing fluid extended cracks stepwise without BD or AE bursts.
2 The difference observed in crack extension can be explained by comparing the stress intensity factors at the crack tip. In water and oil injections, the stress intensity factor of mode I at a crack tip increases with crack length; in other words, a crack never stops once it starts to extend. On the other hand, in the case of pressurization via a urethane sleeve for the fracturing fluid not to be allowed to invade the induced crack, the stress intensity factor decreases with crack length; that is, a crack never extends without an additional increase in applied load.
3 The observation of the generated cracks by the naked eye on the lateral surfaces as well as on the cores subsequently taken by co-axially drilling around the borehole (i.e., overcoring)

elucidated that viscous oil tends to generate thick and planar cracks with few branches. In contrast, water generates thin and wavelike cracks with many secondary branches.

4 The flat jack pressures in the X-direction, P_X, normal to the surface of the crack induced in the Y-direction, showed some sudden increases corresponding to the specimen BD. The magnitudes of the pressure increase in oil injection were much larger than those in water injection. The strain increases measured with BDs by strain gauges glued along the induced cracks on the top surface were also larger in oil injection than in water injection. These measuring results, corresponding to the crack observation by the naked eye, indicate that viscous oil tends to generate a crack having a wider aperture than water.
5 In water injection, the crack extending direction changes easily not only with the respective BDs but also within a period of each BD. In contrast, in oil injection, it changes neither with the respective BDs nor within a period of each BD. The difference could be attributed to the difference in viscosity between water and oil. That is, viscous oil fracturing tends to generate thicker and planar cracks, ignoring various defects like microcracks in the specimen; in contrast, water generates thinner and wavelike cracks affected by the defects.
6 These observations are also consistent with discussions derived from the established fluid-loss equation, which indicates that high-viscosity fracturing fluid results in low fluid-loss velocity. In the case of oil injection, the leak-off from the generated fracture plane should be low because the viscosity is high. Consequently, the pressure in the generated fracture easily increased with the viscous oil injection and resulted in the extension of thick and planar cracks. In contrast, in the case of water injection, because the leak-off from the fracture was high due to the low viscosity, thin and wavelike cracks having many secondary branches were generated with the leak-off of the slick water.
7 Although only around seven large AE events in each experiment were examined using the fault plane solutions, the shear-type mechanism seemed dominant during water injection and pressurization via a urethane sleeve; in contrast, tensile failure was dominant during oil injection.
8 With the help of Hill's model for volcanic earthquake swarms, shear-type fractures with water injection are considered to be induced by connecting tips of pre-existing defects lying in the direction of the maximum compressive stress along a rift plane of the granite specimen. In contrast, tensile-type fracture dominant in oil injection is considered to be induced by the intrusion of the viscous oil into the defects themselves. On the other hand, since the AE events induced by pressurization via a urethane sleeve were caused by compressive stresses without the effects of fluid injection and manifested normal quadrant-type distributions of P-wave initial motions, thereby implying shear fracture, they may correspond to usual seismic events.

From these results, it can be concluded that viscous oil tends to generate thick and planar cracks with few branches under domineering tensile-type fracturing; in contrast, water tends to generate thin and wavelike cracks with many branches under domineering shear-type fracturing. These differences were most likely caused by a structure-sensitive property of the granite specimen, depending on the viscosity of the fracturing fluid. Since the phenomenon would not be specific to the granite alone, the experimental results suggest the strong possibility of controlling patterns and geometry of induced cracks by changing the viscosity of the fracturing fluid in field stimulation treatments.

3.3 New crack generation by viscous fluid injection in a specimen with a pre-existing crack

3.3.1 *Introduction*

Zoback et al. (1977) showed in their laboratory experiment using a rock specimen with a pre-existing crack that injection of a large viscosity drilling mud can generate a new crack in the maximum compressive direction without the influence of the pre-existing crack. The experimental results help understand the viscous fracturing fluid's effect in HF. However, they did not clarify the process of the new crack generation, although they observed the new crack after the experiments and discussed BD pressures dependent on flow rates.

Thus, to clarify the process of a new crack generation in a specimen with a pre-existing crack, we monitored AE and located their hypocenters in a laboratory experiment similar to the experiment by Zoback et al. (1977).

3.3.2 *Experimental method*

3.3.2.1 Specimen and outline of the experimental procedure

The loading system, the same as the one shown in Figure 3.2 for the experiments described in Section 3.2, was used for this experiment. As a specimen, a 19 cm cubic block of Chinese Nukame granodiorite with a 2 cm diameter hole shown in Figure 3.28 was employed. This specimen was taken from the same block as that from which the specimen, S, used in the experiment in Section 2.2, was taken. The granodiorite's rift plane was oriented horizontally to correspond to the *XY*-plane in the Cartesian coordinate system. P-wave velocities along the specimen's *X*-, *Y*-, and *Z*-directions were 4.57, 4.52, and 4.26 km/s, respectively. The granodiorite was produced in China, and its average grain size was 0.5 mm, which is considerably smaller than that of usual granite and granodiorite. The specimen was taken from the same block as the one labeled "Specimen S" in Section 2.2 of Chapter 2, and its photomicrographs with plane polarized light and with crossed polarized light were already shown in Figure 2.1d.

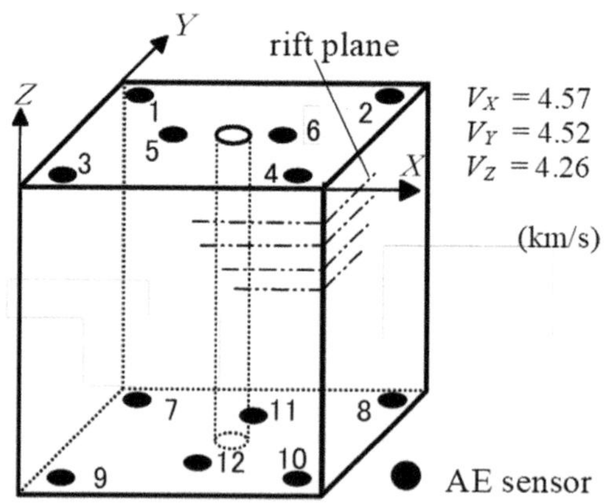

Figure 3.28 Specimen, setting positions of AE sensors, and coordinate system (Nakayama et al., 2002).

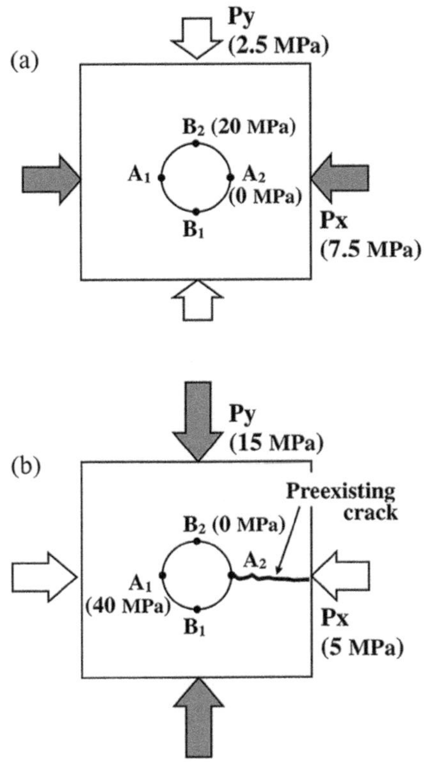

Figure 3.29 Illustration of the experimental procedure. The hole diameter is shown as larger than that of the real scale. (a) Stress condition at the water injection. (b) Stress condition at the viscous oil injection in the specimen having a crack generated by the water injection (Ishida et al., 2003).

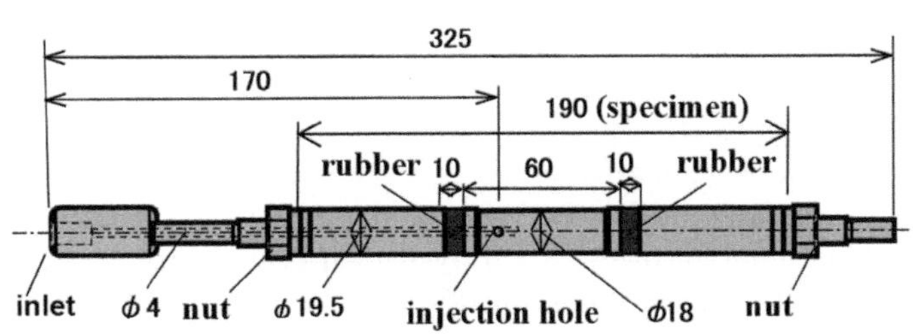

Figure 3.30 Mechanical-type double packer used to inject the water and the viscous oil into the hole (unit: mm). (Slightly Modified Nakayama et al., 2002.)

In the experiment, as shown in Figure 3.29, at first, water was injected into a hole bored in the center of the specimen on which biaxial confining pressures were applied, and a crack was generated parallel to the direction of the maximum confining pressure. After that, regarding the generated crack as a pre-existing joint, the maximum confining pressure direction was turned to 90 degrees, and viscous oil having 80 times larger viscosity than that of water was injected into the hole to generate a new crack parallel to the direction of the turned maximum confining pressure.

3.3.2.2 Water injection

At the water injection of the first procedure, as shown in Figure 3.29a, two horizontal confining pressures, 7.5 MPa in the *X*-direction and 2.5 MPa in the *Y*-direction, were applied to the specimen through flat jacks. In the *Z*-direction, no confining pressure was applied. Under the biaxial confining pressures, in the *XY*-plane, from the solution by Kirsch (1898), the tangential compressive stress on the hole wall has the maximum value of 20 MPa ($=3Px-Py$) at the two points, B_1 and B_2, whose *Y*-coordinates represent the minimum and the maximum, and the minimum value of 0 MPa ($=3Py-Px$) arises at the two points, A_1 and A_2, whose *X*-coordinates represent the maximum and the minimum (refer to Section 1.3). Therefore, when the hole wall is pressurized by water injection, the maximum tensile stress should arise at the two points, A_1 and A_2, on the two-dimensional section of the hole wall, and thus, it should be easiest to initiate cracks at the two points and to extend in the X-direction, that is along the *XZ*-plane in the three dimension.

A mechanical-type double packer having a diameter of 18 mm and a total length of 325 mm, shown in Figure 3.30, was used. By screwing the pair of nuts coming out of the specimen, the two 10 mm long rubber units on both sides of the pressuring section were compressed axially and expanded laterally, and as a result, the 60 mm long pressurizing section was tightly sealed. A screw pump was connected to the inlet of the packer 170 mm distant from the injection hole, as shown in Figure 3.30. Through this packer, the screw pump injected tap water having a viscosity of 1 mPa·s (1 mPa·s $=1.0\times10^{-3}$ Pa·s $=1.0\times10^{-2}$ dyn·s/cm^2 $=1$ cP) at a constant flow rate of 0.25 cm^3/s, by rotating the handle of the screw pump at 1.5 revolutions per second. After the pressurization, with BD pressure, a crack was generated in the *X*-direction as expected. It was confirmed by the naked eye's observation on the specimen's top and bottom surfaces.

3.3.2.3 Viscous oil injection

Regarding the crack generated by the water injection as a pre-existing joint, after changing the direction of the maximum confining pressure to 90 degrees so as to be perpendicular to the generated crack surface, oil having a viscosity of 80 mPa·s (transmission oil for a car, Mahha Super Hypoid Multi, GL-5 80W-90, Fuji Kosan Co., Ltd.) that is 80 times larger than that of water) was injected into the hole. As shown in Figure 3.29b, in this case, the maximum confining pressure, 15 MPa, in the *Y*-direction and the minimum confining pressure, 5 MPa, in the *X*-direction were applied to the specimen. Both confining pressure magnitudes were twice as large as those in the water injection. Under the confining pressures, if the pre-existing joint was ignored, the solution by Kirsch (1898) indicates that the tangential compressive stress on the hole wall has a maximum value of 40 MPa ($=3Py-Px$) at the two points, A_1 and A_2, and the minimum value of 0 MPa ($=3Px-Py$) at the two points, B_1 and B_2 (refer to Section 1.3). Therefore, without hole pressure, the mouth of the pre-existing joint on the hole wall at point A_2 should close under a compressive stress of 40 MPa, if the crack surfaces neither caused relative displacements nor gaps due to mineral particles detached from the surfaces. Both confining pressures were increased to twice as large as those in the water injection to obtain more stable experimental conditions because the results by Zoback et al. (1977) and our preliminary experiments suggested that the BD pressure using viscous fluid becomes larger than that using water. In other words, even if a larger pressure is applied than that in the water injection in a hole, it is ensured that the mouth of the pre-existing crack at the point, A_1 or A_2, on the hole wall keeps closing without opening.

In the viscous oil injection, replacing the screw pump used for the water injection, a hand hydraulic pump was used to cover a small amount of the oil leakage along the pre-existing crack by supplying a sufficient flow rate. The hand hydraulic pump injected the viscous oil into the hole at a flow rate of around 8 cm^3/s, which was 32 times larger than that of water injected by the screw pump. The hand pump was connected to the inlet of the packer at a distance of 170 mm from the injection hole of the packer, as shown in Figure 3.30, without removing water in the packer to avoid disturbing the crack induced by the water injection. As a result, some amount of water remained in the packer at the beginning of the viscous oil injection.

3.3.2.4 *AE and pressure measurements*

Magnitudes of the injected hole pressures and confining flat jack pressures were measured and recorded every 0.01 s through transducers.

To monitor AE for the experiment, we newly introduced the AE data recording and analyzing apparatus, "Type-01 (a custom-made item)" supplied by Iwatsu Electric Co., Ltd., which had 12 channels. Thus, 12 sensors were glued on the top and bottom surfaces of the specimen to detect AE signals. The sensors have a resonance frequency of 150 kHz and a cylindrical shape of 1.74 cm in diameter and 1.63 cm in length (Type R-15, Physical Acoustic Corporation). As shown in Figure 3.31, the detected AE signals were amplified 40 dB in the pre-amplifier and 30 dB in the main amplifier; thus, they were amplified 70 dB in total. The amplified signal of each channel was counted by the ringdown method with a threshold level of 250 mV and recorded every 0.01 ms.

It is necessary to record AE wave data to locate an AE hypocenter and analyze its focal mechanism. To record AE wave data efficiently and sufficiently, only when one of the following three pairs of AE sensors, Nos. 1 and 7, or Nos. 4 and 10, or Nos. 5 and 11, both detect signals exceeding a threshold level, a wave memory is triggered through the trigger unit. Every channel of the wave memory has a record length of 2 kwords and 8 bit resolution in amplitude. In the wave memory, the signal was digitized using a sampling interval of 0.2 μs.

A dead time of 4.4 ms just after recording each AE event was set to prevent the hard disk from recording too much noise caused by "ringing", which is the vibration following a large AE event. The dead time of 4.4 ms was about 11 times larger than the duration of each recorded AE event. AE events continue to be recorded without stopping, unless the number of AE events reaches 16,384 to fill up the storage capacity of the wave memory. All wave data were stored in DRAM (Dynamic Random Access Memory), and they were transferred and recorded in PD (Phase Change Rewritable Disk) having 650 MB with the AE count rates for each channel and the pressure data.

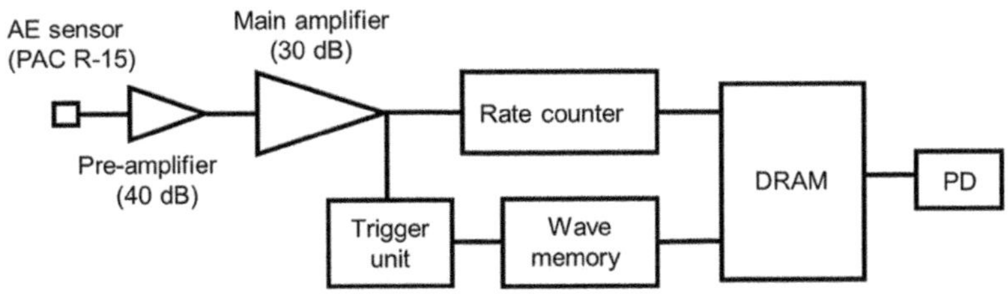

Figure 3.31 AE monitoring system.

3.3.3 *Results and discussion*

3.3.3.1 *Pressures and AE count rates*

Figures 3.32–3.33 show changes in the pressures and AE count rates in the water and the viscous oil injection. In these figures, the solid bold lines indicate the injected fluid pressures, P_o, while the fine solid lines indicate the flat jack pressures applying in the *X*- and the *Y*-directions, P_X and P_Y, as the confining pressures. The bar diagrams of both figures indicate AE count rates per 0.01-s interval, using a semilogarithmic scale. The AE rates shown in the figures are the average rates of all twelve channels originally counted for each channel.

In the water injection, as shown in Figure 3.32, the injected water pressure suddenly decreased after showing a BD pressure, 22.5 MPa. The bar diagram indicates that AE bursts occur with the BD pressure. As shown in the next section, at this time, a crack was generated from the injection hole to a lateral surface along the *X*-direction. Small jags along the pressure curve larger than 10 MPa were due to pressure decreases when changing from one hand to the other hand to rotate the handle of the screw pump, since the handle became harder to rotate with the pressure increase.

In the viscous oil injection shown in Figure 3.33, after recording four small peaks of the injected oil pressure indicated by the arrows, a large pressure drop was induced by generating a new crack in the *Y*-direction. The magnitudes of the four peak pressures, 35.0, 34.5, 33.5, and 33.0 MPa, gradually decreased with the elapsed time. The packer was not removed from the hole after the water injection to avoid relative displacements between the crack surfaces, particularly at its mouth on the hole wall. Thus, when the viscous oil injection was started by changing a pipe from water to oil at the inlet of the packer, around 6 cm^3 of water remained in the packer. As shown in the following section, because the AE hypocenters accompanied by the first, second, and third pressure drops were located along the pre-existing joint, these pressure drops were most likely caused by seepage of the water left in the packer into the pre-existing joint. On the other hand, the fourth large pressure drop was most likely induced by the generation of a new

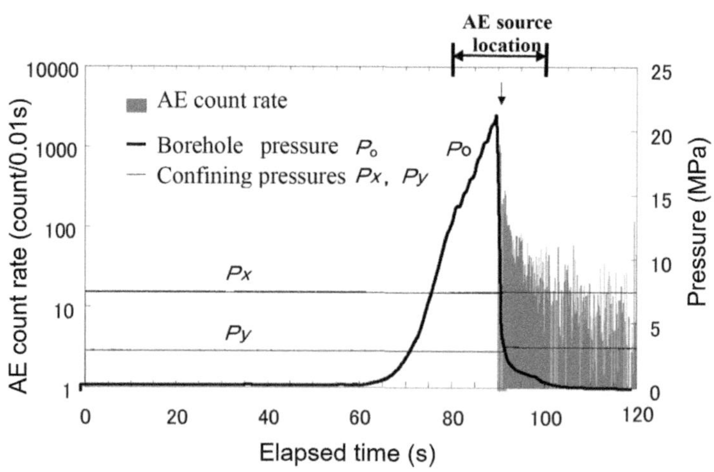

Figure 3.32 Changes in AE count rate, borehole pressure, and confining pressures in the water injection. (Slightly Modified Nakayama et al., 2002.)

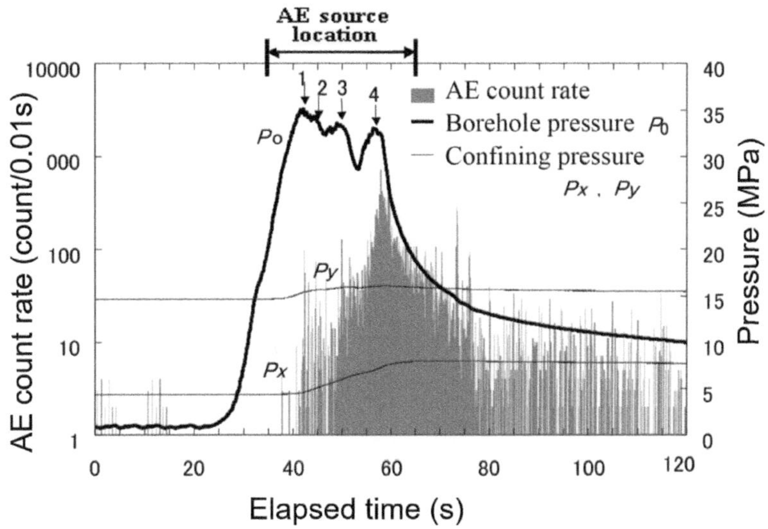

Figure 3.33 Changes in AE count rate, borehole pressure, and confining pressures in the viscous oil injection. (Slightly Modified Nakayama et al., 2002.)

crack, because the AE hypocenters accompanied by the pressure drop were located along the newly generated crack in the *Y*-direction.

3.3.3.2 *Observed cracks and located AE hypocenters*

We observed induced cracks on the specimen surfaces and showed them on unfolded planes of the specimen surfaces in Figure 3.34. The broken and the solid lines indicate cracks generated by the water injection and the viscous oil injection, respectively. AE hypocenters were located by considering the anisotropy of P-wave velocity in the specimen (Rothman et al., 1974). Accuracy of the location was expected to fall within around 15 mm because only AE events satisfying the following conditions were located; P-wave initial motions could be detected by more than five sensors, and the standard deviation of differences between the observed arrival time and back-calculated arrival time was within 3 μs. The accuracy was confirmed by locating the points using signals induced by dropping a steel ball on a specimen surface.

Projections on the *XY*-, *XZ*-, and *YZ*-plane of 236 located AE hypocenters satisfying the above conditions out of 1,264 AE events recorded during the elapsed time from 80 to 100 s in the water injection are shown in Figure 3.35. The *XY*-plane projection shown in Figure 3.35a indicates that the AE hypocenters are distributed from the hole to the lateral surface along the *X*-direction, corresponding to the crack designated by the broken line in Figure 3.34.

Figure 3.36 shows 180 located AE hypocenters out of 2,600 AE events recorded during the elapsed time from 35 to 65 s in the viscous oil injection. The *XY*-plane projection shown in Figure 3.36a indicates that the AE hypocenters are distributed from the hole to the two lateral surfaces along not only the *Y*-direction but also the *X*-direction. Because the AE hypocenter distribution along the *Y*-direction corresponds to the crack designated by the solid line in Figure 3.34, the AE events are most likely accompanied by the new crack generated by the

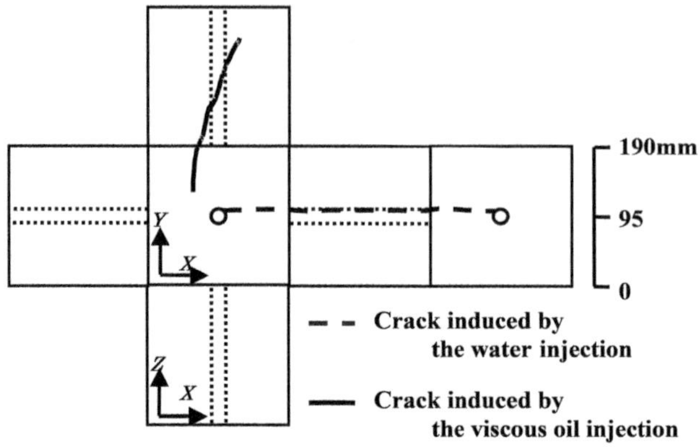

Figure 3.34 Cracks observed on the specimen surfaces (Nakayama et al., 2002).

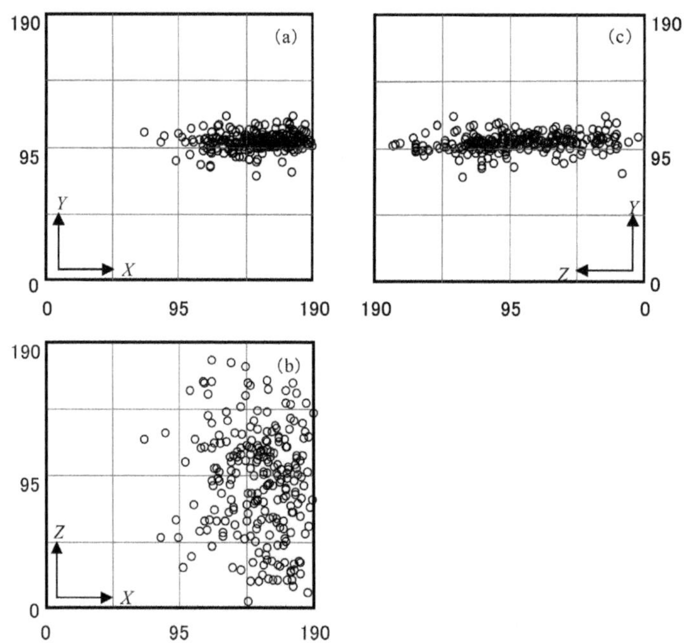

Figure 3.35 Projections of located AE sources induced by the water injection. (a) *XY*-plane, (b) *XZ*-plane, and (c) *YZ*-plane (Nakayama et al., 2002).

viscous oil injection. In contrast, because the AE hypocenter distribution along the *X*-direction in Figure 3.36a corresponds to that in Figure 3.35a and with the crack designated by the broken line in Figure 3.34, the AE events most likely occurred along the pre-existing joint generated by the water injection.

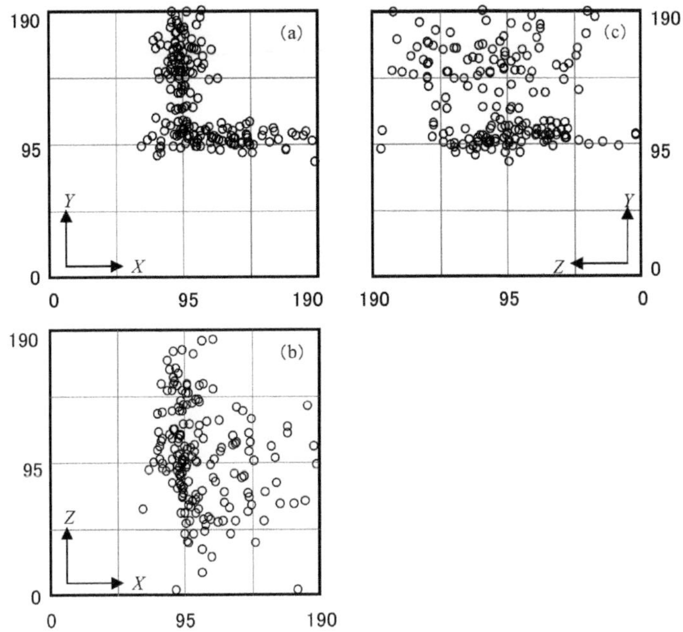

Figure 3.36 Projections of located AE sources induced by the viscous oil injection. (a) *XY*-plane, (b) *XZ*-plane, and (c) *YZ*-plane (Nakayama et al., 2002).

3.3.3.3 *Changes in the AE hypocenter distribution along the elapsed time*

Changes in the AE hypocenter distribution along the elapsed time help to understand the crack generation during the injections. Figure 3.37a–c shows the *X*-, *Y*-, and *Z*-coordinates of AE hypocenters along the elapsed time from 80 to 100 s in the water injection. Broken lines in Figure 3.37a–b indicate that $X=95$ and $Y=95$ mm, respectively, the center coordinates of the injection hole, while those in Figure 3.37c indicate that $Z=65$ and 125 mm, the upper and lower ends of the pressurizing section along the injection hole, respectively. On the respective right sides of Figure 3.37a–c, the crack observed on the specimen surfaces is shown. Figure 3.37d shows the injected water pressure and the AE count rates along the enlarged lateral axis of the elapsed time, using the same data shown in Figure 3.32. In this figure, the arrows (a–c) indicate the occurrence time of AE events whose focal mechanism will be discussed later.

The temporal change of the *X*-coordinates of located AE hypocenters shown in Figure 3.37a indicates that AE hypocenters distribute from the broken line to the plus direction of the *X*-axis and reach 190 mm corresponding to the lateral surface of the specimen. On the other hand, as shown in Figure 3.37b, the *Y*-coordinates concentrate along the broken line. The *Z*-coordinates distribute mainly in and around the band between the two broken lines at 6.5 and 12 mm, corresponding to the section of the pressurizing section, as shown in Figure 3.37c. Figure 3.37d shows that the AE events start to occur at the BD pressure. These figures indicate that a crack was initiated at the BD and extended from the injected hole to the plus *X*- and *Z*-directions, and it reached the lateral surface as well as the top and bottom surfaces of the specimen.

Figure 3.38 shows the data during the oil injection into the same specimen after making the crack and changing the direction of the maximum pressure from parallel to perpendicular to

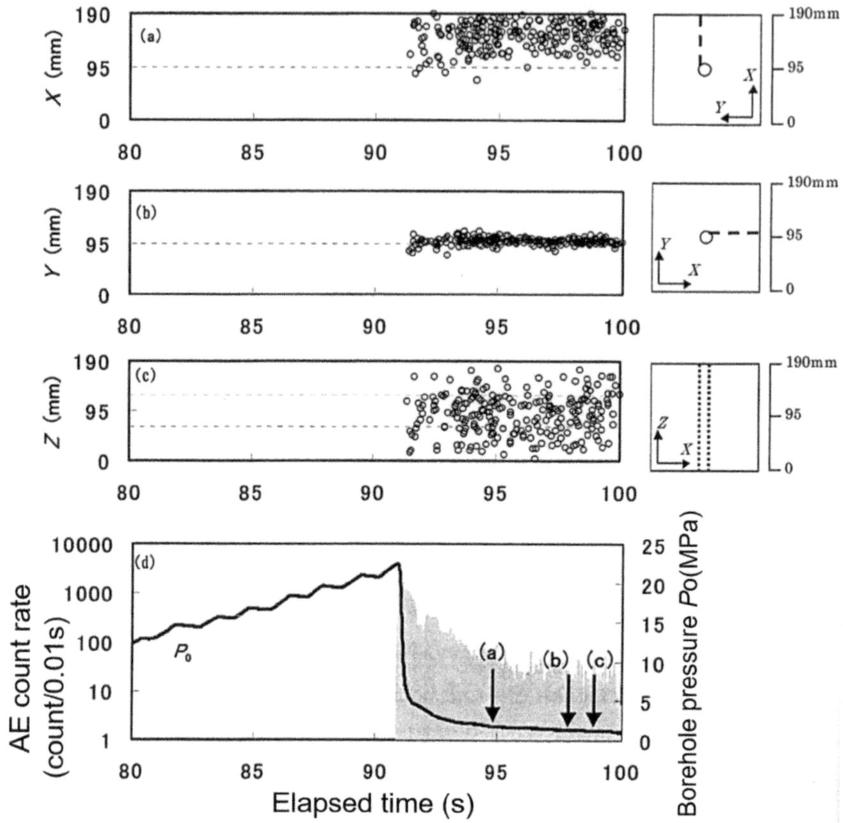

Figure 3.37 Temporal change of X-, Y-, and Z-coordinates of AE hypocenters in water injection in comparison with the change of borehole pressure. (Slightly Modified Nakayama et al., 2002.)

the crack surface. Figure 3.38a–b shows a clear change in AE hypocenter distributions around 55 s in the elapsed time. Before the time of 55 s, the *Y*-coordinates of AE hypocenters distribute within a narrow band along the broken line, while the *X*-coordinates distribute in a wide range of the plus direction from the broken line. This indicates that the AE events occurred along the pre-existing crack induced by the previous HF experiment using water. On the other hand, after a time of 55 s, the *Y*-coordinates distribute in a wide range of the plus direction from the broken line, while the *X*-coordinates distribute within a relatively narrow band along the broken line. This indicates that a new crack was induced in the *Y*-direction. As described at the end of the previous section, "3.3.3.1 Pressures and AE count rates", water of around 6 cm^3 remained in the packer at the start of viscous oil injection. From the comparison of the pressure change shown in Figure 3.38d to the movements of the *X*- and *Y*-coordinates of located AE hypocenters shown in Figure 3.38a–b, the water remaining in the packer most likely seeped into the pre-existing crack little by little with the pressure drops just after the peak pressures indicated by the arrows from No. 1 through No. 3 in Figure 3.38d. And, from the fact that the magnitude of the third pressure drop and the AE count rates with the drop are larger than those at the first and the second, a larger amount of water than that at the first and the second pressure drop most likely seeped

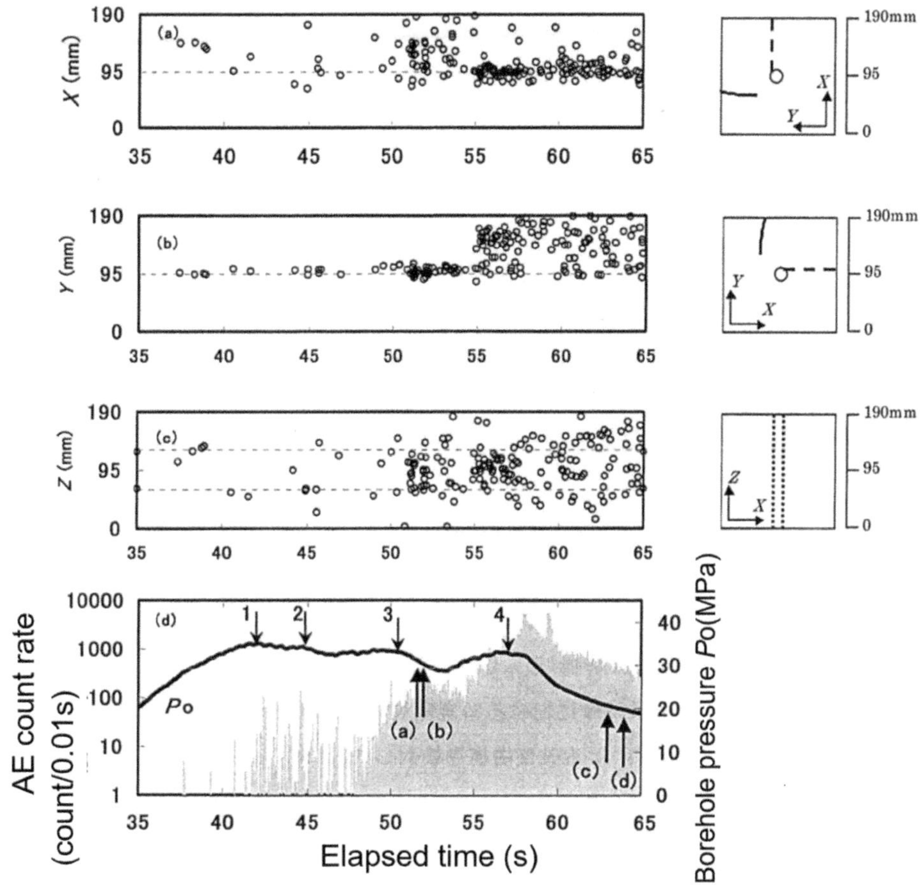

Figure 3.38 Temporal change of X-, Y-, and Z-coordinates of AE hypocenters in the viscous oil injection in comparison with the change of borehole pressure. (Slightly Modified Nakayama et al., 2002.)

into the pre-existing crack. As a result, almost all the water remaining in the packer probably seeped out, and the viscous oil filled the packer. Consequently, the new crack was induced in the maximum compressive direction at the fourth pressure drop.

3.3.3.4 Fault plane solutions of AE events

In the experiments, we obtained fault plane solutions for the located AE events having polarities of P-wave initial motions that were clearly distinguished between compression and dilatation at seven or more sensors.

More than 10 AE events satisfying the criteria existed in the water injection, while more than 20 events were in the viscous oil injection. Figures 3.39–3.40 show a few typical fault plane solutions obtained for these events for water and viscous oil injection, respectively. In these figures, the open or closed circles on the stereo net indicate the sensor position (azimuth and an angle of emergence from a hypocenter to a sensor) that recorded a compressional or a dilatational P-wave initial motion, respectively. On the polarity of a sensor, before the experiment, all sensors were glued on the bottom surface of the specimen, and a steel ball was dropped on the upper surface of it. The calibration test confirmed that the downward trace corresponds to the compression, whereas the upper, to the dilatation. (Please refer to Section 1.4.1 in Chapter 1 and Kasahara, 1981, on the fault plane solution.)

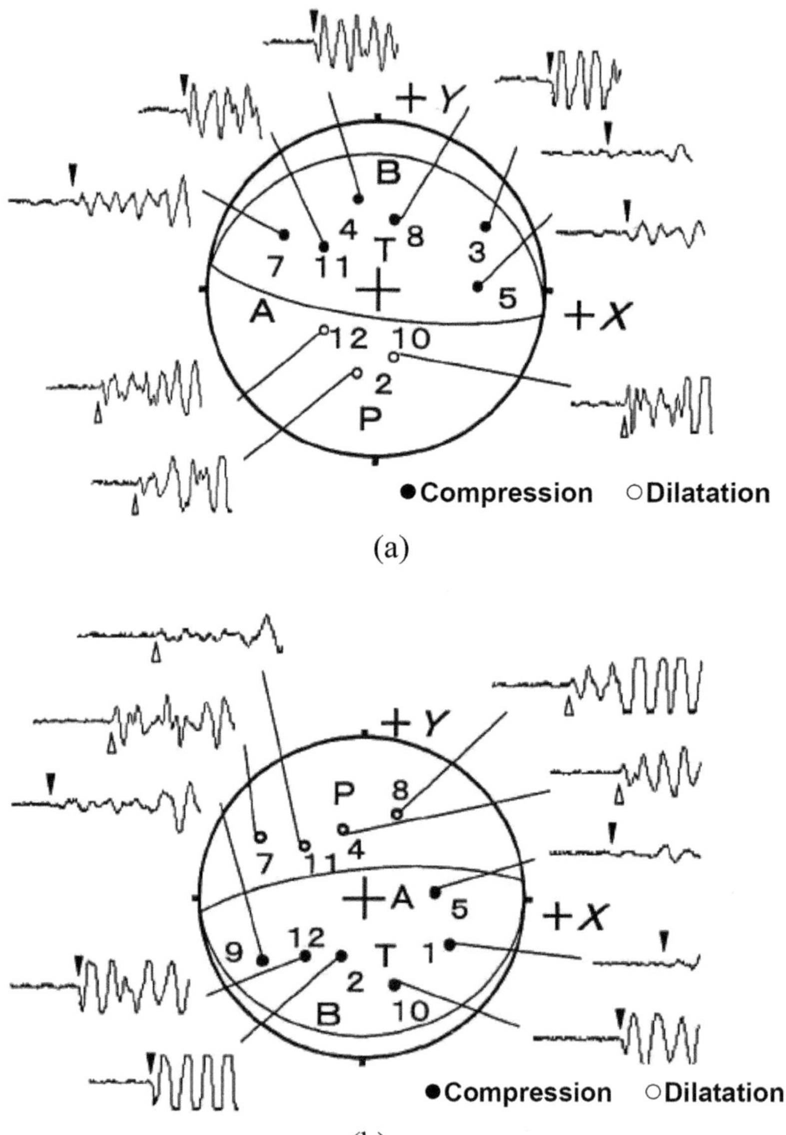

Figure 3.39 Fault plane solutions of AE events induced by water fracturing. Hypocenter coordinates are shown in a unit of mm. (Nakayama et al., 2002). (a) Occurrence time 94.80 s. Hypocenter coordinates (153, 96, 111). (b) Occurrence time 97.50 s. Hypocenter coordinates (143, 100, 74). (c) Occurrence time 98.90 s. Hypocenter coordinates (164, 102, 61).

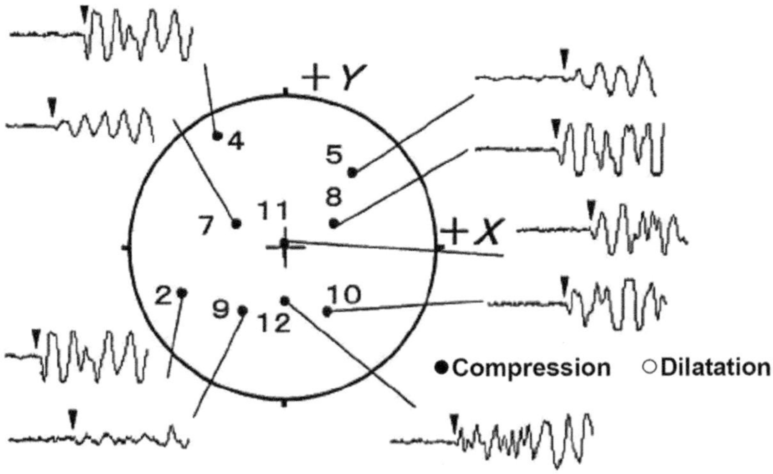

(c)

Figure 3.39 (Continued) Fault plane solutions of AE events induced by water fracturing. Hypocenter coordinates are shown in a unit of mm. (Nakayama et al., 2002). (a) Occurrence time 94.80 s. Hypocenter coordinates (153, 96, 111). (b) Occurrence time 97.50 s. Hypocenter coordinates (143, 100, 74). (c) Occurrence time 98.90 s. Hypocenter coordinates (164, 102, 61).

3.3.3.4.1 WATER INJECTION

Figure 3.39a–c shows three typical fault plane solutions of AE events recorded during the water injection. Arrows on the pressure record with the labels (a–c) in Figure 3.37d indicate the recorded time of the respective AE events.

The AE event whose solution is shown in Figure 3.39a occurred at the location (X, Y, Z)=(153, 96, 111), on the crack induced by the water injection that extends from the hole to the Y-direction. The figure indicates that sensors observed both compressional and dilatational P-wave initial motions, and they showed quadrant distribution suggesting shear fracture. Either of the nodal planes, A and B, that divide the two regions, where the closed and open circles distribute respectively, corresponds to a real fracture plane that induced the AE event. Because the AE event occurred on the crack on the XZ-plane extending along the $+X$-direction from the hole, the nodal plane, A, is most likely the real fracture plane. The letter positions, P and T, indicate the directions of the maximum and minimum compressive stresses that induced the fracture. However, the direction of the P-axis does not agree with the X-direction in which the maximum confining pressure applies, and it rather lies between the directions of the intermediate and the minimum confining pressures, Y and Z.

Figure 3.39b also suggests the shear fracturing mechanism, and the directions of the two nodal planes are similar to those of Figure 3.39a. However, the directions of the P- and T-axes are replaced by each other in comparison with those in Figure 3.39a. Most of the other AE events for which we obtained fault plane solutions suggested shear fracturing mechanisms similar to those shown in Figure 3.39a–b.

However, for a few of them, all sensors received compressional P-wave initial motions suggesting tensile fracturing mechanisms, as shown in Figure 3.39c. Such tensile fracturing is most likely induced by water intruding into a crack and pressurizing and expanding the crack, as shown in Figure 1.2 in Chapter 1. As discussed in the previous section, 3.2, the shear fracturing

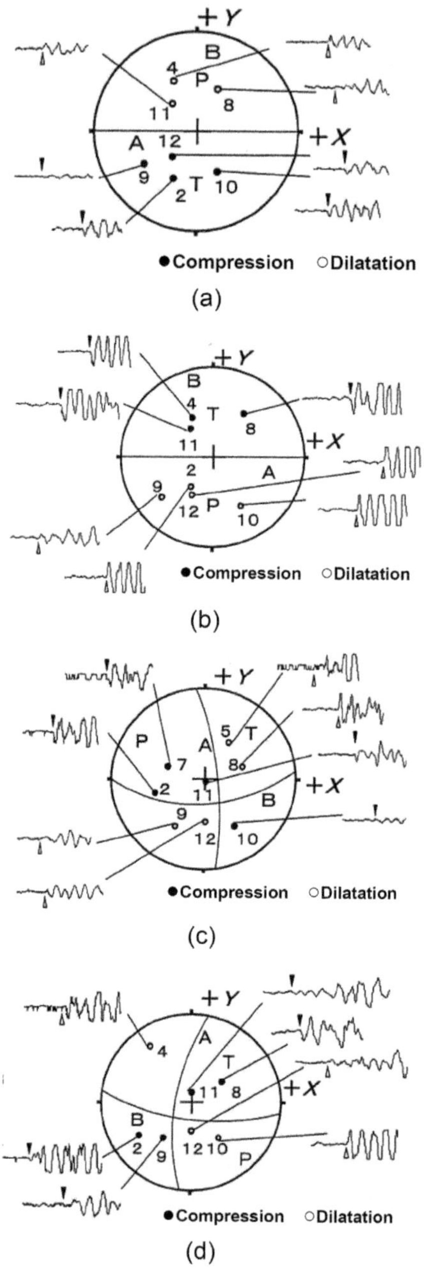

Figure 3.40 Fault plane solutions of AE events induced by viscous oil fracturing. Hypocenter coordinates are shown in a unit of mm. (Nakayama et al., 2002). (a) AE event induced on the pre-existing crack. Occurrence time 51.18 s. Hypocenter coordinates (130, 95, 107). (b) AE event induced on the pre-existing crack. Occurrence time 51.19 s. Hypocenter coordinates (124, 105, 73). (c) AE event induced on the new crack. Occurrence time 63.23 s. Hypocenter coordinates (93, 146, 114). (d) AE event induced on the new crack. Occurrence time 63.98 s. Hypocenter coordinates (95, 120, 136).

mechanism is dominant in water injection; however, the directions of the maximum and minimum compressive stresses that induced the fracture do not always correspond to those of the confining pressure, and in addition, they are often replaced by each other at every AE event. Microfracture accompanying the AE events is likely induced by a shear slip along a weak plane pre-existing in a specimen with the intrusion of water. The microfractures connect each other to make a macroscopic fracture extending in the direction of the maximum confining pressure, although the direction of each microfracture accompanying an AE event does not correspond to the direction of the maximum confining pressure. A similar macroscopic crack extension composed of many small jagged planes is also observed in a three-point-bending test (Kao et al., 2011) as discussed in Section 2.5.2, "Mechanism of shear fracturing", in the previous chapter and Section 3.2.3.6, "Fault plane solutions of AE events", in this chapter.

3.3.3.4.2 VISCOUS OIL INJECTION

Figure 3.40a–d shows four typical fault plane solutions of AE events recorded during the viscous oil injection. Figure 3.40a–b shows those of AE events that occurred after the third BD as indicated by arrows labeled (a–b) on the pressure record in Figure 3.38d. In this duration, almost all AE events occurred on the crack induced by the previous water injection that extends along the $+X$-direction from the injection hole, as shown in Figure 3.38a–c. The hypocenter coordinates of these events, (130, 95, 107) and (124, 105, 73), also indicate the locations on the crack. On the fault plane solutions of these AE events, two nodal planes, A and B, dividing the two regions where closed and open circles distribute, respectively, correspond to the diameter and the outer circle. Because the strike and dip of the nodal plane, A, corresponding to a diameter of the stereo net agree with those of the crack along the XZ-plane induced by the previous water injection, the nodal plane, A, is most likely the real fracture plane that accompanied the AE events. These fault plane solutions are quite similar to those induced by the water injection shown in Figure 3.39a–b, in the direction of the nodal plane, A, considered as the real fracture plane, and at the point suggesting shear fracturing. Most of the fault plane solutions for the other AE events induced in this duration, from 50 to 55 s in the elapsed time, when AE hypocenters distributed along the $+X$-direction are similar to those shown in Figure 3.40a–b.

Figure 3.40c–d shows the fault plane solutions of the AE events that occurred after the fourth BD, as indicated by arrows labeled (c–d) on the pressure record in Figure 3.38d. In this duration, almost all AE events occurred on the new crack induced by the viscous oil injection that extends along the $+Y$-direction from the injection hole, as shown in Figure 3.38a–c. The hypocenter coordinates of these events, (93, 146, 114) and (95, 120, 136), also indicate the locations on the crack. Figure 3.40c–d indicates that two nodal planes, A and B, divide the two regions where closed and open circles distribute, suggesting shear fracture, similar to Figure 3.40a–b. Because the real fracture planes that accompanied the AE events should lie in the direction close to the YZ-plane corresponding to the new macroscopic fracture plane that extends along the $+Y$-direction from the injection hole, the nodal plane, A, most likely corresponds to the real fracture plane in the both fault plane solutions.

Most of the fault plane solutions for the other AE events induced after 55 s, when the new crack extended along the $+Y$-direction, are similar to those shown in Figure 3.40c–d.

In the HF experiment using the same viscous oil as fracturing fluid for a specimen of Kurokami-Jima granite with no pre-existing crack, tensile fracture dominated as discussed in Section 3.2.3.6, "Fault plane solutions of AE events", showing Figure 3.25 in this chapter. However, in this experiment, the AE events with fault plane solutions shown in Figure 3.40a–b occurred after the third BD on the pre-existing crack induced by the previous water injection. We can understand the shear AE events if they were induced by slippage of the pre-existing crack due to an invasion of fracturing fluid into the crack. In contrast to these AE events, the AE events that occurred on the new crack induced by the viscous oil injection showed that the shear fracture mechanism, as shown in Figure 3.40c–d, is inconsistent with the tensile dominant fracture mechanism of the previous experiments discussed in Section 3.2.3.6., shown in Figure 3.25. In addition, the grain size of Chinese Nukame granodiorite used in this experiment is much smaller than that of Kurokami-Jima granite, as shown in Photo 2.1 in Section 2.2.1.1 in the previous chapter. The smaller grain size restrains shear fracture and rather enhances tensile fracture because tensile fracture dominates in HF in granite having smaller grains like Chinese Nukame granodiorite, whereas shear fracture dominates in those having larger grains, as elucidated in Section 2.2, "HF experiments on granite specimens with different grain sizes", in the previous chapter. Therefore, the injected viscous oil was most likely contaminated with water remaining in the packer or in the crack, and the water probably induced the AE events, showing the mechanism of shear fracture.

3.3.4 *Concluding remarks*

The following remarkable features were observed in the HF experiments in the cubic granite specimens:

1 Regarding a crack generated previously by water injection as a pre-existing joint, in the 19-cm cubic granite specimen on which the maximum compressive pressures were applied in the direction normal to the pre-existing joint, the viscous oil having 80 times larger viscosity than that of water was injected. As a result, a new crack was generated in the direction normal to that of the pre-existing crack almost without the influence of the pre-existing crack.
2 In the viscous oil injection, a small pressure drop was repeated three times, before the fourth large pressure drop was recorded generating the new crack. AE sources accompanied by the three pressure drops were distributed along the pre-existing joint, whereas those accompanied by the fourth large pressure drop were distributed along the newly generated crack. From the fact, the three pressure drops before the large pressure drop were most likely caused by seepage of the water remaining in the packer into the pre-existing crack because some amount of water remained in the packer when the hand pump for oil injection was connected to the packer, replacing the screw pump for water injection.
3 We compared fault plane solutions obtained for AE events that occurred after the third BD to those after the fourth BD. Although the former events were located along the pre-existing joint, whereas the latter were along the newly generated crack, they all indicated a shear fracturing mechanism. On the other hand, a tensile mechanism dominated in an HF experiment for a granite specimen without a pre-existing joint using the same viscous oil, whereas shear dominated using water, as shown in Section 3.2.3.6, "Fault plane solutions of AE events", in this chapter. This discrepancy between the two experiments can be explained by the following: the injected viscous oil was most likely contaminated with water remaining in the packer or in the crack, and the water probably induced the AE events, showing the mechanism of shear fracture.
4 The experimental results suggest that we can generate a new crack without the influence of a pre-existing joint in a rock mass by HF operation using viscous fracturing fluid. However, as well as in a field HF operation, water should remain in the drilled hole when changing fracturing fluid from water to viscous fluid. Thus, the similar fracturing process observed in the experiment would most likely also occur in field operations due to contamination of injected viscous fluid with water.

3.4 Conclusions

In this chapter, the effects of fluid on a crack extension in HF, and the different features of crack extension and fracturing mechanism depending on the viscosity of fracturing fluid, were elucidated in the laboratory experiments. Second, the availability of viscous fracturing fluid for HF to generate a new crack in a rock mass with a pre-existing joint was experimentally examined. As a result, we can conclude the following:

1 When using fracturing fluid like water and viscous oil, a crack extends rapidly without stopping once it starts to extend by invading into a crack and applying the fluid pressure at a crack tip. This effect of fracturing fluid was clearly elucidated by comparing the fracturing by expanding the urethane sleeve without using any fracturing fluid; a crack never extends without an additional increase in applied load in the urethane sleeve fracturing. The effect of fracturing fluid on a crack extension in HF is consistent with the theory on the stress intensity factors at the crack tip; the stress intensity factor of mode I at a crack tip increases with crack length when the fluid pressure is applied at a crack tip, whereas the stress intensity factor decreases with crack length when the fracturing fluid is not allowed to invade the fracture.
2 The observation of the cracks generated by viscous oil and water injections elucidated that viscous oil tends to generate thicker and planar cracks with few branches having wider apertures than water. The difference in crack features is consistent with that in the fracturing mechanisms deduced from AE events; that is, the shear-type mechanism is dominant during water injection, whereas the tensile-type is dominant during oil injection.
3 The results of the experiment using a specimen with a pre-existing crack suggested that viscous fluid injection can generate a new crack in a rock mass with a pre-existing joint, ignoring its existence. The locations and fracturing mechanisms of monitored AE events suggested that water remaining in the pressurizing system invaded the pre-existing crack, accompanied by a few small pressure drops, and after that, viscous fluid generated a new crack with a large pressure drop. A similar fracturing process would most likely also occur in actual field operations due to contamination of injected viscous fluid with water.
4 The experimental results suggested that viscous fracturing fluid is useful for generating a new crack in a rock mass with a pre-existing joint and generating thicker and planar cracks with few branches having wider apertures than water. They also suggested that slicker fracturing fluid like water, rather than viscous oil, can generate thinner wavelike cracks having more tortuosity and branches. This implies that much slicker fluids like supercritical carbon dioxide and other gases can make cracks advantageous to extract heat or produce shale gas and oil in some cases than conventional water-based fluids.

References

Baria, R. & Green, A. S. P. (1986) Seismicity induced during a viscous stimulation at the Camborne school of mines hot dry rock geothermal energy project in Cornwall, England. In: *Progress in Acoustic Emission III, Institute of Industrial Science, University of Tokyo*, Tokyo, Japan, pp. 407–429.

Boler, F. M., & Spetzler, H. (1986) Radiated seismic energy and strain energy release in laboratory dynamic tensile fracture. *Pure Appl. Geophys.*, 124, 759–772.

Carter, R. D. (1957) Appendix to "*Optimum Fluid Characteristics for Fracturing Extension*". In Howard, G. C., Fast C. R. (eds), *Drilling and Production Practice,* API (American Petroleum Institute), Washington, DC, USA, p. 267.

Chen, Q., Ishida, T., & Mizuta, Y. (1996) Effect of injected water on hydraulic fracturing deduced from acoustic emission monitoring. *Jpn. J. JSCE*, 541(III-35), 173–187. (in Japanese with English abstract)
Crouch, S. L. & Starfield, A. M. (1983) *Boundary Element Methods in Solid Mechanics*. Unwin Hyman, New South Wales, Australia.
Dingwell, D. B. (1995) Viscosity and an elasticity of melts. In: Ahrens, T. J. (ed) *Mineral Physics and Crystallography: a Handbook of Physical Constants*. American Geophysical Union, Washington, DC, USA, pp. 209–217.
Gidley, J. L., Holditch, S. A., Nierode, D. E., & Veatch Jr., R. W. (1989) *Recent Advances in Hydraulic Fracturing*. Society of Petroleum Engineers of AIME, Texas, USA.
Haimson, B. C. (1978) The hydrofracturing stress measuring method and recent field results. *Int. J. Rock Mech. Min. Sci. & Geomech. Abstr.*, 15, 167–178.
Hayashi, K., Motegi, S. & Abe, H. (1988) Characteristics of energy of elastic waves due to sudden growth of subsurface reservoir cracks for geothermal heat extraction, In Yamaguchi, K., Takahashi, H., Niitsuma, H. (eds) *Progress in Acoustic Emission*. The Japanese Society for NDI, Tokyo, Japan, pp. 147–152.
Howard, G. C., and Fast, C. R. (1957) Optimum fluid characteristics for fracturing extension. In: Howard, G. C., Fast C. R. (eds), *Drilling and Production Practice, API (American Petroleum Institute)*, Washington, DC, USA, pp. 261–270.
Howard, G. C. and Fast, C. R. (1970) *Hydraulic Fracturing*. Society of Petroleum Engineers of AIME, Texas, USA.
Hubbert, M. K. & Willis, D. G. (1957) Mechanics of hydraulic fracturing. *Pet. Trans. Am. Soc. Min. Eng.*, 210, 153–168.
Ishida, T. (1999) *An Introduction to Acoustic Emission of Rock*. Nagoya, Japan, Kinmiraisha. (in Japanese)
Ishida, T., Chen, Q., & Mizuta, Y. (1996) Effect of fluid viscosity in hydraulic fracturing deduced from acoustic emission monitoring. *Jpn. J. JSCE*, 547(III-36), 183–198. (in Japanese with English abstract)
Ishida, T., Chen. Q., & Mizuta, Y. (1997) Effect of fluid viscosity in hydraulic fracturing deduced from acoustic emission monitoring. *Pure Appl. Geophys.*, 150, 627–646.
Ishida, T., Chen. Q., Mizuta, Y. & Roegiers, J-C. (1998) Influence of fluid viscosity on the hydraulically induced crack geometry. *Int. J. Rock Mech. Min. Sci.*, 35(4/5), 460–462. (Paper No.30)
Ishida, T., Chen, Q., Mizuta, Y. & Roegiers, J-C. (2004) Influence of fluid viscosity on the hydraulic fracturing mechanism. *Trans. ASME, J. Energy Res. Technol.*, 126, 190–200.
Ishida, T., Nakayama, Y., Ohike, Y. & Mizuta, Y. (2003) Acoustic emission monitoring of a new crack generation by viscous fluid injection in a granitic specimen having a preexisting joint. In: Saito, T. & Murata, S. (eds) *Environmental Rock Engineering: Proceedings of the 1st Kyoto International Symposium on Underground Environment—Role of Geo-technology to the Underground Environment*, 17–18 March 2003, Kyoto, Japan. Swets & Zeitlinger Publishers/ A.A. Balkema Publishers, Lisse, pp. 179–184.
Kao, C.S., Carvalho, F. C. S. & Labuz, J. F. (2011) Micromechanisms of fracture from acoustic emission. *Int. J. Rock Mech. Min. Sci.*, 48, 666–673.
Kasahara, K. (1981) *Earthquake Mechanics*. Cambridge University Press, Cambridge, p. 248.
Kirsch, C. (1898) Die Theorie der Elastizität und die Bedürfnisse der Festigkeitslehre. *Z. VDI*, 42, 797–807. (in German).
Matsunaga, I., Kobayashi, H., Sasaki, S. & Ishida, T. (1993) Studying hydraulic fracturing mechanism by laboratory experiments with acoustic emission monitoring. *Int. J. Rock Mech. Min. Sci. & Geomech. Abstr.*, 30, 909–912.
Maxwell, S. C. & Cipolla, C. (2011) What does microseismicity tell us about hydraulic fracturing? *Presented at Society of Petroleum Engineers Annual Technical Conference and Exhibition*, Denver, Colorado, USA, Paper No. SPE 146932.
Mizuta, Y., Sano, O., Ogino, S. & Katoh, H. (1987) Three dimensional stress determination by hydraulic fracturing for underground excavation design. *Int. J. Rock Mech. Min. Sci. & Geomech. Abstr.*, 24, 15–29.
Nakayama, Y., Ishida, T., Ohike, Y. & Mizuta, Y. (2002) Acoustic emission monitoring of a new crack generation by viscous fluid injection in a granitic specimen having a preexisting joint. *J. Jpn. Soc. Civ. Eng.*, 715(III-60), 345–356.

Rothman, R. L., Greenfield, R. J. & Hardy, Jr. H.R. (1974) Errors in hypocenter location due to velocity anisotropy. *Bull. Seismol. Soc. Am.*, 64, 1933–1996.

Sasaki, S. (1995) A study on characteristics and source mechanism of acoustic emission induced by hydraulic fracturing. Ph. D. thesis, Faculty of Science, Tohoku University, Sendai. (in Japanese).

Shimizu, H., Murata, S. & Ishida, T. (2011) The distinct element analysis for hydraulic fracturing in hard rock considering fluid viscosity and particle size distribution. *Int. J. Rock Mech. Min. Sci.*, 48, 712–727. https://doi.org/10.1016/j.ijrmms.2011.04.013

Talebi, S. & Cornet, F. H. (1987) Analysis of the microseismicity induced by a fluid injection in a granitic rock mass. *Geophys. Res. Lett.*, 14, 227–230.

Zoback, M. D. & Pollard, D. D. (1978) Hydraulic fracture propagation and the interpretation of pressure time records for in-situ stress determination. In: *Proceedings of 19th US Rock Mechanics Symposium, Mackay School of Mines*, Reno, Nevada, pp. 14–22.

Zoback, M. D., Rummel, F., Jung, R. & Raleigh, C. B. (1977) Laboratory hydraulic fracturing experiments in intact and pre-fractured rock. *Int. J. Rock Mech. Min. Sci. & Geomech. Abstr.*, 14, 49–58.

Chapter 4

Numerical simulation using the distinct element method for hydraulic fracturing considering fluid viscosity and particle size distribution

4.1 Introduction

To better understand the HF (hydraulic fracturing) mechanism, a considerable amount of research has been carried out in the past few decades. According to conventional theory, HF is formed by tensile crack generation (Hubbert and Willis, 1957). On the other hand, shear-type mechanisms were observed in most of the acoustic emission (AE) events recorded during the laboratory and field HF experiments (Baria and Green, 1986; Talebi and Cornet, 1987; Falls et al., 1992; Urbancic et al., 1999), as described in the previous chapters. To give a rational explanation for the disagreement between conventional theory and AE monitoring, and to better understand the HF mechanism, various numerical analysis techniques have been developed. The Finite Element Method (FEM) and the Boundary Element Method (BEM) have been commonly used to simulate HF in complex three-dimensional structures (Shah et al., 1997; Vandamme and Curran, 1989) Al-Busaidi et al. (2005) simulated HF in granite by using the distinct element method (DEM), and the results were compared with the AE data from the experiment. However, the simulation results showed that the disagreement mentioned above was not solved successfully. Recently, Ouchi et al. (2015) applied a peridynamics approach, and Yoshioka et al. (2023) applied a successive energy minimization approach to simulate HF. Despite these efforts, the mechanism of HF propagation has not been sufficiently clarified.

As described in Chapter 2, Matsunaga et al. (1993) conducted HF experiments for various rocks and acrylic resin, and found that rock texture, such as grain size, affects the HF mechanism. Ishida et al. (2004) extended this work, and the HF experiments were conducted for four different types of granitic rock specimens having different grain sizes to investigate the influence of grain size on induced crack geometry and fracturing mechanisms (Ishida et al., 2000a,b; Ishida, 2001). The fault plane solutions of AE events indicated that the dominant micro-fracturing mechanism becomes tensile rather than shear with the decrease of the grain size. Their experimental results indicate that the texture of rock, like the grain size of granitic rocks, considerably affects the geometry, surface roughness, and microcracking mechanism of hydraulically induced cracks.

In Chapter 3, HF experiments were conducted using water, viscous oil, and a urethane sleeve without injection fluid in the same types of granitic rock specimens (Ishida et al., 2004). In HF using water and viscous oil, the BD (breakdown) pressure (which is defined as a peak pressure immediately before it drops suddenly) is recorded and a crack rapidly extends to a specimen surface, while in the urethane sleeve fracturing, the BD pressure is not recorded and an additional increase of applied load is needed to extend a crack. The difference suggests that the important effect of injected fluid in HF is the wedge effect, in which the fluid intrudes into a

DOI: 10.1201/9781003282631-4

crack and applies the fluid pressure to the crack tip by opening the crack, which can be explained by the stress intensity factor calculated from the linear fracture mechanics. When comparing the source mechanisms of AE events induced by low-viscosity water and higher-viscosity oil, the shear-type mechanisms are dominant when low-viscosity fluid is injected, and both shear and tensile type mechanisms are observed when high-viscosity fluid is injected.

Shimizu et al. (2010, 2011) discussed the influence of the grain size distribution and the fluid viscosity that was observed in the experiments reported in Chapters 2 and 3. A fluid flow algorithm that can consider the fluid viscosity and permeability is introduced into the DEM program. They performed a series of simulations for HF in hard rock using the flow-coupled DEM code and successfully discussed the experimental results. In this chapter, by introducing their DEM analyses and discussion (Shimizu et al., 2010, 2011), we try to give insights to explain rationally the disagreement between conventional theory and the AE monitoring results.

4.2 Simulation methodology

4.2.1 Formulation of mechanics of bonded particles

The DEM for granular materials was originally developed by Cundall and Strack (1979). They developed a well-known commercially available DEM code (particle flow code (PFC)) (Potyondy and Cundall, 2004). In this chapter, the two-dimensional DEM was employed, and we used our own DEM code that we had written to study HF. Since thorough details of the fundamental DEM algorithm can be found in Cundall and Strack (1979) and Potyondy and Cundall (2004), we focus mainly on the primary differences between the DEM code used in this chapter and the parallel-bond model in the two-dimensional particle flow code (Potyondy and Cundall, 2004).

Although the DEM is one of the numerical techniques based on the discontinuum model, it can also be applied to the continuum by introducing bonds between particles. In the two-dimensional DEM, the intact rock is modeled as a dense packing of small, rigid circular particles. Neighboring particles are bonded together at their contact points with normal, shear, and rotational springs and interact with each other as shown in Figure 4.1.

The increments of normal force f_n, the tangential force f_s, and the moment f_θ can be calculated from the relative motion of the bonded particles, and are given as

$$f_n = k_n\left(dn_j - dn_i\right) \tag{4.1}$$

$$f_s = k_s\left\{ds_j - ds_i - 1/2\left(d\theta_j - d\theta\right)_i\right\} \tag{4.2}$$

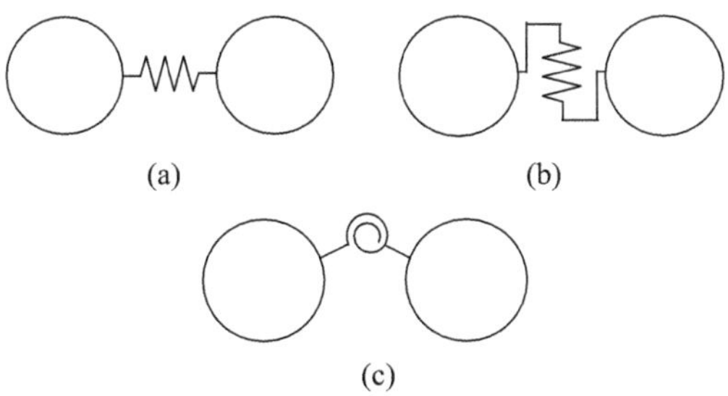

Figure 4.1 Three spring models between two bonded particles (Shimizu et al., 2010). (a) Normal spring. (b) Shear spring. (c) Rotational spring.

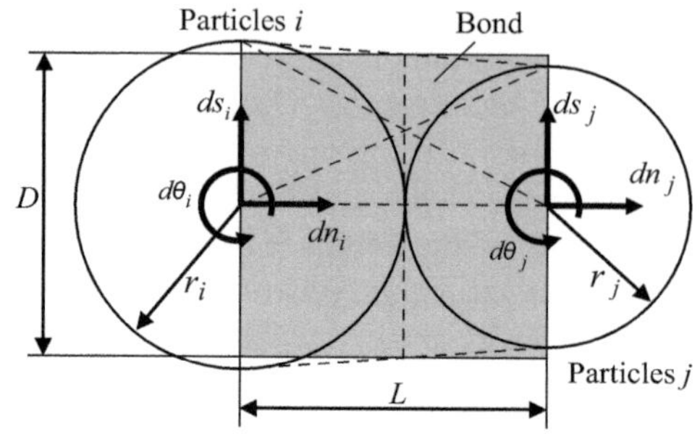

Figure 4.2 Bonded particles model (Shimizu et al., 2010, 2011).

$$f_\theta = k_\theta \left(d\theta_j - d\theta_i \right) \tag{4.3}$$

where k_n, k_s, and k_θ are the stiffnesses of normal, shear, and rotational springs, respectively; d_n, d_s, and $d\theta$ are normal and shear displacements and rotation of particles, respectively; and r_i and r_j are the radii of the bonded particles. A bond between the particles is presented schematically as a gray rectangle in Figure 4.2, where D and L are the bond diameter and the bond length, respectively. D is obtained from the harmonic mean of the radii of two particles. L and D are given by

$$L = r_i + r_j \tag{4.4}$$

$$D = \frac{4r_i r_j}{\left(r_i + r_j \right)} \tag{4.5}$$

Because the DEM is formulated as a fully dynamic system, small amounts of viscous damping are necessary to provide dissipation of high-frequency vibration. Unless contact damping is introduced, the assemblies cannot reach equilibrium. Contact damping operates on the relative velocities at the contacts and is represented by dashpots acting in the normal and shear directions at the contact points.

Because the simulation of laboratory rock tests, such as the uniaxial compression test, requires quasi-static loading, the coefficients of viscous contact damping are determined to provide critical viscous damping that approximates quasi-static loading. The coefficients of viscous contact damping in both normal and shear directions, c_n and c_s, are given by the following equations:

$$c_n = 2\sqrt{m_{ij} k_n} \tag{4.6}$$

$$c_s = 2\sqrt{k_s / k_n} \tag{4.7}$$

where m_{ij} is given by the mass of two particles m_i and m_j, as follows:

$$m_{ij} = \frac{2m_i m_j}{m_i + m_j} \tag{4.8}$$

If the stiffnesses of the springs, k_n, k_s, and k_θ, are set as tuning parameters treated independently, a significant effort will be required to determine appropriate values for them. Therefore, the stiffnesses of the normal and rotational springs, k_n and k_θ, are calculated using beam theory, and the stiffness of shear springs k_s is calculated by multiplying the stiffness of the normal spring k_n and a constant stiffness ratio α. Thus, the stiffnesses of the springs are given by the following equations:

$$k_n = \frac{E_p A}{L} \tag{4.9}$$

$$k_s = \alpha k_s \tag{4.10}$$

$$k_\theta = \frac{E_p I}{L} \tag{4.11}$$

where A is the cross-sectional area of the bond, and I is the moment of inertia of the bond. E_p is Young's modulus of particles and bonds. The moment of inertia, I, depends on the shape of the cross-section, and a rectangular cross-section is assumed in this study.

Young's modulus E_p assigned to the particles and the stiffness ratio α are microscopic parameters. These values are different from Young's modulus and Poisson's ratio of the rocks obtained from the laboratory experiments and simulation of the uniaxial compression tests.

The normal stress σ and shear stress τ acting on the cross-section of the bond are calculated using the following equations. The positive stress and strain mean compression:

$$\sigma = \frac{f_n}{D} \tag{4.12}$$

$$\tau = \frac{f_s}{D} \tag{4.13}$$

4.2.2 *Microcrack generation*

When σ exceeds the strength of the normal spring σ_c or τ exceeds the strength of the shear spring τ_c, the bond breaks, and three springs are removed from the model altogether. The criteria for bond breaks are summarized as follows. They imply that the normal spring breaks only by tension, whereas compression does not cause the bond breaks.

Bond break criterion 1: $|\sigma| \geq \sigma_c$ and $\sigma < 0$ (tensile stress)
Bond break criterion 2: $|\tau| \geq \tau_c$

Each bond breakage represents the generation of microcracks. A microcrack is generated at the contact point between two particles. A crack length is assumed to be the same as the bond diameter D, and its direction is perpendicular to the line joining the two centers of particles.

In the parallel-bond model developed by Potyondy and Cundall (2004), the moment acting on the parallel-bond (which is expressed as an elastic beam) contributes to the normal stress acting on the particles. This means that the bond breakage is judged by the maximum tensile stress acting on the cross-section of the assumed elastic beam of their model. On the other hand, in this study, because the spring is introduced to restrict the rotation of the particles and is used only to calculate the moment acting on the particles, the normal stress calculated by Eq. (4.12) does not include the moment of the elastic beam. This means that the bond breakage in the model presented here is judged by the average normal stress acting on the cross-section of the assumed elastic beam. Moreover, this procedure makes it possible to calibrate the tensile strength σ_c and the shear strength τ_c almost independently. This is the difference in the mechanism of particle bonding between the parallel-bond model proposed by Potyondy and Cundall (2004) and the model presented in this chapter.

4.2.3 *Classification of crack modes*

In the AE measurement during the laboratory experiment, the AE hypocenter can be calculated from the arrival time of the *P*-wave first motion, and the source mechanism of AE events is determined from the spatial distribution of the *P*-wave first motion polarities as shown in the previous chapters, 1 through 3. For tensile cracks, all sensors detect the compressional *P*-wave first motion. On the other hand, for shear cracks, both compressional and dilatational *P*-wave first motions are detected. This polarity of the *P*-wave first motion will depend on the stress state at the crack generation. Therefore, in this study, the crack modes can be classified using the shear–tensile stress ratio $|\tau/\sigma|$ regardless of broken spring type (normal or shear springs) as follows:

Crack classification criterion 1: $|\tau/\sigma| \leq 1$ and $\sigma < 0$ (tensile stress) → tensile crack.
Crack classification criterion 2: $|\tau/\sigma| > 1$ and $\sigma < 0$ (tensile stress) → shear crack.
Crack classification criterion 3: $\sigma > 0$ (compressive stress) →shear crack.

When a microcrack is generated, the strain energy stored in both the normal and shear springs at the contact point is released. This produces a force imbalance, and the subsequent stress redistribution induces an AE event. The magnitude of the AE event is related to the kinetic energy generated in the model, which propagates outward from the bond breakage points. Although the strain energy at the contact point does not equal the magnitude of the AE event, the magnitude of the AE event would strongly relate to the released strain energy. For this reason, the strain energy E_k calculated using Eq. (4.14) is assumed to be the energy corresponding to the magnitude of the AE event:

$$E_k = \frac{f_n^2}{2k_n} + \frac{f_s^2}{2k_s} \tag{4.14}$$

4.2.4 *Fluid flow algorithm*

The fluid flow algorithm (Al-Busaidi et al., 2005; Kiyama et al., 1994; Bruno, 1994; Thallak et al., 1991) that can consider the fluid viscosity and permeability is introduced into the DEM program to simulate the HF. In the fluid flow algorithm, as shown in Figure 4.3, the aperture between the adjoining particles is supposed to be a flow channel, and a series of enclosed domains is created by connecting the centers of adjoining particles.

As shown in Figure 4.4, each channel is supposed to be a set of parallel plates with some aperture, and the Poiseuille equation models the fluid flow in the channel. Therefore, the volumetric laminar flow rate is given by the following equation (Zimmerman and Bodvarsson, 1996):

$$Q = \frac{w^3}{12\mu}\frac{\Delta P}{L_p} \tag{4.15}$$

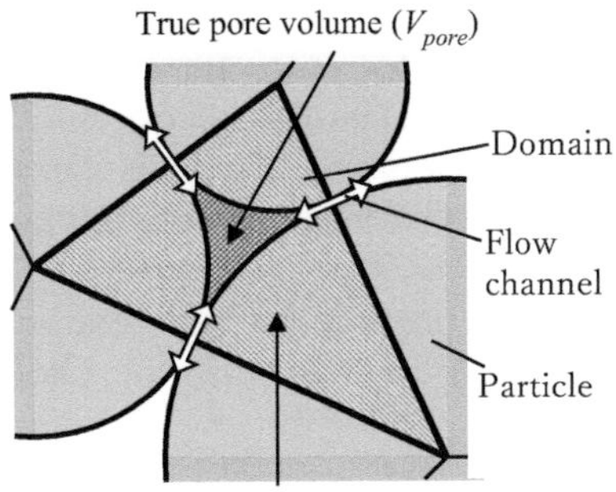

Figure 4.3 Channel–domain model (Shimizu et al., 2010, 2011).

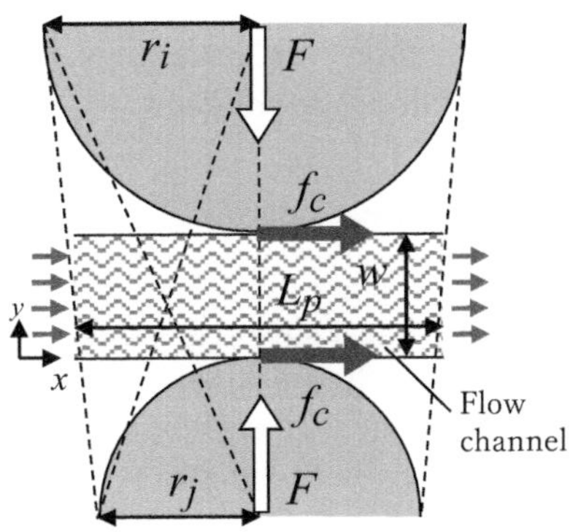

Figure 4.4 Fluid flow in a channel (Shimizu et al., 2010, 2011).

where w is the aperture, L_p is the length of the channel, ΔP is the difference in pressure across a channel, and μ is the viscosity of the fluid. Because the model is two-dimensional, an out-of-plane thickness is supposed to be a unit. L_p is assumed to be obtained from the harmonic mean of the radius of two particles r_i and r_j, and given by

$$L_p = \frac{4r_i r_j}{r_i + r_j} \tag{4.16}$$

According to Eq. (4.15), fluid flow never occurs when the two particles are in contact ($w = 0$). To avoid this, the w will be given by Eq. (4.17) relating to the compressive normal force F at the contact:

$$w = \frac{w_0 F_0}{F + F_0} \tag{4.17}$$

where w_0 is assumed to be the initial aperture for particles that are just touching, and F_0 is the normal force at which the channel aperture decreases to half of its initial aperture. Because the flow rate Q in Eq. (4.15) is the microscopic flow rate in one flow channel and the assembly of many flow channels expresses the fluid flow in a rock model, the entire rock model cannot be calculated directly from Eq. (4.15). Therefore, the value of w_0 is determined as the permeability of the entire rock model obtained by simulating the permeability test corresponds to the value of an actual specimen. This implies that tuning the value of w_0 can adjust the permeability of the entire rock model, and particle displacement updates the porosity and permeability of the rock model.

Each domain accumulates the fluid pressure acting on the surface of surrounding particles, and the fluid flow calculation updates the fluid pressure. The following continuity equation gives a change of fluid pressure dP:

$$dP = \frac{K_f}{V_r}\left(\sum Qdt - dV_r\right) \tag{4.18}$$

where $\sum Q$ is the total flow rate for a one-time step from the surrounding channels, dt is the duration in the one-time step, K_f is the fluid bulk modulus, V_r is the apparent volume of the domain, and dV_r is the change of the volume in the domain.

4.2.5 *Fluid flow and fluid pressure*

Following the fluid flow algorithm by Al-Busaidi et al. (2005), the fluid pressure of each domain acts on particles. In addition to this, the shear stress caused by fluid flow in the channel has been newly introduced in our own code. The fluid flow in the channel is supposed to be the two-dimensional Poiseuille flow. Therefore, the laminar flow between two parallel plates extending in the x-direction, as shown schematically in Figure 4.4, will now be considered. The plates are at the planes $y=0$ and $y=w$, and the flow is in the x-direction; hence, there is no velocity component in the y-direction. The velocity distribution u for laminar flow between parallel plates is a function of y only, and given by

$$u = \frac{w^2}{2\mu}\left\{\frac{y}{w} - \left(\frac{y}{w}\right)^2\right\}\frac{\Delta P}{L_p} \tag{4.19}$$

Viscous fluid flow along a solid boundary will induce a shear stress on that boundary. The shear stress at a surface element parallel to a plate, at the point y, is given by

$$\tau_f = \mu\frac{du}{dy} = \left(\frac{w}{2} - y\right)\frac{\Delta P}{L_p} \tag{4.20}$$

In particular, the wall shear stress is given by

$$\tau_{f0} = \frac{w}{2}\frac{\Delta P}{L_p} \tag{4.21}$$

Therefore, total force acting on a plate is given by

$$f_c = L_p\tau_{f0} = \frac{w}{2}\Delta P \tag{4.22}$$

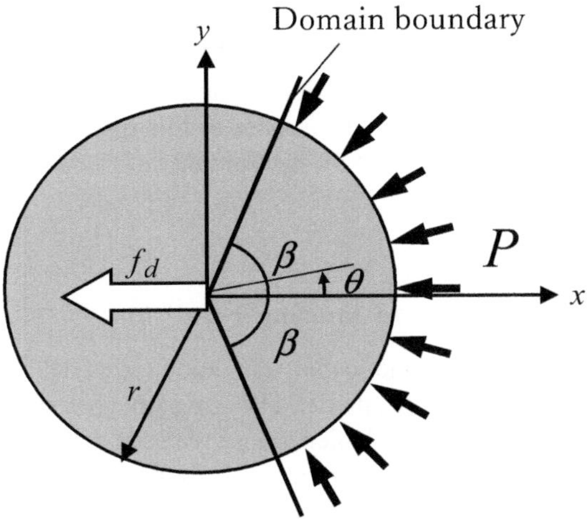

Figure 4.5 Fluid pressure acting on the particle (Shimizu et al., 2011).

where L_p is the length of the channel. As shown in Figure 4.4, the total force f_c is given to particles as a shear force that acts on the surface of two particles that form channels.

Each domain accumulates the fluid pressure, and the fluid pressure acts on the surface of surrounding particles as shown in Figure 4.5. When fluid pressure P acts on a particle whose radius is r, Eq. (4.23) gives the total force f_d that acts on the center of a particle:

$$f_d = \int_{-\beta}^{\beta} P \cos\theta r d\theta \tag{4.23}$$

where β is the corner half-angle of a domain.

As a result, by introducing the fluid flow algorithm, the shear stress caused by fluid flow and the fluid pressure accumulated in each domain act on particles.

4.2.6 *Saturation*

The fluid flow algorithm presented by Al-Busaidi et al. (2005) supposed that the entire model is always filled with fluid (consider only saturated condition). However, the specimen might not always be saturated in actual laboratory experiments. Such a condition with different saturations (saturated or partially saturated conditions) might influence the simulation results. Therefore, to consider the partially saturated conditions, fluid flow algorithm is further improved, and the saturation factor in each domain is introduced. The saturation factor in each domain is defined as

$$S_t = \frac{V_f}{V_r \varphi} \tag{4.24}$$

where V_r is the domain volume as shown in Figure 4.3, V_f is the volume of fluid in the domain, and φ is the supposed porosity of the model. Because the assembly of simple circular particles expresses the DEM model, it is difficult to accurately reproduce an actual rock's porosity. Therefore, the supposed pore volume of the domain obtained from the entire volume of the domain,

V_r, and supposed porosity, φ, are used in Eq. (4.24) instead of using the true pore volume of the domain, V_{pore}. For $1 < S_t$, the domain is filled with fluid, while for $1 > S_t$, the domain is partially saturated. When the partially saturated condition is considered, fluid pressure is supposed to be the same value as the atmospheric pressure (0 MPa in this research) in the partially saturated domain and to increase only after the domain is saturated.

4.3 Simulation condition

4.3.1 *Rock specimen model and loading condition*

Figure 4.6a illustrates the rock model and loading condition for the HF. The assembly of particles bonded with each other expresses the rock model. The size of the rock specimen is 200 mm in width and 200 mm in height. A borehole for fluid injection is created at the center of the rock model. The diameter of the borehole is 20 mm. The rock model is surrounded by the four confining walls. The left and under walls are fixed, and the right and upper walls can move to apply the constant confining pressure. Two confining pressures, 10 MPa in the x-direction and 5 MPa in the y-direction, were applied to the rock model. The frictional force was not applied between the model and the confining walls. The rock models used in the simulations were made according to the following procedures.

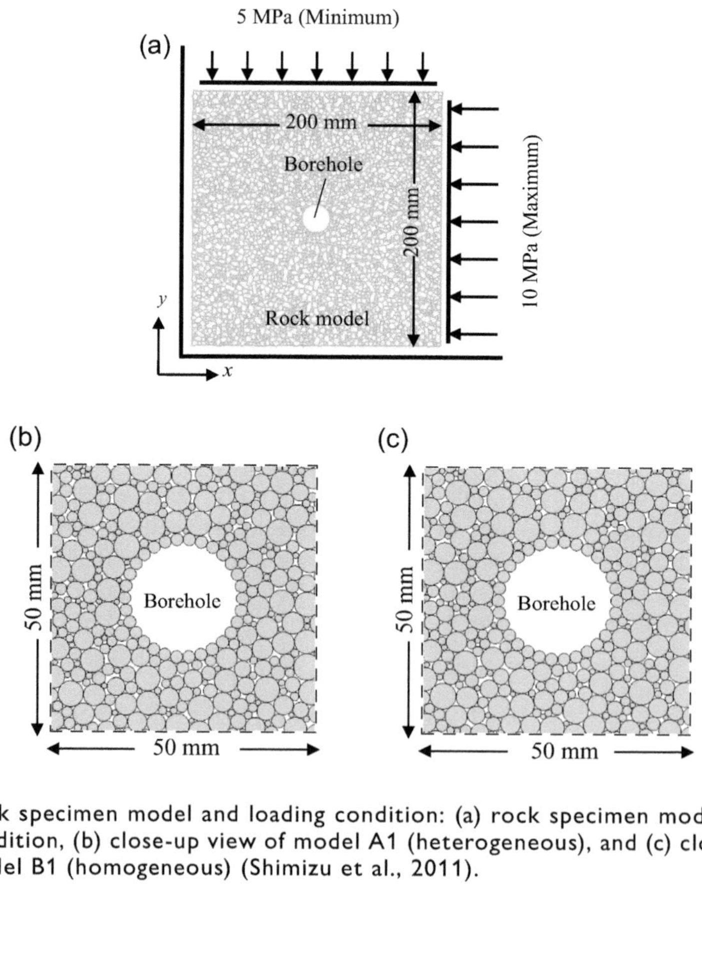

Figure 4.6 Rock specimen model and loading condition: (a) rock specimen model and loading condition, (b) close-up view of model A1 (heterogeneous), and (c) close-up view of model B1 (homogeneous) (Shimizu et al., 2011).

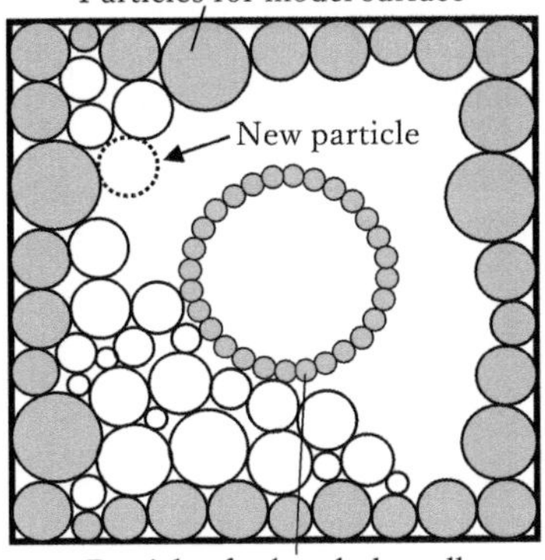

Figure 4.7 Particle packing method (Shimizu et al., 2010, 2011).

At first, particles were aligned along one edge of the model as shown in Figure 4.7. The particle radius was selected following a uniform distribution between the maximum and minimum radius using random numbers. In addition, particles that have the same radius are arranged in a circular form at the center of the model to form the inner wall of the borehole. During this process, the edge of the model and the surface of the borehole can be smoothed, and unnecessary stress concentration that originates in the model geometry can be avoided. After arraying particles along every side of the model and the inner wall of the borehole, the inside of the model is filled with particles. The new particle radius was selected following a uniform distribution between the maximum and minimum radius using random numbers and arranged to contact the pre-existing particles with at least three contact points. This process is repeated until a new particle with a minimum radius cannot be arranged.

In this study, two types of rock models with different ranges of particle radius were considered to investigate the influence of the particle size distribution on the HF behavior. For each type, three rock models were generated, and in total, six stochastic particle models were created. Models A1–A3 have the same particle size distribution with different random particle packings, and are called "heterogeneous models". The number of particles for a heterogeneous model is about 6,500, with the particle radius ranging from 0.5 to 2.5 mm. On the other hand, models B1–B3 are called "homogeneous models". The number of particles for a homogeneous model is about 18,000, with the particle radius ranging from 0.5 to 1.0 mm. Figure 4.8 shows the particle size cumulative curve for heterogeneous models and homogeneous models. Figure 4.6b and c illustrates the close-up view of the heterogeneous model A1 and the homogeneous model B1, respectively. Although the microscopic parameters such as Young's modulus of particles and strength of the spring are the same through a rock model, the stiffness of the bonding springs and the stress acting on the bonds are given as a function of the particle radius by Eqs. (4.9–4.13). Therefore, the stress distribution in the rock model is irregularly distributed according to the distribution of the particle radius. Models A1–A3 with wide ranges of particle radius become more heterogeneous than models B1–B3. For this reason, models A1–A3 are supposed to be heterogeneous models, while models B1–B3 are supposed to be homogeneous models. In addition, a low-viscosity fluid (0.1 mPa · s) and a high-viscosity fluid (100 MPa · s) are used as the fracturing fluid to investigate the influence of the fluid viscosity on HF behavior. Fracturing fluid was injected at a constant pressurizing rate.

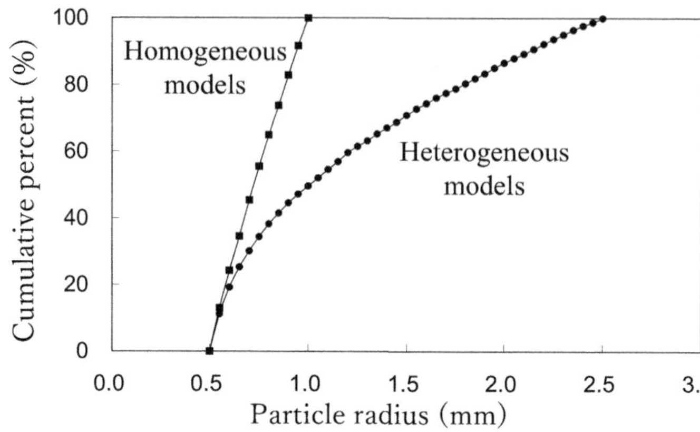

Figure 4.8 Particle size cumulative curves of the heterogeneous models (models A1–A3) and the homogeneous models (models B1–B3) (Shimizu et al., 2011).

4.3.2 Calibration

Preliminary simulations of uniaxial compression and Brazilian tests calibrated the microscopic mechanical parameters used in this simulation. The microscopic parameters in this study were calibrated using the macroscopic mechanical properties of Kurokamijima granite. The confining wall is assumed to be steel. The microscopic mechanical parameters used in this simulation and the calibration results are shown in Table 4.1. The macroscopic parameters of both rock models show good agreement with the experimental results.

As mentioned above, because the permeability of the rock model cannot be determined directly, the value of w_0 is calibrated as the permeability of the entire rock model obtained by simulating the permeability test. The permeability test was conducted using a rock model with a width of 200 mm and a height of 100 mm. Under the initial condition, all domains in the rock model are fully saturated. The fluid flow was established through the model by maintaining the fluid pressure P_{in} on the left-hand side of the model at 0.2 MPa and the fluid pressure P_{out} on the right-hand side of the model at 0.1 MPa. This pressure difference causes fluid flow only through the existing network of flow channels. Fluid flow does not occur through the upper and lower sides of the model. The simulation of the permeability test is continued until the inflow Q_{in} equals outflow Q_{out} and a steady flow state is achieved ($Q = Q_{in} \approx Q_{out}$).

Supposing the rock model as an isotropic medium and according to Darcy's law, the steady flow rate Q_{steady} is given as

$$Q_{steady} = \frac{kH\left(P_{in} - P_{out}\right)}{\mu W} \tag{4.25}$$

where μ and k are the viscosity of the fluid and the macroscopic permeability of the rock model, respectively. Therefore, the permeability of the rock model can be given by

$$k = \frac{Q_{steady}\mu W}{H\left(P_{in} - P_{out}\right)} \tag{4.26}$$

Table 4.1 Rock model properties and input parameters (Shimizu et al., 2011)

Microscopic parameters	*Heterogeneous*	*Homogeneous*
Number of particles	6,407	18,505
Maximum particle radius	2.5 mm	1.0 mm
Minimum particle radius	0.5 mm	0.5 mm
Porosity of the model	12.5%	18.5%
Width	200 mm	200 mm
Height	200 mm	200 mm
Borehole radius	10 mm	10 mm
Particle density	2,620 kg/m^3	2,620 kg/m^3
Young's modulus of the wall (E_w)	200 GPa	200 GPa
Friction coefficient of the wall ($\tan\varphi_w$)	0.0	0.0
Poisson's ratio of the wall (v_w)	0.3	0.3
Young's modulus of the particle (E_p)	66 GPa	84 GPa
Friction coefficient of the particle ($\tan\varphi_p$)	0.5	0.5
Poisson's ratio of the particle (v_p)	0.25	0.25
Shear/normal spring stiffness ratio (α)	0.55	0.7
Shear strength of bonding (τ_c)	157.5 MPa	249 MPa
Tensile strength of bonding (σ_c)	15.8 MPa	21 MPa
Assumed porosity of the model (po)	0.2%	0.2%
Initial saturation (S_t)	10%	10%
Initial aperture (w_0)	7.028 × 10^{-7} m	5.775 × 10^{-7} m
Bulk modulus of the fracturing fluid (K_f)	2.0 Gpa	
Fluid viscosity for the low-viscosity fluid (μ)	0.1 mPa·s	
Fluid viscosity for the high-viscosity fluid (μ)	100 mPa·s	

Calibration results	*Experiment*	*Simulation*	
	Kurokamishima-granite	*Heterogeneous*	*Homogeneous*
unconfined compressive strength (UCS) of the rock model (MPa)	200.0	199.5	199.9
Tensile strength (MPa)	10.0	10.2	10.2
Young's modulus (GPa)	70.0	70.3	70.2
Poisson's ratio	0.250	0.254	0.248
Permeability (m^2)		1.0 × 10^{-17}	1.0 × 10^{-17}

4.3.3 Calculation procedure for the stress distribution in the DEM model

In the DEM, a particle is in contact with many others, and the contact forces, such as normal and tangential forces acting on each contact point, are calculated individually. These contact forces are in various directions according to the direction of contact. Therefore, to discuss the stress state in the DEM model, the contact forces acting on each contact point should be converted into the stress distribution. The stress acting on a particle is calculated by the following procedure in this research.

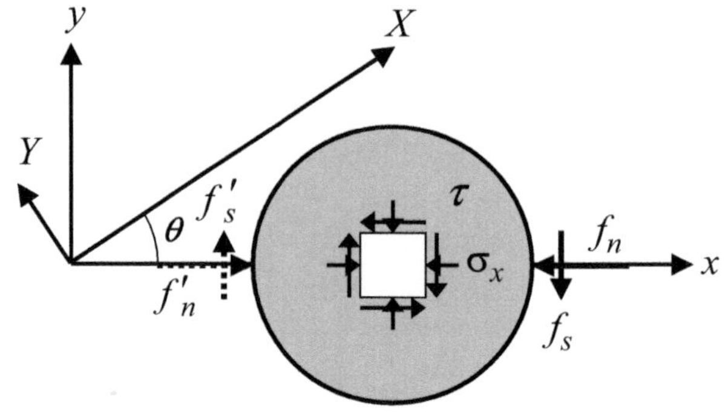

Figure 4.9 Normal and shear stress acting on a particle (Shimizu et al., 2011).

As shown in Figure 4.9, the normal and tangential contact forces f_n and f_s are acting on a particle, and the imaginary forces f_n' and f_s' are supposed. The two sets of forces, f_n' and f_n, f_s' and f_s, are of the same magnitude and opposite in direction, respectively. The normal force f_n is parallel to the x-axis. Assume that these forces are acting on a small area as shown in Figure 4.9, and are in equilibrium. The stresses that act on the small area are given by

$$\sigma_x = \frac{f_n}{D} \tag{4.27}$$

$$\sigma_y = 0 \tag{4.28}$$

$$\tau = \frac{f_s}{D} \tag{4.29}$$

where D is the contact width, given by Eq. (4.5).

Then, the normal and shear stresses, σ_x, σ_y, and τ_{xy}, are obtained at the center of the particle, expressed concerning the local coordinates x and y. As shown in Figure 4.9, the new global coordinates X and Y will now be considered, and the new stresses, σ_x', σ_y', and τ_{xy}' expressed in the global coordinates X and Y are given as follows (Jaeger et al., 2007):

$$\sigma_x' = \frac{\sigma_x + \sigma_y}{2} + \frac{\sigma_x - \sigma_y}{2}\cos 2\theta + \tau_{xy}\sin 2\theta \tag{4.30}$$

$$\sigma_y' = \frac{\sigma_x + \sigma_y}{2} - \frac{\sigma_x - \sigma_y}{2}\cos 2\theta - \tau_{xy}\sin 2\theta \tag{4.31}$$

$$\tau_{xy}' = -\frac{\sigma_x - \sigma_y}{2}\sin 2\theta + \tau_{xy}\cos 2\theta \tag{4.32}$$

where θ is the rotation angle from the local coordinates x and y, with counterclockwise angles taken as positive.

These calculations are applied to all contact points of each particle, and the total stresses, σ_x, σ_y and τ_{xy}, are obtained from the accumulation of the stress calculated by Eqs. (4.30–4.32) for each contact point. Because the imaginary forces f_n' and f_s' are supposed, the total stresses are divided by two. The total stresses, σ_X, σ_Y, and τ_{XY}, are given by

$$\sigma_X = \frac{\sum \sigma'_{xi}}{2} \tag{4.33}$$

$$\sigma_Y = \frac{\sum \sigma'_{yi}}{2} \tag{4.34}$$

$$\tau_{XY} = \frac{\sum \tau'_{xyi}}{2} \tag{4.35}$$

As a result, the stresses that act on each particle are obtained. However, these are discrete data. To investigate the stress distribution's tendency and character in detail, the continuation of the obtained discrete data (interpolation) is necessary. In this research, Inverse Distance Weighting was used as a technique for interpolation. Inverse Distance Weighting is one of the most commonly used techniques for interpolation, and is based on the assumption that the interpolating surface is a weighted average of the discrete data, and the weight assigned to each discrete data point diminishes as the distance from the interpolation point to the data point increases. The simplest form of Inverse Distance Weighting is called "Shepard's method" (Shepard, 1968). The equation used is the following:

$$Z_j = \frac{\sum_{i=1}^{n} Z_i d_{ij}^{-m}}{\sum_{i=1}^{n} d_{ij}^{-m}} \tag{4.36}$$

where Z_j denotes an interpolated (arbitrary) value, Z_i is the discrete data ($i = 1, 2, \ldots\ldots, n$), d_{ij} is the distance from the data point to the interpolation point, n is the total number of data points within the maximum distance from the interpolation point, and m is the positive real number, called the power parameter, and controls how the weighting factors drop off as the distance from the reference point increases. For $0 < m < 1$, the interpolated value Z_j has smooth peaks over the interpolated points, whereas for $m > 1$, the peaks become sharp. Here, the maximum distance from the interpolation point to the data point is four times as large as the maximum particle radius in the DEM simulation, and the power parameter that $m = 1$ is employed in this research.

4.4 Summary of the simulation results

As mentioned in Section 4.3.1, the rock models A1–A3 are heterogeneous models, whereas the rock models B1–B3 are homogeneous models. The low-viscosity fluid (0.1 mPa · s) and the high-viscosity fluid (100 mPa · s) were used as fracturing fluids. As a result, 12 cases of HF simulations with different combinations of rock model and fracturing fluid were performed. In all cases, two confining pressures, 10 MPa in the x-direction and 5 MPa in the y-direction, were applied to the rock model.

Simulation results, such as crack initiation pressure, BD pressure, and the number of microcracks generated during the simulation, are shown in Table 4.2. In this chapter, the case using model A1 and low-viscosity fluid is called "case A1—low", and the other cases are named in the

Table 4.2 Summary of the simulation results (Shimizu et al., 2011)

	Crack initiation pressure (MPa)	*BD pressure (MPa)*	*The number of microcracks*	
			Tensile crack (%)	*Shear crack (%)*
Case A1-low	19.73	21.40	151 (93.8)	10 (6.2)
Case A1-high	23.84	26.09	144 (94.1)	9 (5.9)
Case A2-low	18.40	18.89	159 (94.6)	9 (5.4)
Case A2-high	21.14	37.41	260 (90.9)	26 (9.1)
Case A3-low	20.91	20.96	131 (95.6)	6 (4.4)
Case A3-high	23.39	27.20	154 (92.2)	13 (7.8)
Case B1-low	20.26	21.34	261 (98.5)	4 (1.5)
Case B1-high	23.84	40.42	258 (91.2)	25 (8.8)
Case B2-low	20.02	22.32	236 (97.1)	7 (2.9)
Case B2-high	24.74	39.46	319 (90.9)	32 (9.1)
Case B3-low	17.32	20.21	314 (99.0)	3 (1.0)
Case B3-high	23.79	37.95	247 (90.8)	25 (9.2)

same manner. As shown in Table 4.2, when the high-viscosity fluid was used, the crack initiation pressure and the BD pressure became higher than those with low-viscosity fluid, regardless of the rock model.

Because the number of particles of a homogeneous model is larger than that of a heterogeneous model, the number of bonds between particles of a homogeneous model is also larger than that of a heterogeneous model. As a result, the total number of microcracks becomes large when a homogeneous model is used.

In all cases, the percentage of tensile crack generation is 90% or more. The conventional theory presumes that the HF fracture is created by a tensile crack that extends along the direction of maximum compressive principal stress (Hubbert and Willis, 1957). From this, even though a few shear cracks were generated, tensile cracks were dominant in all cases, as expected in theory. However, this result does not agree with the result obtained from the laboratory AE measurement experiments, as shown in Section 3.2 in this book. This disagreement between the simulation results and the experimental results will be discussed in Section 4.7.1.

In almost all cases, the percentage of shear crack generation was small when the low-viscosity fluid was used. This tendency is clearly shown in cases using the homogeneous models (models B1–B3). On the other hand, when a heterogeneous model (models A1–A3) was used, the influence of the fluid viscosity on the crack mode became relatively small except for the case A2—high. The shear crack generation in case A2—high will be explained in Section 4.7.2.

In this study, two types of rock models and two types of fracturing fluid were used. When the combination of the type of rock model and the fracturing fluid is the same, similar results were obtained, as shown in Table 4.2. Therefore, the cases A1—low, A1—high, B1—low, and B1—high are considered to be the representative cases in this study. Figure 4.10 illustrates the geometry of the fracture formed in these cases. For all cases, the orientation of the HF fractures was parallel to the direction of maximum compressive principal stress, as expected in theory. These results indicate that the effect of the confining stress was appropriately expressed in the DEM simulations. Figure 4.11a–d shows the location of the x-coordinates of the cracks versus the time step. The crack mode is identified based on the criterion described in Section 4.2.3. A tensile crack is shown with a closed circle and a shear crack with an open circle. Figure 4.11e

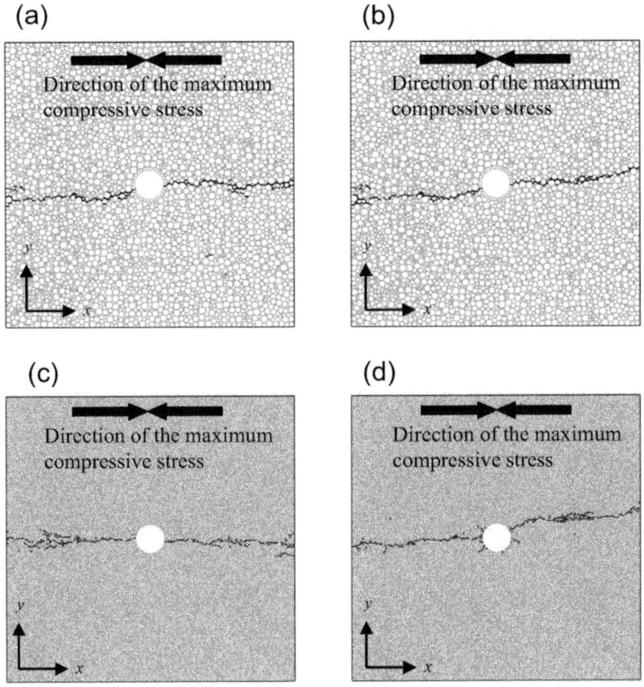

Figure 4.10 Spatial distribution of all the cracks obtained from each case. The solid lines indicate the crack generations: (a) case A1—low, (b) case A1—high, (c) case B1—low, and (d) case B1—high (Shimizu et al., 2011).

shows the evolution of the fluid pressure in the borehole. HF has been initiated before the BD (peak) pressure in all cases. This result shows good agreement with the HF process deduced from AE measurements in the laboratory experiments conducted by some researchers (Zoback et al., 1977; Zhao et al., 1996).

4.5 Crack initiation pressure

As shown in Table 4.2, when the low-viscosity fluid was used, the crack initiation pressure was lower than that with the high-viscosity fluid. For the cases using the low-viscosity fluid, the crack initiation pressure was about 20 MPa. On the other hand, the crack initiation pressure was about 23 MPa for the cases using the high-viscosity fluid. The effect of fluid infiltration and pore pressure gradient around the borehole can explain this result. There is a large number of small pores inside rocks. When the borehole pressure increases with fluid injection, fracturing fluid penetrates the interconnected pores of rock from the borehole wall. The fluid penetration causes an additional pore pressure around the borehole. The pore pressure reduces the effective stress of rock around the borehole and enhances the generation of microcracks (Schmitt and Zoback, 1993). To investigate the effect of fluid infiltration on the tensile rupture of porous rock, Haimson and Fairhurst (1967) and Haimson (1968) developed a sophisticated model theoretically to consider fluid penetration. Moreover, Ito (2008) and Ito and Hayashi (1991) developed a new

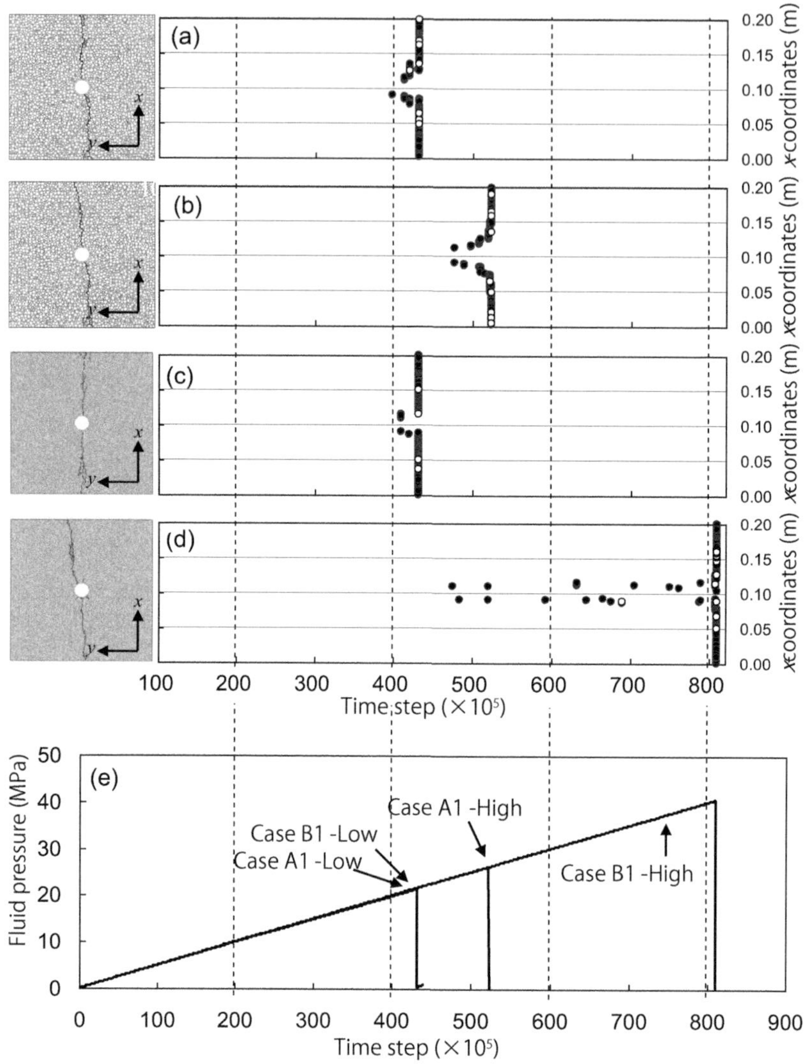

Figure 4.11 Results of the HF simulation. Time–space distribution of cracks generated in each case: (a) case A1—low, (b) case A1—high, (c) case B1—low, (d) case B1—high, and (e) evolution of borehole pressure (Shimizu et al., 2011).

theory based on the point stress criterion. This theory supposes that fracture initiation occurs when the maximum tensile effective stress first reaches the tensile strength of a rock at a point not on the wellbore surface but inside the rock.

Figure 4.12 shows a close-up view of the rock model around the borehole for each case at the time step just before the crack initiation. The solid lines indicate the fluid penetration (saturated) area, and each domain's shade indicates the fluid pressure. As mentioned in Section 4.2.5, the saturated area is judged by the saturation factor, St, calculated by Eq. (4.24). For $St > 1$, the domain is supposed to be saturated. When the low-viscosity fluid was used, the fracturing fluid

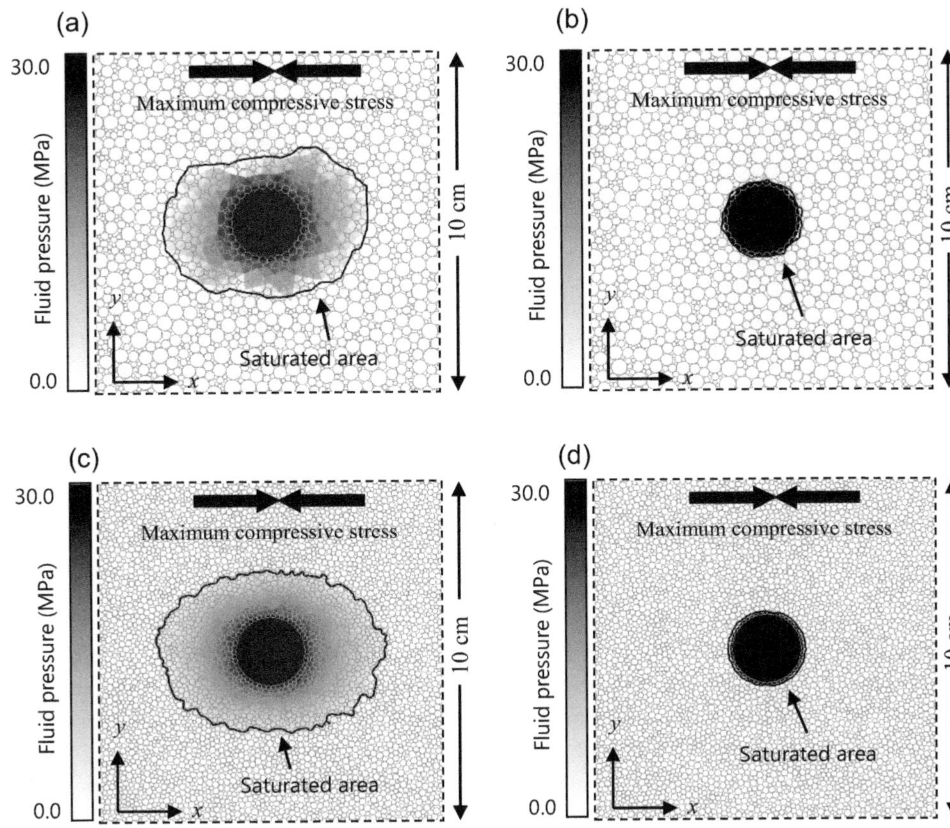

Figure 4.12 Close-up view of the rock model around the borehole in each case at the time step just before the crack initiation. The solid lines indicate the fluid penetration (saturated) area, and the shade of each domain indicates the fluid pressure: (a) case A1—low, (b) case A1—high, (c) case B1—low, and (d) case B1—high (Shimizu et al., 2011).

widely infiltrated the rock model from the borehole wall and increased fluid pressure around the borehole. According to Eq. (4.17), the compressive normal force at the contact point affects the flow rate between particles. When the compressive normal force increases, the aperture of the channel, w, decreases. As a result, because the maximum confining pressure is 10 MPa in the x-direction, fluid flow perpendicular to the x-axis decreases. Thus, the fluid saturation area is not a circle but an oval shape. On the other hand, fracturing fluid did not infiltrate into the rock model when high-viscosity fluid was used.

Figure 4.13a shows the spatial distribution of the maximum principal stress at the initial step. Figure 4.13b–c shows the spatial distribution of the maximum principal stress in cases B1—low and B1—high at the time step just before the crack initiation, respectively. The calculation procedure for the stress distribution in the DEM model is described in Section 4.3.3. The distribution of the principal stresses is calculated based on the forces acting on the particles. In short, Figure 4.13 shows the distribution of the effective stress.

At the initial step (borehole pressure is 0 MPa), the maximum principal stress exhibits mainly compressive stress in the entire model and is the highest at the borehole surface across the

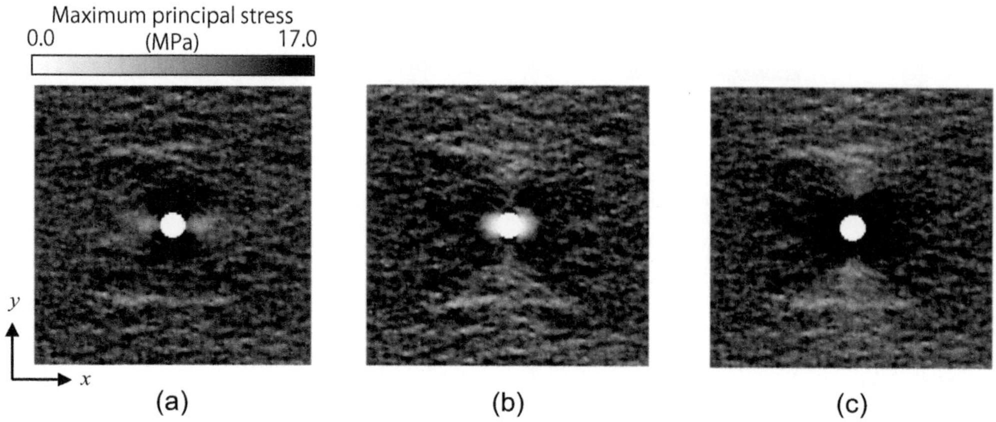

Figure 4.13 Spatial distribution of the magnitude of the maximum principal stress at the initial step and at the time step just before the crack initiation: (a) initial step (fluid pressure: 0 MPa); (b) case B1—low (fluid pressure: 20.26 MPa); (c) case B1—high (fluid pressure: 23.84 MPa) (Shimizu et al., 2011).

y-axis, as shown in Figure 4.13a. This agrees with conventional elastic theory. However, as shown in Figure 4.13b, the maximum principal stress around the injection hole decreased when the low-viscosity fluid was used. On the other hand, when the high-viscosity fluid was used, such a decrease in effective stress was not observed, and the maximum principal stress around the borehole increased due to the borehole pressure, as shown in Figure 4.13c. This result indicates that a decrease in the crack initiation pressure in cases that use low-viscosity fluid is caused by the decrease in effective stress due to the rise of fluid (pore) pressure around the borehole.

In actual HF experiments, it is difficult to observe the infiltration behavior of fluid and the change of stress distribution due to the fluid injection directly. On the contrary, the effect of fluid infiltration on the tensile rupture of permeable rock can be successfully reproduced by the coupled fluid flow and the DEM. However, several problems remain to be investigated. Schmitt and Zoback (1992) discussed the infiltration and failure problem of HF in detail and suggested that the pore pressure in a low porosity specimen under rapid strain conditions is diminished due to elastic pore volume expansion. This results in a higher effective confining stress near the borehole surface and requires a higher borehole pressure at fracture initiation. Brace and Martin (1968) called this mechanism "dilatancy hardening". In contrast, as mentioned in Section 4.2.6, fluid pressure in the partially saturated domain is supposed to be 0 MPa in the simulation. Thus, the dilatancy hardening explained above is not considered accurate in this study. For further investigation into the effect of the fluid infiltration, improvement of the DEM code will be needed in the future.

4.6 BD pressure

As shown in Table 4.2, when the high-viscosity fluid was used, the BD pressure was markedly higher than that with the low-viscosity fluid. This result agrees well with the theory (Zoback et al., 1977; Zoback and Pollard, 1978), and the comparison between the simulation result of HF by the DEM and the fracturing process led by Linear Elastic Fracture Mechanics (LEFM) can explain the result. In addition, the results also agree well with the results of laboratory experiments using supercritical carbon dioxide and liquid state carbon dioxide, water, and viscous oil that will be shown in the next chapter, Chapter 5, although the experimental results described in Chapter 3 that used only the two fracturing fluids, water and viscous oil, could not show clearly the result.

Generally speaking, LEFM is often used as an analytical technique for propagating the fracture in rock-like materials. LEFM is a theory that supposes application to the continuum. On the other hand, the DEM is basically a discontinuity analysis technique for granular materials. Therefore, the breakage of individual bonds that connect particles in the DEM cannot be directly compared with LEFM. However, by supposing the connection of individual microcracks in the DEM as a propagation of one macroscopic fracture, the simulation result by the DEM can be interpreted by LEFM. Such research has already been reported and shows that the fracture strength calculated from a macroscopic fracture in the DEM agrees well with the one calculated from LEFM (Potyondy and Cundall, 2004; Huang, 1999; Moon et al., 2007).

Figure 4.14a–b shows the fluid pressure acting on the inside of HF fractures during low-viscosity and high-viscosity fluid injection, respectively. The solid lines indicate the crack generations, and the shade of each domain indicates the fluid pressure. As shown in Figure 4.14a,

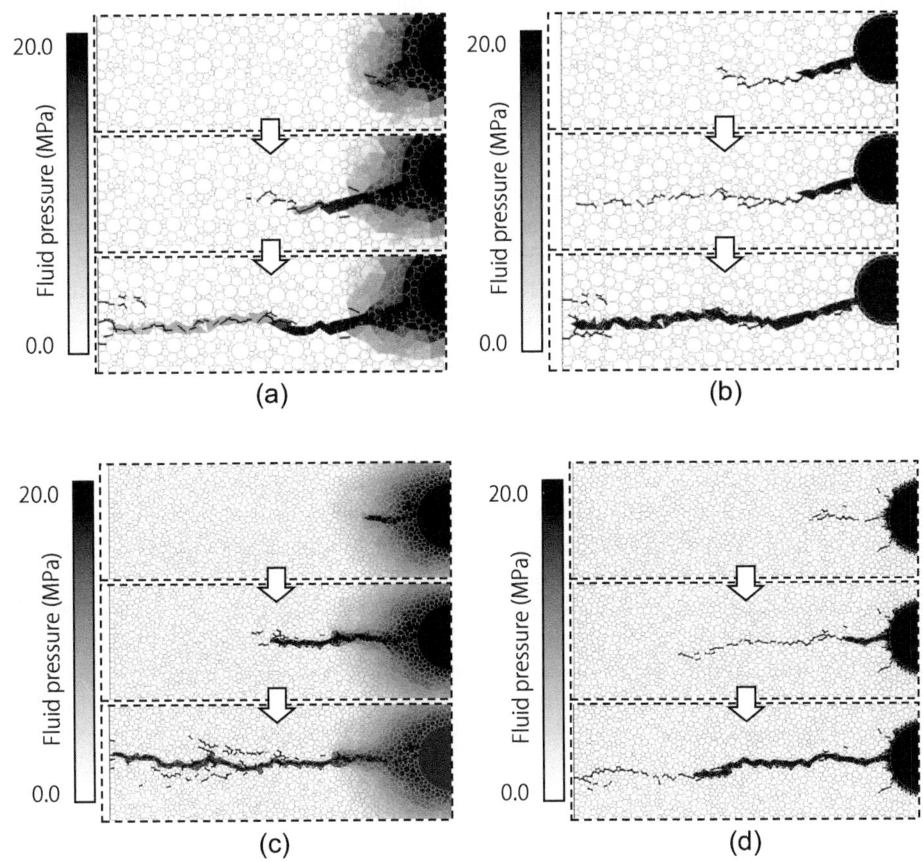

Figure 4.14 Fracture propagation and fluid infiltration behavior. The solid lines indicate the crack generations, and the shade of each domain indicates the fluid pressure: (a) case A1—low, (b) case A1—high, (c) case B1—low, and (d) case B1—high (Shimizu et al., 2011).

when a low-viscosity fluid was used, the fluid infiltrated into the fracture immediately, and the fluid pressure was applied throughout the fracture surface. On the other hand, when high-viscosity fluid was used, only the fracture elongated first. Then, the fluid infiltrated slowly into the fracture, and fluid pressure was applied only to a part of the fracture surface.

Newman (1971) theoretically derived stress intensity formulas for the abovementioned cases, and Zoback and Pollard (1978) also showed similar formulas. As shown in Figure 3.23 in the previous chapter, in the case that pressure acts all over the fracture surface, the stress intensity factor monotonically increases with crack length. Thus, the fracture propagates continuously without stopping once it is initiated. On the other hand, in the case that pressure acts only in the borehole, the stress intensity factor slowly decreases with crack length. Therefore, when high-viscosity fluid was used as the fracturing fluid, the fracture never extended without additional pressure. Thus, the simulation result of HF by the DEM showed good agreement with the fracturing process explained by LEFM, and the effect of fluid viscosity on the BD pressure was discussed.

As shown in Table 4.2, the crack initiation pressures of heterogeneous and homogeneous models were almost the same when the same fracturing fluid was used. As mentioned in the previous section, the macroscopic stress state around the borehole strongly affects crack initiation pressure, and the macroscopic parameters of both rock models are the same (see Table 4.1). Therefore, this result seems reasonable. However, as shown in Table 4.2, the BD pressure in the cases using a homogeneous model was remarkably higher than that in the cases using a heterogeneous model when the high-viscosity fluid was used.

This result can be explained by the difference in crack generation processes. The propagation of fracture during HF is affected by the stress intensity factor at the fracture tips, as discussed above. This fact implies that the BD pressure is strongly affected by the microscopic properties around the fracture tip. In this study, the strength of the springs for homogeneous models is larger than that for heterogeneous models, although the other macroscopic parameters of both rock models are the same. As shown in Section 2.2 in the previous chapter, the laboratory HF experiment conducted by Ishida et al. (2000a,b) revealed that the BD pressure decreased with increasing grain size of the rock specimen, and they gave possible explanations for this result as follows. In an actual rock specimen, there are many pre-existing flaws such as pores, microcracks, and grain boundaries, and these grain-scale discontinuities induce complex macroscopic behaviors. When a rock specimen contains a lot of large grains, the flaws between the large grains become large, and the flaws become triggers for the propagation of the fractures. As a result, BD pressure decreases with increasing grain size.

For these reasons, BD pressures in the cases using a homogeneous model became remarkably high. Meanwhile, when the low-viscosity fluid is used and the fluid pressure acts all over the fracture surface, the stress intensity factor monotonically increases with crack length as mentioned above. In this case, the effect of the microscopic properties would not be essential since the fracture propagates continuously without stopping once it is initiated.

4.7 Geometry and microcracking mechanism of HF

4.7.1 *HF in the heterogeneous model*

As shown in Figure 4.10a–b, the geometry of the induced fracture in cases A1—low and A1—high seems almost the same. However, the tendency of the microcrack generation was different due to the difference in fluid viscosity. To quantify the geometry of the fracture, the total crack length and the average crack aperture are introduced. The total crack length is defined as the sum of the length of all microcracks. A crack length is supposed to be the same as the bond diameter D that is given by Eq. (4.5). Because the model width is 200 mm and the borehole diameter is 20 mm, the total crack length becomes 180 mm when the fracture is completely straight without branches. Consequently, a large total crack length means that the geometry of the fracture is more complex. The average crack aperture is calculated at the time step when the fracturing fluid reaches the model boundary. The total crack lengths in A1—low and A1—high were 199 and 186 mm, respectively. On the other hand, the average crack apertures in A1—low and A1—high were 0.090 and 0.181 mm, respectively. These results indicate that the fracture in case A1—low is thin and wavelike with many secondary branches, and that the fracture in case A1—high is thick and planar with few branches.

Figure 4.15a–b shows the spatial distribution of all the microcracks generated during the HF simulations in cases A1—low and A1—high, respectively. The closed circle indicates a location of a tensile crack, and the open circle indicates that of a shear crack. The diameters of the circle correspond to the respective magnitudes of energy obtained by Eq. (4.14). As shown in Figure 4.15a, when the low-viscosity fluid was used, the energy emitted from the shear crack was larger than that from the tensile crack. It is theoretically indicated that the energy emitted from a tensile crack is smaller than that from a shear crack (Hayashi and Motegi, 1988). The simulation results are consistent with the theory. When the low-viscosity fluid was used, shear

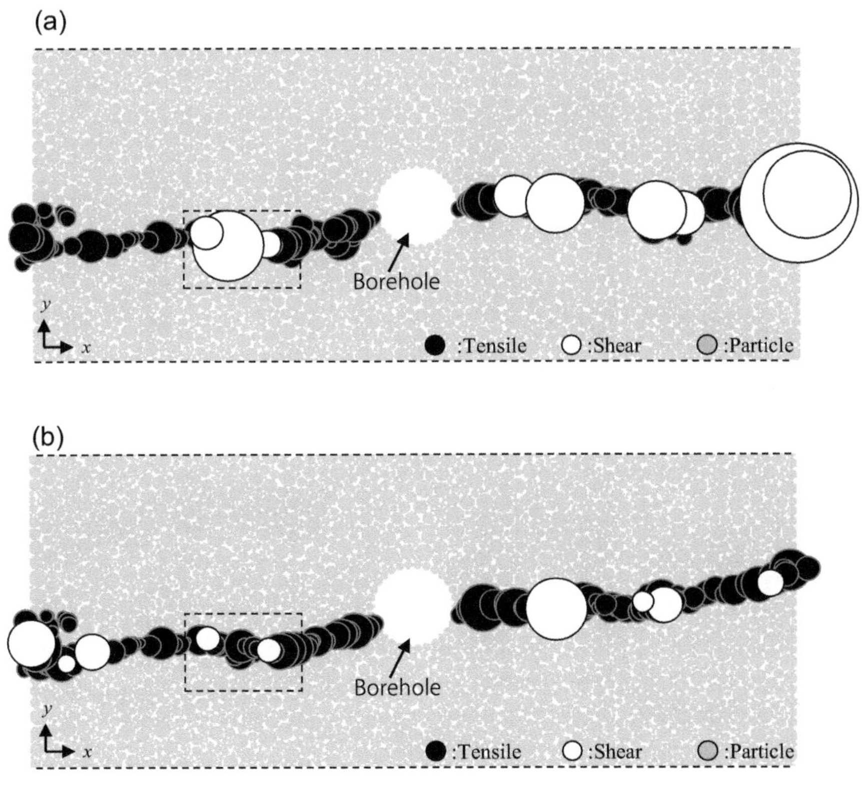

Figure 4.15 Crack types and magnitude of energy emitted from the cracks in (a) cases A1—low and (b) A1—high (Shimizu et al., 2011).

cracks emitting significantly large amounts of energy were generated in the region enclosed by a dotted line square in Figure 4.15a. On the other hand, when the high-viscosity fluid was used, the energy emitted from the shear crack was relatively small, as shown in Figure 4.15b, and such a shear crack emitting large amounts of energy was not generated at the same position (see the dotted line square in Figure 4.15b) even though the same rock model was used. The difference in the fracturing process due to the difference in fluid viscosity caused this result.

Figure 4.16a illustrates a close-up view of the fracture propagation and fluid filtration behavior in case A1—low. Tensile and shear cracks are shown as thick solid lines and open ellipses. When low-viscosity fluid was used, the fracture propagated in the direction of maximum compressive stress, and the fluid immediately infiltrated into the fracture. At this time, the fracture

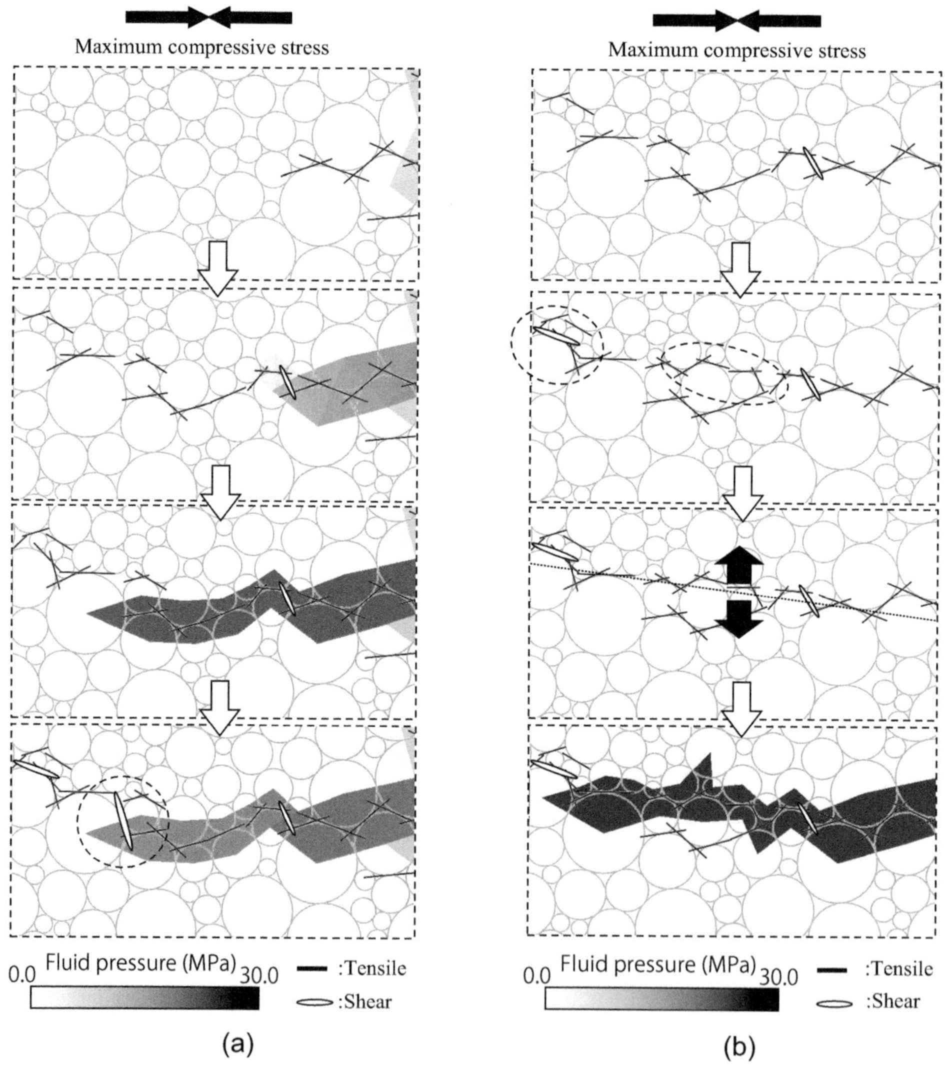

Figure 4.16 Close-up view of the fracture propagation and fluid filtration behavior: (a) case A1—low and (b) case A1—high (Shimizu et al., 2011).

was mainly formed by tensile failure, whereas shear cracks were few. However, when a large particle existed on the course of the fracture growth, propagation of the fracture was obstructed, and the fluid pressure was applied throughout the fracture surface. Beyond the obstructing particle, tensile cracks were generated in front of the fracture tip due to the fluid pressure acting on the fracture surface because the tensile strength of the bond was small compared with the shear strength. Finally, when the fluid pressure acting on the fracture tip sufficiently increased, shear cracks emitting significantly large amounts of energy were formed to connect these tensile cracks. This fracturing process is similar to Hill's model (Hill, 1977) as shown in Figure 2.30, which was originally proposed for volcanic earthquake swarms as explained in Section 2.5.2.

On the other hand, Figure 4.16b illustrates a close-up view of the fracture propagation and fluid infiltration behavior in case A1—high. When high-viscosity fluid was used, the fracture propagated first without fluid infiltration. At this time, the microcrack geometry was similar to that in case A1—low. However, the fracturing fluid did not immediately infiltrate into the fracture because the fluid viscosity was high. The fluid pressure was applied to the borehole and only a part of the fracture surface. As mentioned in Section 4.3.1, the microscopic parameters given to the bonds between particles, such as Young's modulus of particles and the strength of the spring, are constant in this research. However, since the stiffness of the bonding springs and the stress acting on the bonds are given as a function of the particle radius by Eqs. (4.9–4.13), the strain energy given by Eq. (4.14) becomes a function of the particle radius. Hence, the magnitude of the energy emitted along with microcrack generation was irregularly distributed according to the distribution of the particle radius of the rock model. The microcracks are likely to be generated between small particles, and the energy emitted from such microcracks becomes small.

As borehole pressure increased, many microcracks were generated between small particles without infiltration of the high-viscosity fluid. In this case, the number of shear cracks depends on the particle arrangement because the shear cracks are generated along the grain boundaries that are diagonally across the direction of maximum confining stress. When these microcracks were connected to form one straight fracture, the fracture width gradually increased due to the borehole pressure. Finally, high-viscosity fluid infiltrated into the existing fracture when the fracture width was sufficiently increased. As a result, the fracture became thick planar with few branches when high-viscosity fluid was used, and the shear crack emitting large amounts of energy observed in case A1—low was not generated in case A1—high. The microscopic observation in laboratory experiments by Matsunaga et al. (1993) and Ishida et al. (2004) showed that HF fractures were mainly located at the grain boundary. They also pointed out that the fractures induced by high-viscosity oil injection could be observed very clearly because they had large widths, whereas fractures induced by low-viscosity water injection could only be detected after careful and close observation because they had only extremely thin widths. In addition, high-viscosity oil tends to generate thick planar fractures with few branches, while low-viscosity water tends to generate thin and wavelike cracks with many secondary branches. By considering the particles in the DEM as the mineral grains, the fracture propagation process in the DEM simulations mentioned above gave a rational explanation for the HF behavior observed in the experiment.

As shown in Table 4.2, tensile cracks were dominant in all cases, although a few shear cracks were generated. This result does not agree with the result obtained from the laboratory AE measurement experiments conducted by Ishida et al. (2004), as shown in the previous chapter, Chapter 3. Recorded AE waveforms in their HF experiment indicated that the shear-type fracturing seemed dominant in low-viscosity water injection, and both tensile and shear AE were recorded in high-viscosity oil injection. This disagreement between the results obtained from the simulation and experiment can be explained by considering the energy of microcracks.

As mentioned above, when a low-viscosity fluid is used as the fracturing fluid, the low-viscosity fluid can easily infiltrate into the pores, defects, and microcracks, and the HF fracture has tortuous paths following the boundary of the mineral grain. When the fluid pressure on the fracture tip increases sufficiently, shear cracks emitting large amounts of energy are formed to connect pre-existing microcracks. Because the tensile strength of rock is small compared with the compressive strength, the energy emitted from a tensile crack is small compared with that from a shear crack. Such a small AE is easily buried in noise and difficult to measure in the experiments. As shown in Chapter 3, only a few percent of the located AE events showed clearly the first motions of the P-wave and were possible to obtain reliable mechanism solutions, even though several hundred AE sources were located during HF experiments conducted by Ishida et al. (2004). Hence, only for about five events in each specimen, their fracturing mechanisms could be examined based on the polarities of P-wave first motions (Ishida et al., 2001, 2004). Therefore, when a low-viscosity fluid was used, the shear-type AE with large energy was dominantly observed in AE measurement experiments conducted by Ishida et al. (2004).

On the other hand, when the high-viscosity fluid is used, fracturing fluid cannot infiltrate into the microcracks, and a straight fracture is formed according to the stress state in the rock due to the increase in the borehole pressure. The fracturing fluid will infiltrate along the existing straight fracture after enough opening of the fracture due to the borehole pressure. Although a few shear cracks are generated along the grain boundaries that diagonally cross the direction of maximum confining stress, the energy emitted from these shear cracks is small, as shown in Figure 4.15b, because such microcracks are likely to be generated between small particles. Therefore, very few shear cracks emitting large amounts of energy are observed when the high-viscosity fluid is used. In addition, when the high-viscosity fluid infiltrates into the fracture, fluid pressure becomes very high. Thus, in an actual HF, highly pressurized fluid causes large fracture openings, and tensile AE with large energy may be generated. As a result, both tensile and shear AE were observed during the injection of high-viscosity fluid in the AE monitoring conducted by Ishida et al. (2004), as shown in Chapter 3.

4.7.2 HF in the homogeneous model

Figure 4.17 shows the spatial distribution of all the microcracks and the magnitude of energy in cases B1—low and B1—high, respectively. By comparing Figure 4.15 with Figure 4.17, it is found that the energy emitted from the microcrack generation in model B1 was relatively small compared with that in model A1. This tendency agrees well with the results of the uniaxial experiments conducted by Eberhardt et al. (1999) and the laboratory HF experiment by Ishida (2001). Their experimental results showed that the number of detected AE events decreased markedly with decreasing grain size. As mentioned in Section 4.7.1, the energy emitted along with microcrack generation is irregularly distributed following the distribution of the particle radius of the rock model. The energy emitted from a microcrack generated between small particles becomes small. As a result, the energy emitted from the microcrack generation in the homogeneous models consisting of small particles was relatively small compared with that in the heterogeneous models, including many large particles.

Because the fracture path in case B1—low was markedly different from that in case B1—high as shown in Figure 4.10c–d, the geometry of the fracture could not be compared simply by the total crack length and the average crack aperture. However, Figure 4.17a clearly shows

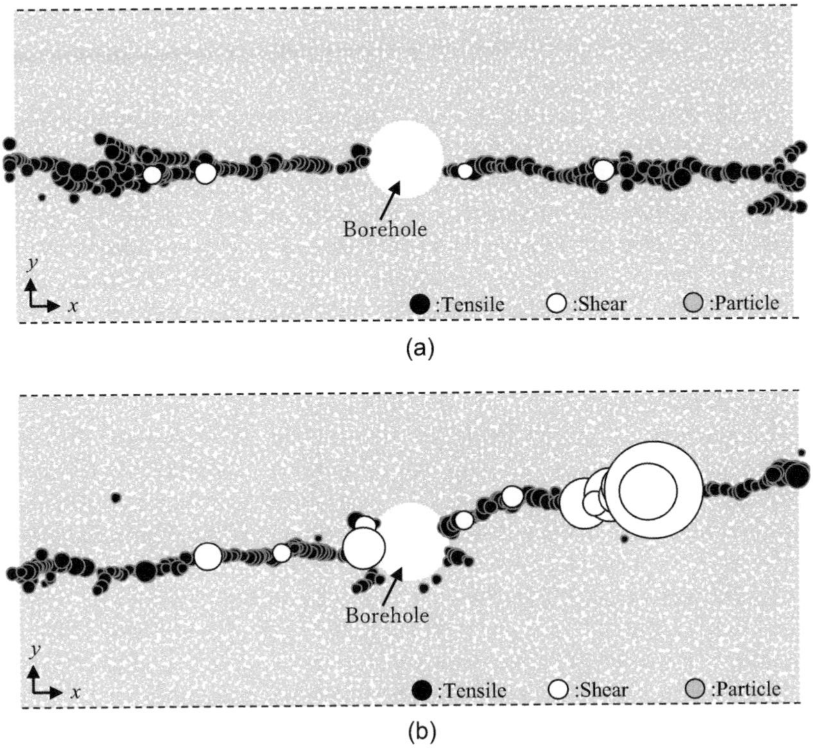

Figure 4.17 Crack types and magnitude of energy emitted from the cracks in cases B1—low and B1—high (homogeneous model). The diameter of the circle corresponds to the magnitude of crack energy: (a) case B1—low and (b) case B1—high (Shimizu et al., 2011).

that many small branching cracks were generated when the low-viscosity fluid was used. On the other hand, as shown in Figure 4.17b, such a branching crack was not generated when the high-viscosity fluid was used. This result supports the consideration related to the influence of the viscosity of the fracturing fluid discussed in the previous section, Section 4.7.1.

As shown in Table 4.2, the percentage of shear crack generation was very small when a homogeneous model and a low-viscosity fluid were used (cases B1—low, B2—low, and B3—low). This tendency agrees with the experimental results. The fault plane solutions of AE implied shear-type fracturing in the specimens with large grains, whereas they implied tensile fracturing in the specimens with small grains (Matsunaga et al., 1993; Ishida, 2001). In addition, Sasaki et al. (1988) reported that all recorded AE events indicated tensile fracturing mechanisms in HF in an acrylic resin block, as shown in Section 2.3.2. The acrylic resin is impermeable and could be considered a homogeneous material that consists of extremely fine grains. These results indicate that, with decreasing grain size, the dominant micro-fracturing mechanism becomes tensile rather than shear. The particle size distribution of the model can explain this result. As mentioned in the previous section, Section 4.7.1, when the particle radius was widely distributed and relatively large particles were contained in the rock model, shear cracks were generated along the grain boundaries that cross diagonally the direction of maximum confining stress. On the other hand, when all particles in the rock model are small, the HF fracture can develop straight in the direction of the maximum confining stress. Therefore, the number of shear cracks is very small, and shear cracks that emit substantial energy are difficult to generate in a homogeneous rock model, as shown in Figure 4.17a.

However, as shown in Figure 4.17b, when high-viscosity fluid was used, many shear cracks that emit large amounts of energy were generated even in the homogeneous model. Figure 4.18 shows a close-up view of the time–space distribution of cracks generated in case B1—high from time steps 807–811($\times 10^5$). This figure indicates that the fracture did not develop smoothly but developed stepwise. The generation of shear cracks led to such a stepwise development of the fracture. However, the number of shear cracks generated during the fracture propagation was very few, and the energy emitted from such a shear crack was small. The region enclosed with a dotted circle in Figure 4.18 indicates that many shear cracks emitting significantly large amounts of energy were generated when the fluid infiltrated the fracture after the fracture development.

As mentioned above, shear cracks are difficult to generate in the homogeneous model. However, the rock model is not perfectly homogeneous even when all particles in the rock model are very small, because the rock model is expressed by the assembly of particles. The HF fracture is microscopically tortuous due to the particle arrangement, and a few shear cracks are generated when the fracture grows diagonally across the direction of maximum confining stress. At this time, a higher fluid pressure is required to extend the fracture because a part of the confining stress and the borehole pressure act on the fracture tip as the shear stress. Thus, the fracture develops stepwise, accompanied by the shear crack generation. In the actual AE measurement during HF using low-viscosity water and high-viscosity oil injection, AE events spread from the borehole throughout the specimen, within short periods corresponding to the respective pressure drops. In particular, when the fracturing fluid was not allowed to penetrate the fracture, AE sources spread stepwise (Ishida et al., 2004) as shown in Chapter 3. The DEM simulation results agree well with these AE behaviors.

When the fracture was microscopically tortuous and there existed a region where the fracture width was narrowed locally, a higher fluid pressure was required so that a high-viscosity fluid could go through such a narrow region. Consequently, BD pressure became remarkably high, and many new microcracks, including the shear cracks, were generated due to the impact of fluid infiltration as shown in the dotted circle in Figure 4.18. Such large shear cracks were not seen in cases A1—high and A3—high, although the high-viscosity fluid was used. Because the

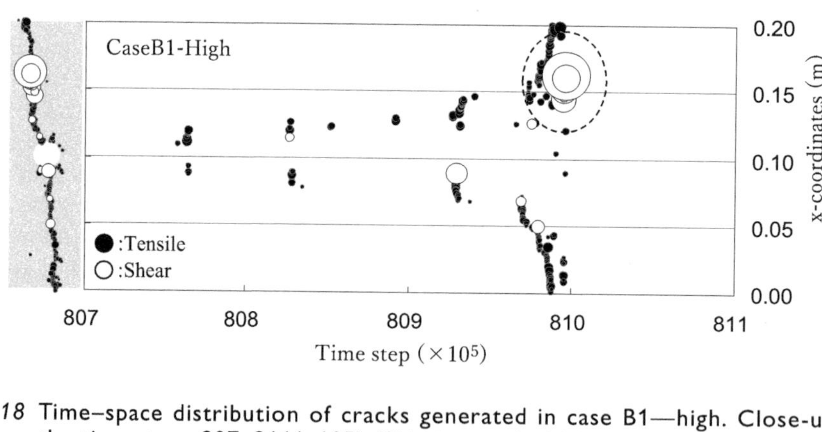

Figure 4.18 Time–space distribution of cracks generated in case B1—high. Close-up view of the time steps 807–811(×105) (Shimizu et al., 2011).

particle number of heterogeneous models was about 1/3 of that of homogeneous models, the region where the fracture width was narrowed partially, as mentioned above, did not appear in these cases. However, a phenomenon similar to case B1—high possibly occurs in cases using the heterogeneous model with different particle arrangements, as in case A2—high. These shear cracks emit significantly large amounts of energy, and AE that occurs from such large shear cracks would be observed even in field-scale AE measurement as reported by Baria and Green (1986) and Talebi and Cornet (1987). To validate this phenomenon, more detailed analysis, such as field-scale simulations, would be required.

4.8 Conclusions

A series of simulations for HF in hard rock was performed by using the flow-coupled DEM code to discuss the influence of the fluid viscosity and the particle size distribution. The simulation results show good agreement with the actual experimental results, including the AE measurement data shown in Chapter 3.

In this Chapter, 12 cases of HF simulations with different combinations of the rock model and fracturing fluid were performed. For all cases, the orientation of HF fractures is parallel to the direction of maximum compressive principal stress. This result indicates that the effect of the confining stress was appropriately expressed in the DEM simulations.

Crack initiation pressure and BD pressure for the low-viscosity fluid were lower than those for the high-viscosity fluid. Fracturing fluid easily penetrates through the interconnected pores into a rock from the borehole wall, and the fluid penetration causes pore pressure to increase around the borehole. Such an increase in pore pressure reduces the effective stress around the borehole and induces microcracking. Consequently, the crack initiation pressure with the low-viscosity fluid was lower than that with the high-viscosity fluid.

The low-viscosity fluid can easily infiltrate into the fracture, and the fluid pressure was applied throughout the fracture surface. Hence, the fracture propagates continuously without stopping once it is initiated because the stress intensity factor at the fracture tip monotonically increases with crack length. On the other hand, when the high-viscosity fluid is used, fluid pressure is applied only to a part of the fracture surface. As a result, the fracture never extends without additional pressure. For this reason, when the high-viscosity fluid was used, the BD pressure was markedly higher than that with the low-viscosity fluid.

When the fluid pressure acting on the fracture tip increases sufficiently, shear cracks that emit large amounts of energy are generated to connect pre-existing microcracks, and the HF fracture has tortuous paths according to the boundary of the mineral grain. The energy released from a tensile crack becomes small because the tensile strength of the rock is small. A small AE is easily buried in the noise and difficult to measure in an experiment. As a result, in AE measurement experiments, the shear type of AE with large energy is dominantly observed, although the tensile cracks are dominantly generated in the simulation. On the other hand, when the high-viscosity fluid is used, the fracturing fluid will infiltrate along the existing straight fracture after the fracture opens enough due to the borehole pressure. As a result, a thick and planar fracture with few branches was generated, and very few shear cracks emitting large amounts of energy were observed. Consequently, both tensile and shear AE can be observed during the injection of high-viscosity fluid in actual AE monitoring.

When the particle radius is widely distributed and relatively large particles are contained in the rock model, shear cracks are generated along the grain boundaries that diagonally cross the direction of maximum confining stress. On the other hand, when all particles in the rock model are small, the HF fracture can develop straight in the direction of the maximum confining stress. Consequently, the number of shear cracks is very small, and shear cracks that emit large amounts of energy are difficult to generate in a homogeneous rock model.

In actual HF experiments, it is difficult to observe the fluid infiltration behavior and change in stress distribution due to the fluid injection directly. On the contrary, the DEM can directly represent grain-scale microstructural features of rock, such as pre-existing flaws, pores, microcracks, and grain boundaries, by considering each grain as a DEM particle without any complicated constitutive laws. The HF process and the effect of fluid infiltration on the tensile rupture of permeable rock can be successfully reproduced and discussed in detail by the coupled fluid flow and the DEM. This suggests that the DEM model may be a strong tool for understanding the fracture behavior of permeable rock.

References

Al-Busaidi, A., Hazzard, J. F. & Young, R. P. (2005) Distinct element modeling of hydraulically fractured Lac du Bonnet granite. *J. Geophys. Res.*, 110, B06302.

Baria, R. & Green, A. S. P. (1986) Seismicity induced during a viscous stimulation at the Camborne School of Mines Hot Dry Rock geothermal project in Cornwall, England. In: *Progress in Acoustic Emission*, vol III, *Institute of Industrial Science, University of Tokyo*, Tokyo, Japan, pp. 407–429.

Brace, W. F. & Martin, III R. J. (1968) A test of the law of effective stress for crystalline rocks of flow porosity. *Int J. Rock Mech. Min. Sci. Geomech. Abstr.*, 5, 415–426.

Bruno, M. S. (1994) Micromechanics of stress-induced permeability anisotropy and damage in sedimentary rock. *Mech Mater.*, 18, 31–48.

Cundall, P. A. & Strack, O. D. L. (1979) A discrete numerical model for granular assemblies. *Geotechnique*, 29, 47–65.

Eberhardt, E., Stimpson, B. & Stead, D. (1999) Effect of grain size on the initiation and propagation thresholds of stress-induced brittle fractures. *Rock Mech. Rock Eng.*, 32, 81–99.

Falls, S. D., Young, R. P., Carlson, S. R. & Chow, T (1992) Ultrasonic tomography and acoustic emission in hydraulically fractured Lac du Bonnet grey granite. *J. Geophys. Res.*, 97, 6867–6884.

Haimson, B. C. (1968) Hydraulic fracturing in porous and nonporous rock and its potential for determining in situ stresses at great depth. Ph.D. thesis, University of Minnesota, Minneapolis, USA.

Haimson, B. C. & Fairhurst. C. (1967) Initiation and extension of hydraulic fracture in rocks. *Soc., Petrol, Eng., J.*, 7, 310–318.

Hayashi, K. & Motegi, S. (1988) Characteristics of energy of elastic waves due to sudden growth of subsurface reservoir cracks for geothermal heatex traction. In: *Progress in Acoustic Emission*, vol IV, pp. 265–272.

Hill, D. (1977) A model for earthquake swarms. *J. Geophys. Res.*, 82, 1347–1352.

Huang, H. (1999) Discrete element modeling of tool-rock interaction. Ph.D. thesis, University of Minnesota, Minneapolis, USA.

Hubbert, M. K. & Willis, D. G. (1957) Mechanics of hydraulic fracturing. *Petrol. Trans. Am. Soc. Min. Eng.*, 210, 153–168.

Ishida, T. (2001) Acoustic emission monitoring of hydraulic fracturing in laboratory and field. *Constr. Build. Mater.*, 15, 283–15, 295.

Ishida T., Chen, Q., Mizuta, Y. & Roegiers J. C. (2004) Influence of fluid viscosity on the hydraulic fracturing mechanism. *J. Energy Resour. Technol.*, 126, 190–200.

Ishida, T, Mizuta, Y., Matsunaga, I., Sasaki, S. & Chen, Q. (2000a) Effect of grain size in granitic rock on crack extension in hydraulic fracturing. In: Girard, J., Liebman, M., Breeds, C., & Doe, T. (eds) *Proceedings of the fourth North American Rock Mechanics Symposium*, Balkema, Rotterdam, Netherland, pp. 1105–1111.

Ishida, T., Sasaki, S., Matsunaga, I., Chen Q. & Mizuta, Y. (2000b) Effect of grain size in granitic rocks on hydraulic fracturing mechanism. In: *Trends in Rock Mechanics, Geotechnical Special Publications*, vol 102, ASCE Press, Reston, VA, USA, pp. 128–139.

Ito, T. (2008) Effect of pore pressure gradient on fracture initiation in fluid saturated porous media. *Eng., Fract., Mech.*, 75, 1753–1762.

Ito, T. & Hayashi, K. (1991) Physical background to the breakdown pressure in hydraulic fracturing tectonic stress measurements/*Int. J. Rock Mech. Min. Sci.*, 28, 285–293.

Jaeger, J. C., Cook, N. G. W. & Zimmerman, R. W. (2007) *Fundamentals of Rock Mechanics*. 4th ed. Wiley-Blackwell, Oxford.

Kiyama, H., Nishimura, T. & Fujimura, H. (1994) An advanced distinct element model coupling with pore water. *J. Geotech. Eng., Jpn. Soc. Civil Eng.*, 499, 31–39 (in Japanese with English Abstract)

Matsunaga, I., Kobayashi, H., Sasaki, S. & Ishida, T. (1993) Studying hydraulic fracturing mechanism by laboratory experiments with acoustic emission monitoring. *Int. J. Rock Mech. Min. Sci. & Geomech. Abstr.*, 30, 909–912.

Moon, T., Nakagawa, M. & Berger, J. (2007) Measurement of fracture toughness using the distinct element method. *Int. J. Rock Mech. Min. Sci.*, 44, 449–456.

Newman, J. C. (1971) *An Improved Method of Collocation for the Stress Analysis of Cracked Plates with Various Shaped Boundaries*. NASA Tech. Note, D-6373, NASA, Washington, DC, USA.

Ouchi, H., Amit Katiyar, A., Jason York, J, Foster, J. T. & Sharma M. M. (2015) A fully coupled porous flow and geomechanics model for fluid driven cracks: a peridynamics approach. *Compu. Mech.*, 55, 561–576. doi.0.1007/s00466-015-1123-8

Potyondy, D. O. & Cundall, P. A. (2004) A bonded-particle model for rock. *Int. J. Rock Mech. Min. Sci.*, 41, 1329–1364.

Sasaki, S., Matsunaga, I., Kobayashi, H. & Ishida, T. (1988) *Mechanism of Fracturing Inferred from Acoustic Emission*. CRIEPI (Central Research Institute of Electric Power Industry), Abiko, Chiba, Japan, U88034 (in Japanese with English abstracts).

Schmitt, D. R. & Zoback, M. D. (1992) Diminished pore pressure in low porosity crystalline rock under tensional failure: apparent elastic dilatant strengthening. *J. Geophys. Res.*, 97, 273–288.

Schmitt, D. R. & Zoback, M. D. (1993) Infiltration effects in the tensile rupture of thin walled cylinders of glass and granite: implications for the hydraulic fracturing breakdown equation. *Int. J. Rock Mech. Min. Sci.*, 30, 289–303.

Shah, K. R., Carter, B. J. & Ingraffea, A. R. (1997) Hydraulic fracturing simulation in a parallel computing environment. *Int. J. Rock Mech. Min. Sci.*, 34, 282.

Shepard, D. (1968) A two-dimensional interpolation function for irregularly-spaced data. In: *Proceedings of the 1968 ACM National Conference*, New York, USA, pp. 517–524.

Shimizu, H., Murata, S. & Ishida, T. (2010) The distinct element analysis for hydraulic fracturing with fluid having different viscosity. *J. MMIJ*, 126, 124–130. (in Japanese with English abstracts)

Shimizu, H., Murata, S. & Ishida, T. (2011) The distinct element analysis for hydraulic fracturing in hard rock considering fluid viscosity and particle size distribution. *Int. J. Rock Mech. Min. Sci.*, 48(5), 712–727. doi: 10.1016/j.ijrmms.2011.04.013

Talebi, S. & Cornet, F. H. (1987) Analysis of the microseismicity induced by a fluid injection in a granitic rock mass. *Geophys. Res. Lett.*, 14, 227–230.

Thallak, S., Rothenburg, L. & Dusseault, M. (1991) Simulation of multiple hydraulic fractures in a discrete element system. In: *Proceedings of the 32nd US Rock Mechanics Symposium*, Norman, Oklahom, USA, pp. 271–280.

Urbancic, T. I., Shumila, V., Rutledge, J. T. & Zinno, R. J. (1999) Determining hydraulic fracture behavior using microseismicity. In: *Proceedings of the 37th US Rock Mechanics*, Vail, Colorado, USA, pp. 991–997.

Vandamme, L. & Curran, J. H. (1989) A three dimensional hydraulic fracturing simulator. *Int. J. Numer. Methods Eng.*, 28, 909–927.

Yoshioka, K., Katou, M., Tamura, K., Arima, Y., Ito, Y., Chen, Y. & Ishida, T. (2023). Interactions of hydraulic fractures with grain boundary discontinuities in the near wellbore region. *J. Geophys. Res. Solid Earth*, 128, e2022JB024509. https://doi. org/10.1029/2022JB024509

Zhao, Z., Kim, H. & Haimson, B. (1996) Hydraulic fracturing initiation in granite. In: *Proceedings of the Second North American Rock Mechanics Symposium,* Montréal, Québec, Canada, pp. 1231–1235.

Zimmerman, R.W. & Bodvarsson, G. S. (1996) Hydraulic conductivity of rock fractures, Transp. *Porous Media*, 23, 1–30.

Zoback, M. D. & Pollard, D. D. (1978) Hydraulic fracture propagation and the interpretation of pressure time records for in-situ stress determination. In *Proceedings of the 19th US Rock Mechanics and Symposium,* Stateline, Nevada, USA, 14–22.

Zoback, M. D., Rummel, F., Jung, R. & Raleigh, C. B. (1977) Laboratory hydraulic fracturing experiments in intact and prefractured rock. *Int. J. Rock Mech. Min. Sci. Geomech. Abstr.*, 14, pp. 49–58.

Chapter 5

Carbon dioxide fracturing experiment in laboratory

5.1 Introduction

Through laboratory HF (hydraulic fracturing) experiments, we examined the effect of rock properties on the features of induced cracks in Chapter 2. We also discussed how the viscosity of the fracturing fluids affects them in Chapter 3. In a field HF operation, we cannot choose the rock property, but we can select the viscosity of the fracturing fluid. Thus, the selection of fracturing fluid viscosity is one of the usable methods that can control the crack feature. As suggested at the end of Chapter 3, slicker fluids that include various gases can generate cracks that are advantageous for extracting heat or producing shale gas and oil compared to conventional water-based fluids. Out of many kinds of gases such as CO_2 (carbon dioxide), N_2 (nitrogen), and liquefied petroleum gas, CO_2 is the most interesting to many engineers and researchers from the viewpoint of carbon capture and storage and/or carbon capture, utilization, and storage, suggested by Elsworth et al. (2014), Jacobs (2014), and others.

Carbon capture, utilization, and storage is a promising and feasible method for mitigating the greenhouse effect by decreasing the amount of CO_2 emissions. If this method can offer economic benefits to compensate for the costs of carbon capture and storage, it will be much more eagerly developed and adopted.

Brown (2000) proposed to use CO_2 as a substitute for water to extract heat from an enhanced geothermal system (EGS) while realizing the underground sequestration of CO_2 at the same time. Ueda et al. (2005) suggested that CO_2 injected through a borehole can be dissolved into groundwater, and the CO_2-saturated water reacts with the hot rocks to form calcite and clay minerals at the edge of the migrating plume of CO_2. The impermeable layer of the calcite and the clay minerals can act as a cap rock to retain CO_2. Pruess (2006) compared the flow behavior, chemistry, and thermophysical properties of CO_2 and water for geothermal energy extraction and thus concluded that CO_2 is superior to water as a working fluid for EGS. Mohan et al. (2013) explored the possibility of utilizing CO_2 as an alternative geothermal fluid to extract geothermal heat in arid regions and simultaneously sequester CO_2. Xu et al. (2016) assessed the power generation and CO_2 sequestration ability of the Habanero geothermal field in the Cooper Basin of South Australia and suggested through their simulation that the major sequestration mechanisms are the storage in the fracture-stimulation damage zone.

Injection to extract heat energy in EGS is usually injected into a rock mass more than 3,000 m deep, which satisfies the temperature and pressure conditions to make CO_2 a supercritical (SC) state. The viscosity of SC-CO_2 is 1 or 2 orders of magnitude smaller than that of normal liquid water, and it is still lower than that of liquid CO_2 (L-CO_2) formed at a lower temperature. Although the features of cracks induced by HF with a low-viscosity fluid are expected to be

DOI: 10.1201/9781003282631-5

different from those induced by conventional HF using water, the difference has been examined by only visual inspection of specimens (Kizaki et al., 2013) and numerical simulations (Zhou and Burbey, 2014).

On the other hand, to investigate the effect of fracturing fluid viscosity on the feature of cracks, we conducted HF experiments by injecting water and viscous oil into granite blocks under bi-axial stress (Ishida et al., 2004) as shown in Section 3.2, in addition to this, by injecting SC-CO_2 and L-CO_2 into them under hydrostatic stress conditions (Ishida et al., 2012). Following these preliminary experiments, we conducted HF experiments injecting SC-CO_2, L-CO_2, water, and viscous oil, whose viscosities ranged from 0.051 to 336.6 mPa·s, into granite blocks under an identical true triaxial stress condition (Ishida et al., 2016). In the experiments, we focused on the AE (acoustic emission) source distribution, the AE focal mechanism, and the microstructure of the cracks revealed by using an acrylic resin containing a fluorescent compound imbibed into the material after the experiment. The results consistently showed a tendency that a low-viscosity fluid, such as SC-CO_2, induces more three-dimensional sinuous cracks with more secondary branches than higher-viscosity fluids such as water and viscous oil. In this chapter, the results and discussion of the experiments are described, following the paper of Ishida et al. (2016).

5.2 Experimental method

5.2.1 *Specimens and experimental setup*

We used eight cubes (170 × 170 × 170 mm) with a 20 mm diameter central hole of Kurokami-jima granite. The granite is the same as that used in the previous experiments shown in Sections 2.4 and 3.2. The granite is produced in Kurokami-jima Island in the Seto Inland Sea in Southwestern Japan, and it is commonly known as Kurokami-jima granite. The modal composition of the samples for our experiments was about 40% potassium feldspar, 36% plagioclase, 21% quartz, and 3% micas and accessories. The mean grain size was about 2 mm, and some coarse potassium feldspar grains whose size was more than 5 mm were observed.

To confirm the reproducibility of the experimental results, we used two specimens for each of the fracturing fluids: SC-CO_2, L-CO_2, water, and viscous oil (Table 5.1). The granite specimen's inherent rift plane, explained at the end of Section 2.1, was oriented to correspond to the *YZ* plane in the Cartesian coordinate system (Figure 5.1). As shown in Table 5.1, the average

Table 5.1 Specimen number, fracturing fluid, and P wave velocity (Ishida et al., 2016)

Specimen no.	*Fracturing fluid*	*P wave velocity (km/s)*		
		V_X	V_Y	V_Z
G1115	SC-CO_2	4.25	4.86	4.72
G1202	SC-CO_2	5.16	5.68	5.67
G1213	L-CO_2	4.88	5.32	5.26
G1209	L-CO_2	4.75	5.40	5.33
G1102	Water	4.05	4.86	4.72
G1103	Water	4.25	5.00	4.72
G1114	Oil	4.05	4.72	4.60
G1112	Oil	4.25	4.86	4.60
Average		4.46	5.09	4.95

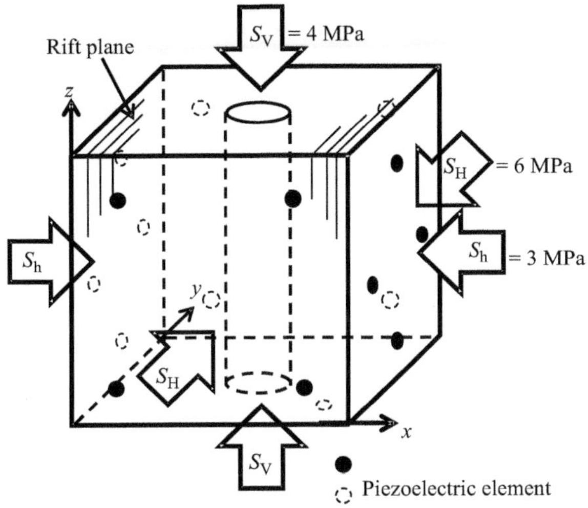

Figure 5.1 Coordinate system and loading condition of the specimen. The filled circles indicate the piezoelectric elements used as AE sensors at open positions, while the open circles with broken lines indicate those at hidden positions behind the specimen (Ishida et al., 2015, 2016).

P wave velocities measured along the *Y* and *Z* directions were around 5.0 km/s, whereas those along the *X* direction normal to the rift plane were around 4.5 km/s. The P wave velocities were measured before the experiments without any confining pressure. Results of Brazilian tests with the disks oriented to make a fracture along the rift plane on 33 cylindrical samples of the same kind of Kurokami-jima granite, measuring 45 mm in diameter and 30 mm in length, showed an average tensile strength of 3.38 MPa with a standard deviation of 0.21 MPa (Ishida et al., 2005). The effective porosity of the granite was 0.84%, where the pore and the bulk volumes were measured with the saturation and the buoyancy methods, respectively (International Society for Rock Mechanics, 1979), and its intrinsic permeability obtained by the air permeability test for hollow cylinder samples was around 9×10^{-19} m^2. The intrinsic permeability corresponds to 1×10^{-11} m/s of hydraulic conductivity, which ranges from 10^{-12} to 10^{-8} m/s for general granitic rock shown by Vutukuri and Katsuyama (1994). To monitor AE events induced by the fluid injection, we glued four cylindrical lead zirconate titanate (PZT) elements 4 mm in diameter and 4 mm thick on each of the four lateral surfaces (a total of 16 elements) of the specimen (Figures 5.1 and 5.2). To apply confining pressures, we covered each lateral surface with a 6 mm thick hard nylon sheet with four 5 mm diameter holes in the positions of the PZT elements. The PZT element wires were guided through grooves in the surface of the sheet, and the sheet was covered again with another 4 mm thick hard nylon sheet. The procedure of specimen preparation for AE monitoring and loading of confining pressures is shown in Figures 5.3 and 5.4.

The reason why we covered each lateral surface with a 6 mm thick hard nylon sheet with four 5 mm diameter holes in the positions of the PZT elements is based on the reflection of our following failure experience. When we applied confining pressures after we encapsulated with plaster the same types of PZT elements glued on specimen surfaces with the specimen, the PZT elements worked well under 1 MPa confining pressure as an AE sensor; however, they didn't work under larger than 1 MPa in the previous experiment (Ishida et al., 2012). It is most likely

Figure 5.2 Photo of the granite specimen of 170 mm cube with a 20 mm diameter center hole.

that the confining pressures were applied also on the PZT elements in the plaster encapsulating the specimen when they were applied to the specimen, a DC voltage induced in the PZT element by the confining pressures saturated the maximum voltage that the AE monitoring system could measure, and, as a result, any high frequency AC signals that the PZT element induced when they detected an AE signal were completely covered by the large DC voltage. By covering each lateral surface with a 6 mm thick hard nylon sheet with four 5 mm diameter holes in the positions of the PZT elements that measures 5 mm in thickness and 4 mm in diameter to prevent from the confining pressures been applied on the PZT elements, the AE signals could be successfully monitored under the confining pressures larger than 1 MPa in the experiments described in this chapter. In addition, we confirmed that the confining pressures were uniformly applied to the specimen surfaces by using "Prescale", which is the film to measure pressure and pressure distribution provided by FUJIFILM Corporation (https://www. fujifilm.com/pl/en/business/inspection/measurement-film/prescale) that was explained in Section 3.2.2.1.

We placed the specimen in a dry condition in a cylindrical pressure cell, except in the case of SC-CO_2 injection, where we filled the cell with hot water to prevent the injected CO_2 from cooling. We inserted four bow spacer blocks between the specimen and the inner wall of the pressure cell and placed a flat jack between the specimen and each spacer block to apply confining pressure in the X and Y directions (Figure 5.5). In the Z direction, we placed flat jacks between

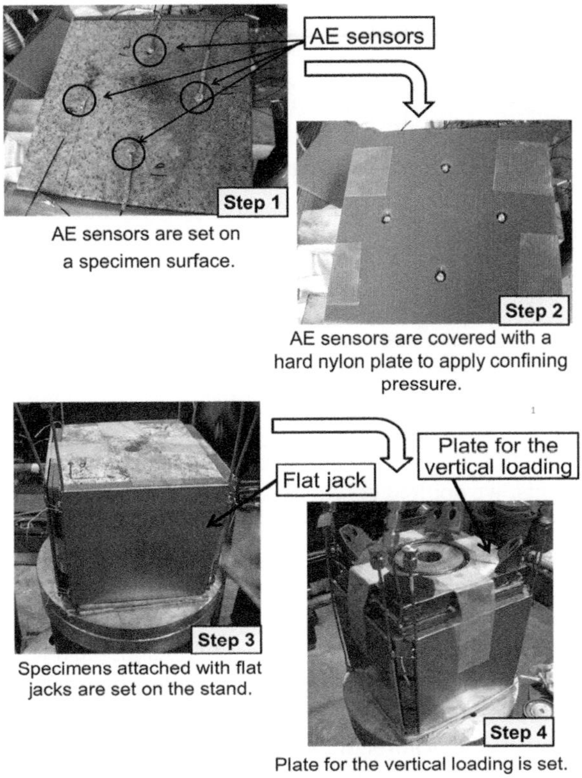

Figure 5.3 Procedure of specimen preparation for AE monitoring and loading of confining pressure.

Figure 5.4 Loading frame to apply confining pressure and the stand on which the specimen is set. The stand is covered and connected with the loading frame when confining pressure is applied to the specimen.

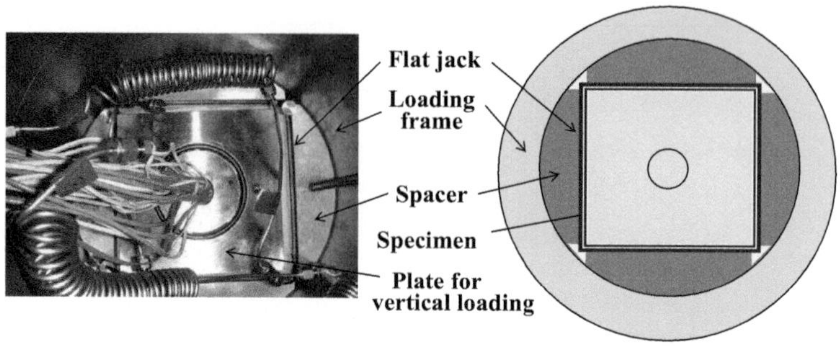

Figure 5.5 (left) Photo and (right) cross-sectional schematic of a specimen, flat jacks, and spacers set in the loading frame. In the photo, the specimen is set under the plate for vertical loading (Ishida et al., 2015, 2016).

the specimen and loading plates supported by the end caps on the top and bottom of the pressure cell. In all the experiments, we applied confining pressures of 3, 6, and 4 MPa in the X, Y, and Z directions, respectively, to provide the deviatoric stress. The stress magnitudes of 3 and 6 MPa were selected following the solution by Kirsch (1898) to keep the compressive stress larger than 3 MPa in the tangential direction all along the hole wall. The minimum tangential stress of 3 MPa induced at the points of the Y direction along the hole wall did not exceed the capacity of the injection pump and pipeline to induce HF; on the other hand, the maximum stress of 15 MPa induced at those of the X direction was sufficiently smaller than that to induce the borehole breakout. The stress magnitude of 4 MPa in the Z direction was selected to be between 3 and 6 MPa in the X and Y directions. Although the confining pressures increase P wave velocities, we ignored the increases in the calculation of AE source location, because they were expected only around 0.1–0.3 km/s from the measurements for granite under confining pressure from 0 to 200 MPa (Sano et al., 1992).

5.2.2 *Method of fluid injection*

Figure 5.6 shows the straddle packer to inject the fracturing fluids, including CO_2. The packer had a 60 mm pressurizing section sealed with two O-rings at each end. Figure 5.7 shows a view of inserting the packer into the central hole of the samples. The pressurizing section was centered along the hole. In all the experiments, we injected the fracturing fluid under the control of the syringe pump to keep the same constant flow rate of 10 mL/min and stopped the injection immediately after HF was induced, which was indicated by a sudden pressure drop. Figure 5.8 shows a view of putting down the loading frame to cover the specimen on the stand and tightening screws to fix the loading frame to the top and bottom loading plates.

Figures 5.9 and 5.10 show the injection system used for the HF experiments. To make SC-CO_2, we fed CO_2 from a bomb to a syringe pump cylinder, which had a capacity of 266 mL. To fill the cylinder as fully as possible, we cooled the CO_2 to keep it in the liquid state by circulating coolant in a cooling unit above the cylinder. The phase diagram in Figure 5.11 shows that CO_2 becomes an SC state when the temperature exceeds 31.0°C and the pressure exceeds 7.38 MPa. After discharging L-CO_2 from the cylinder of the syringe pump at a constant flow rate of 10 mL/min, we heated it with a heater unit. We maintained the temperature at 55°C by bandaging the electric resistance heating ribbon along the pipe connecting the heater unit to the packer (Figures 5.9 and 5.10). In addition, we filled the cell containing the specimen with hot water at

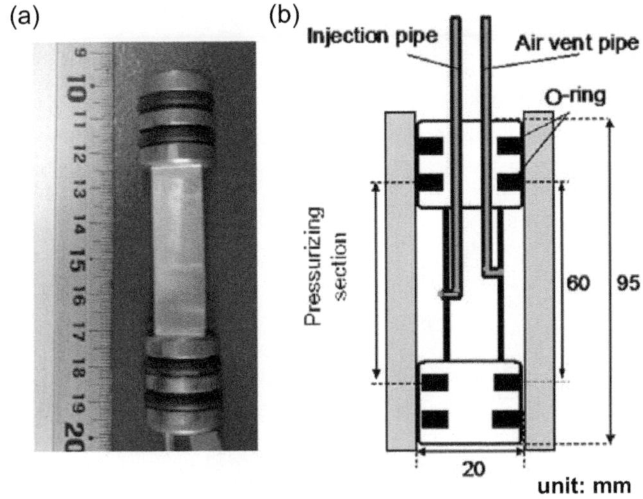

Figure 5.6 Straddle packer used to inject fracturing fluid including CO_2. (a) Photo and (b) cross-sectional schematic of the packer (Ishida et al., 2015, 2016).

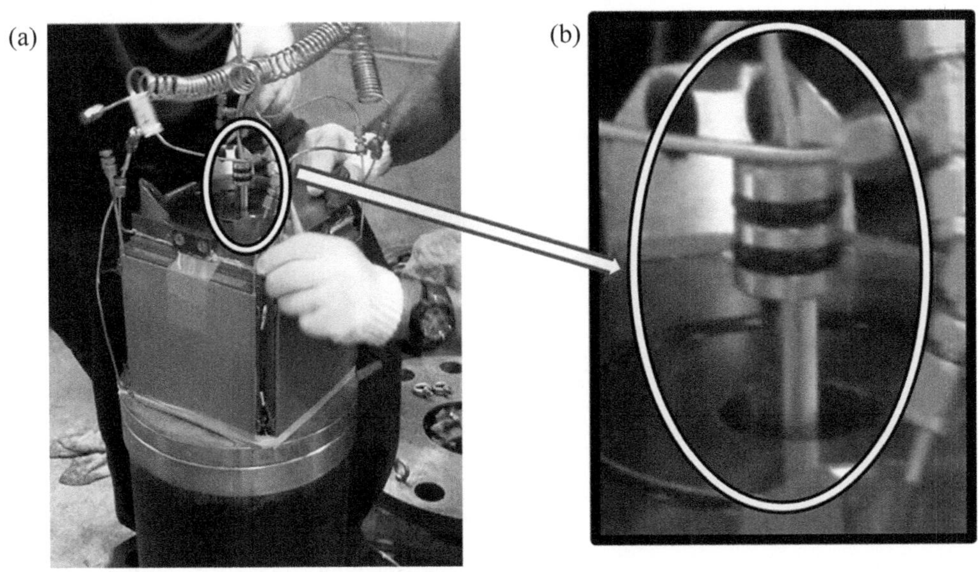

Figure 5.7 (a) View of inserting the straddle packer into the HF hole of the specimen. (b) Enlarged view of the straddle packer in the left view.

a temperature of around 45°C to prevent the injected CO_2 from cooling. Although we did not measure the thermal properties of the Kurokami-jima granite, we believe that the temperature in the specimen did not easily increase within 1 h or so for the SC-CO_2 experiment, even when the outside of the specimen was warmed with hot water and the wall of the injection hole was warmed with SC-CO_2. This expectation comes from our experiences in the heater test in a

Figure 5.8 (left) View of putting down the loading frame to cover the specimen on the stand. (Right) View of tightening screws to fix the loading frame to the top and bottom loading plates.

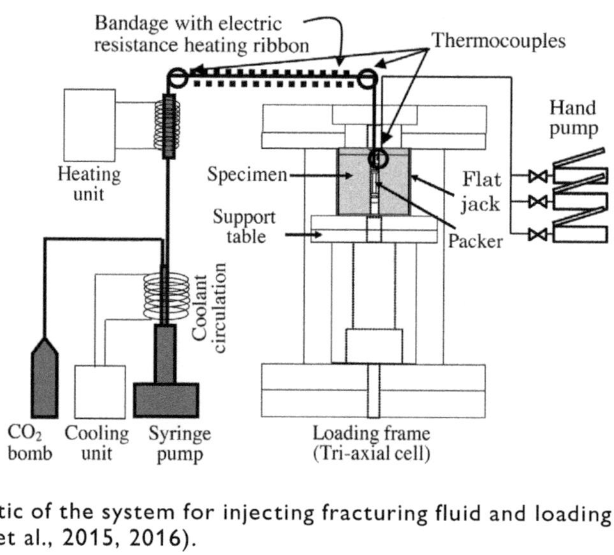

Figure 5.9 Schematic of the system for injecting fracturing fluid and loading confining pressure (Ishida et al., 2015, 2016).

block of Inada granite whose thermal conductivity is 3.3 W/(m·K), specific heat is 0.72 J/(g K), and density is $2.0 \times 10^3 kg/m^3$ (Ishida et al., 1990). A similar tendency was also found in a heater test in a block of Lac du Bonnet granite by Jansen et al. (1993). Thus, during the experiment, it is most likely that the temperature of the specimen only in the vicinity of the surfaces and the hole wall was kept around 40°C, while the temperature in the specimen was kept at

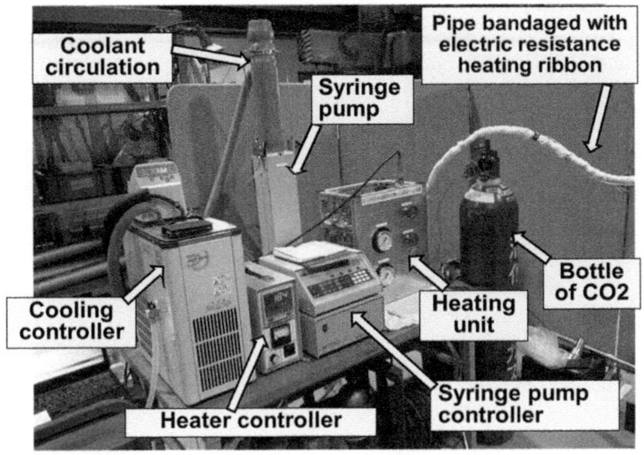

Figure 5.10 View of the actual system for injecting CO_2.

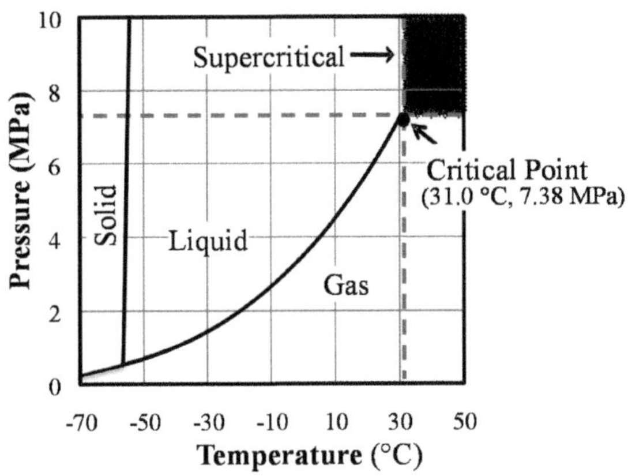

Figure 5.11 Phase diagram of CO_2 (Ishida et al., 2015, 2016).

room temperature, around 20°C, where the specimen was left for more than 1 week before the experiments. Although the specimen was soaked in hot water for around 1 h, the water was not expected to infiltrate around the injection hole owing to the specimen's low permeability, and the HF was induced under almost dry conditions based on our experience from similar previous experiments.

As the other fluids, we used L-CO_2, tap water having a viscosity of 1 mPa·s (1 mPa·s = 1.0×10^{-3} Pa·s = 1.0×10^{-2} dyn·s/cm^2 = 1 cP), and viscous oil that is the transmission oil for a car, having a viscosity of 80 mPa·s (Mahha Super Hypoid Multi, GL-5 80W-90, Fuji Kosan Co. Ltd.). The water and the viscous oil were the same as those used in the experiments described in Chapter 3. HF experiments using these fluids were also conducted under dry conditions at room temperature because we fed the fluids to the packer neither by heating nor by filling the cell with water. The viscosities of these fluids were slightly different from the above values, depending on the room temperature when the experiments were conducted. Figure 5.12 shows the whole view of the experiment.

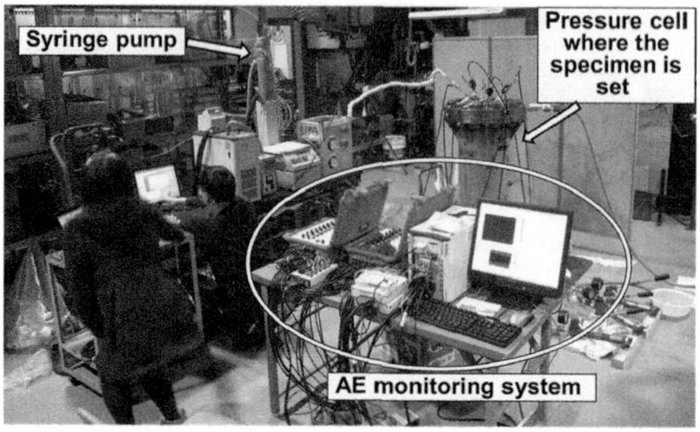

Figure 5.12 Whole view of the experiment.

5.2.3 *Methods for monitoring AE, pressure, and temperature*

The PZT elements glued onto the specimen had a resonance frequency of 300 kHz, and they were covered with an aluminum sheet to avoid electromagnetically induced noise. They were also covered with heat-shrinkable tubes and silicone rubber to provide a waterproof barrier for SC-CO_2 fracturing. We amplified the AE signals by 72–84 dB in total (36 dB in a preamplifier and 36–48 dB in a signal conditioner), processed them with a band-pass filter between 80 kHz and 1 MHz, and recorded them on a hard disk through an analog-to-digital converter (PXI-5105, National Instruments Corp.) under the control of the LabVIEW program. Since the analog-to-digital converter has 16 separate channels, the processed AE signals were digitized with 0.1 μs sampling time and had 2,048 samples in the record length for each event for each sensor. We set a dead time to stop recording for 1 ms immediately after recording an AE event, to prevent the hard disk from recording too much noise caused by "ringing", that is, the vibration following a large AE event. The recording of an AE event was triggered when one of the signals from the 16 AE sensors exceeded 3 V. In addition, we counted the number of AE events per second for each AE sensor when the signal exceeded 3 V.

Every 0.1 s, we measured the injected fluid pressure through transducers (PW-50MPA, Tokyo Sokki Kenkyujo Co., Ltd., with a full scale of 50 MPa and a resolution of 0.1 MPa) set at the top of the packer, and the flat jack pressures in the *X*, *Y*, and *Z* directions through transducers (PW-20MPA with a full scale of 20 MPa and a resolution of 0.04 MPa) set on the connecting pipes of opposing pairs of flat jacks. We measured the temperature changes with T-type thermocouples, which can measure temperatures from −200°C to 400°C with a resolution of 0.5°C, glued to the injection pipe just above the packer in the hole, and between both ends of the section of pipe covered with electric resistance heating ribbon only for SC-CO_2 injection (Figure 5.9).

5.2.4 *Observation of cracks with fluorescent resin*

After the HF experiments, we overcored the tested specimens with a diameter of 68 mm coaxially along the central injection hole (Figure 5.13). The overcored hollow cores were soaked in the thermosetting acrylic resin monomer containing a fluorescent compound under vacuum to allow the resin monomer to penetrate the induced cracks. Then, the resin was heated to fix the resin in the cores before sectioning. The cracks filled with the resin were visualized and observed under ultraviolet light irradiation (Nishiyama and Kusuda, 1994). We made two thin

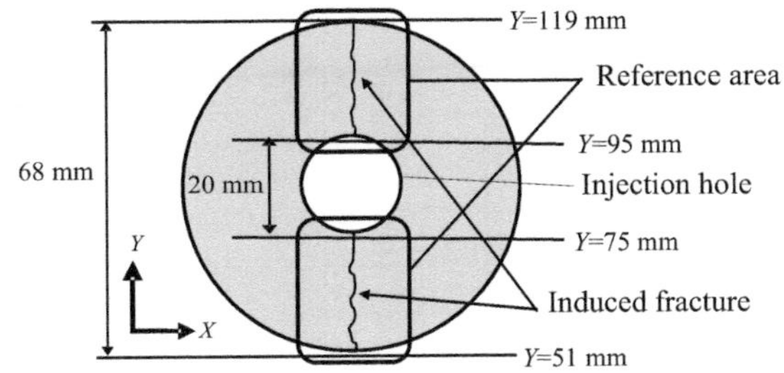

Figure 5.13 Overcored hollow core for microscopic observation and reference areas on a thin section (Ishida et al., 2016).

sections for each fluid, including hydraulically induced fractures on the plane where $Z=85$ mm, corresponding to the center of the pressuring section. No secondary fractures caused by preparing the thin sections were detected because the pretreatment, as mentioned above, was completed before sectioning.

5.3 Results

Table 5.2 summarizes our experiments. Because both specimens for each fluid showed similar trends, the results for only one specimen per fracturing fluid are shown below in Figures 5.14–5.17. These specimens are G1115 for SC-CO_2, G1213 for L-CO_2, G1102 for water, and G1114 for viscous oil, indicated by open circles in Table 5.2.

5.3.1 *Changes in fluid pressure and AE count rate*

Figure 5.14 shows the changes in injected fluid pressure, confining pressure, and AE count rate during the injection of SC-CO_2, L-CO_2, water, and viscous oil into the specimen. The temperature is shown only for SC-CO_2. In the injection of SC-CO_2 and L-CO_2, since we fed CO_2 from the bomb, the bomb pressure of around 5 MPa was applied in the hole immediately after the experiments had started. In contrast, in the injection of water and oil, the pressure gradually increased from zero. In all cases, we shut the return pipe after filling the fluids in the dry empty pressurizing section in the hole, and we kept injecting the fluids at a constant flow rate of 10 mL/min through the experiments. The elapsed time along the lateral axis is taken as zero when the BD (breakdown) pressure, defined as the peak pressure immediately before a large sudden pressure drop, is recorded.

In Figure 5.14a for the case of SC-CO_2 injection, we show the temperature measured with a thermocouple glued on the injection pipe just above the packer in the hole. The temperature most likely indicates the CO_2 temperature, because the temperature always showed the same temperature measured at both ends of the section of pipe covered with electric resistance heating ribbon (see Figure 5.2) even when the sudden temperature change of the CO_2 was induced. Because the temperature shows around 36°C, the test began with CO_2 in a gas state in the borehole, referring to the phase diagram of CO_2 shown in Figure 5.11. The CO_2 should start to become a SC state once the fluid pressure gets closer to the critical pressure, 7.38 MPa. This change could be inferred from the change in the increase rate of the fluid pressure under the constant flow rate, as shown with the arrow in Figure 5.14a. This change was most likely caused by the decrease in fluid compressibility due to the change from the gas to the SC state of CO_2. The fluid pressure

Table 5.2 Summary of experiments (Ishida et al., 2016)

	Specimen no.	*Fracturing fluid*	*Date of experiment*	*Temperature (°C)*	*Viscosity (mPa·s)*	*BD pressure, P_b (MPa)*	*Tensile strength, T (MPa)*[a]	*No. of AE sources*	*L_{av} (mm)*	*Fractal dimension*	*Ratio of compression (%)*	*Tortuosity L/L_0*	*Av. no. of branches*
○	G1115	SC-CO_2	12 Oct. 2012	35.5	0.051	9.10	6.10	118	12.06	2.42	51.30	1.109	2.131
	G1202	SC-CO_2	11 Oct. 2012	38.4	0.053	10.16	7.16	218	11.44	2.40	62.70	1.075	2.083
○	G1213	L-CO_2	10 Dec. 2013	12.8	0.097	11.96	8.96	160	10.78	2.17	54.21	1.120	1.870
	G1209	L-CO_2	16 Jan. 2014	6.0	0.108	11.98	8.98	81	12.52	2.15	54.80	1.102	1.927
○	G1102	Water	14 Sep. 2011	31.4	0.774	12.96	9.96	116	8.15	2.20	56.67	1.095	1.789
	G1103	Water	15 Sep. 2011	30.5	0.788	13.33	10.33	72	9.35	2.42	61.42	–	–
○	G1114	Oil	1 Dec. 2011	11.1	316.0	23.07	20.07	73	7.39	2.01	69.01	1.062	1.704
	G1112	Oil	2 Dec. 2011	10.6	336.6	25.05	22.05	62	8.51	1.96	72.11	1.054	2.029

[a] Tensile strength is obtained from the equation (3), $T = (1+\alpha)P_b - 3$, when $\alpha = 0$. (The unit for *T* and P_b is MPa, and *a* is a coefficient having no dimension.)

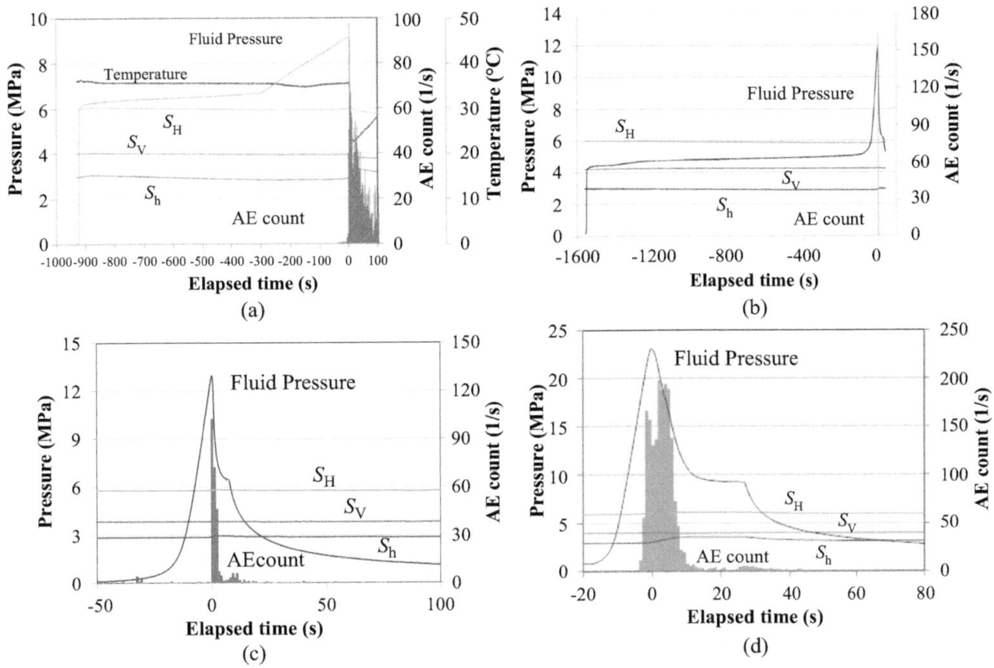

Figure 5.14 Changes in injected fluid pressure, confining pressure, and AE count rate for (a) SC-CO_2, (b) L-CO_2, (c) water, and (d) viscous oil. The temperature change is only shown in Figure 4.5a. The scales of the vertical and horizontal axes differ between plots (Ishida et al., 2015, 2016).

decreased sharply to zero immediately after the BD at 9.10 MPa. The pressure and temperature, 35.5°C, at the BD demonstrate that the HF was induced by CO_2 of a SC state. The pressure decrease, which was caused by leakage through fracturing-induced cracks, most likely changed the CO_2 from a SC to a gas state again. The temperature also decreased to 22.6°C immediately after the BD and came back to 27.9°C at 100 s after the BD. The temperature decrease was probably caused by adiabatic expansion of CO_2, because the measured temperature decrease from 35.2°C to 22.6°C can be induced by 1.15 times larger expansion of the sealed volume of gas-state CO_2 under the assumption of simple adiabatic expansion using Poisson's law. The AE count rate (average of the 16 AE sensors) increased sharply immediately after the BD and AE events continued to occur for several tens of seconds. The confining pressures, S_H, S_V, and S_h (Figure 5.14a), were kept nearly constant at 6, 4, and 3 MPa, respectively, with small fluctuations due to crack growth during BD.

The BD pressures were positively correlated with viscosity: 9.10 MPa for SC-CO_2, 11.96 MPa for L-CO_2, 12.96 MPa for water, and 23.07 MPa for viscous oil.

5.3.2 *AE source locations*

Figure 5.15 shows the locations of AE sources observed during the fluid injections projected onto the three orthogonal planes, *XY*, *YZ*, and *ZX*. All of the AE occurred within 15 s after BD. Waveforms of a typical AE event for the respective four fluids are shown in Figure 5.16.

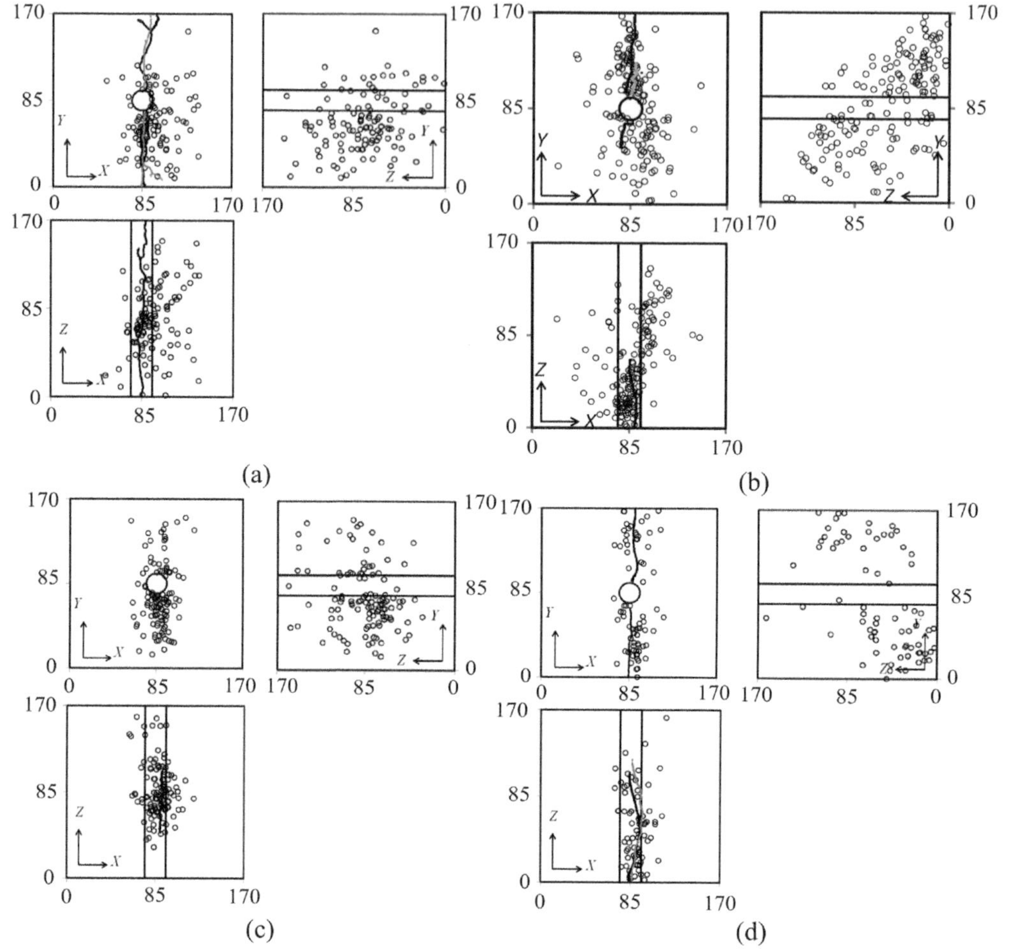

Figure 5.15 AE source distribution and cracks visible on the specimen surface for (a) SC-CO_2, (b) L-CO_2, (c) water, and (d) viscous oil. Dark and light lines show cracks that appeared in different planes facing each other (Ishida et al., 2015, 2016).

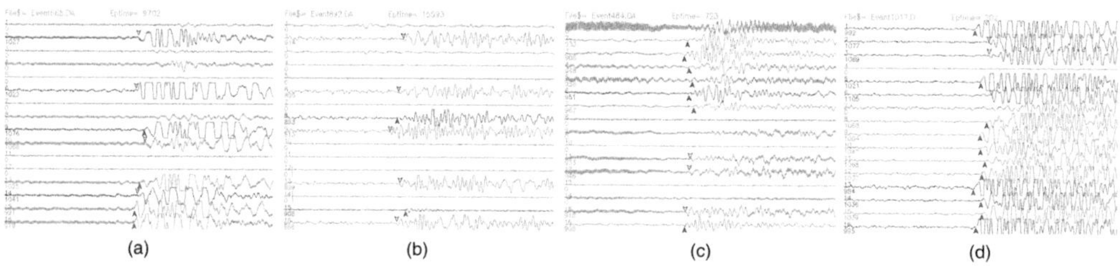

Figure 5.16 Typical AE waveforms induced just after the BD for the respective four fluids. The closed triangles indicate compression in the P wave first motion showing upward trace, while the open triangles indicate dilatation. (a) AE event induced by SC-CO_2 in G1115 at 1.9 s after the BD and the *X*, *Y*, and *Z* coordinates of the source are (78.4, 48.7, 52.6). (b) L-CO_2 in G1213 at 0.2 s and (99.8, 88.7, 88.9). (c) Water in G1102 at 1.6 s and (88.0, 62.4, 73.9). (d) Viscous oil in G1114 at 0.4 s and (80.4, 63.8, 121.6) (Ishida et al., 2016).

As a source location method, we used the iterative method with the least squares principle, which is called Geiger's method (Geiger, 1910, 1912), following the classification by Ge (2003a,b). Considering the anisotropy of the P wave velocity (Rothman et al., 1974), which was measured before the experiment shown in Table 5.1, the AE source locations were determined within an expected accuracy of several millimeters. This was possible because we chose the AE events satisfying the following criteria (Ishida and Sasaki, 2011): (i) five or more P wave arrival times could be read, and (ii) the standard deviation and the maximum of residuals of arrival times were within 3 μs. The number of sources that satisfied these conditions was 118 for SC-CO_2 (Figure 5.15a), 160 for L-CO_2 (Figure 5.15b), 116 for water (Figure 5.15c), and 73 for viscous oil (Figure 5.15d). Figure 5.15 also shows the cracks visible on the specimen surfaces with the located sources. Cracks were visible on the two respective opposite parallel surfaces of the *XY* and *ZX* planes for SC-CO_2 (Figure 5.15a), L-CO_2 (Figure 5.15b), and viscous oil (Figure 5.15d). The cracks visible on the two parallel surfaces are distinguished by using dark lines (near planes; $Z=170$ mm for the *XY* and $Y=0$ mm for the *XZ* planes) and light lines (far planes; $Z=0$ mm for the *XY* and $Y=170$ mm for the *XZ* planes). The other cracks were visible on one plane only, while no visible cracks appeared on the *YZ* planes.

The AE sources were distributed along the cracks, corresponding to the observed surface cracks. However, the AE sources in Figure 5.15a were distributed more widely than those lying along a flat plane; this effect seemed to increase with the decrease in the viscosity of the fracturing fluids. To present the differences quantitatively, the maximum likelihood flat plane for the AE distribution was first estimated for each specimen by minimizing the sum of squares of the distances from a source to the flat plane. After that, the average distance, L_{av}, from a source to the maximum likelihood plane was obtained. For the AE distributions shown in Figure 5.15a–d, the L_{av} values were 12.06 mm for SC-CO_2, 10.78 mm for L-CO_2, 8.15 mm for water, and 7.39 mm for viscous oil.

We also obtained the fractal dimensions (FDs) of the AE distributions with the correlation function method, following Hirata et al. (1987) and Grassberger (1983). Theoretically, the FDs of an infinite number of points distributed on a line, on a plane, and in three-dimensional space are 1, 2, and 3, respectively. However, because the number of distributed points is finite, the FDs obtained by this method tend to be slightly lower than their respective dimensions (Ishida et al., 1998). Despite that, the FD is useful to characterize the features of the AE distribution, and a larger FD suggests more extensive cracking in three dimensions. For the AE distributions shown in Figure 5.15a–d, the FDs for SC-CO_2, L-CO_2, water, and viscous oil were 2.42, 2.17, 2.20, and 2.01, respectively. This suggests that a low-viscosity fracturing fluid such as SC-CO_2 induces cracking in three dimensions rather than in two by, for example, forming sinuous cracks with a greater number of secondary branches.

5.3.3 Induced crack patterns observed with the fluorescence method

The cracks induced by HF in the experiments were subtle in all cases, and they were barely identified by tracing segments where the fracturing fluid leaked on the surfaces of the cubic specimens, although they seemed to be observed more clearly when the more viscous fracturing fluid was used. Even when we overcored with a 68 mm diameter around the 20 mm diameter injection hole coaxially, the overcored hollow core still had substantial strength without breaking into two pieces for almost all of the cases, unlike in the experiments by Schmitt and Zoback (1992, 1993).

Figure 5.17 shows crack propagation patterns observed on the thin sections cut from the hollow cores soaked in the acrylic resin containing a fluorescent compound (Nishiyama and

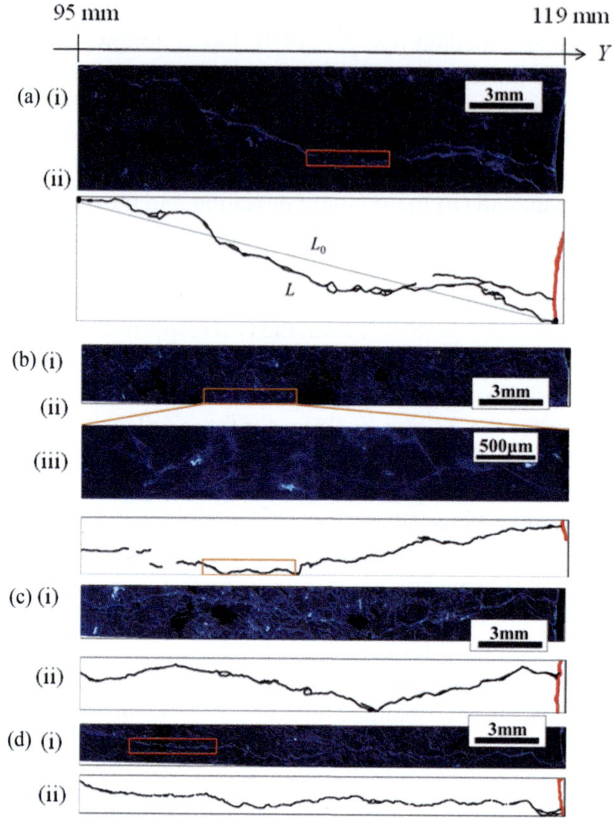

Figure 5.17 Fracture propagation patterns observed on thin sections using the fluorescence method from 95 to 119 mm in the *Y*-coordinate in Figure 5.13. Specimens fractured by (a) SC-CO_2 [G1115], (b) L-CO_2 [G1213], (c) water [G1102], and (d) viscous oil [G1114]. Figure 5.17a(i) and b(i), b(ii), c(i), and d(i) shows the photos taken of the thin section. Figure 5.17a(ii), b(iii), c(ii), and d(ii) shows sketches of the main crack, and the red line on the right side indicates the periphery of the injection hole. In Figure 5.17b, an enlarged photo of the area enclosed by the orange box is also shown to help identify the main crack. The micro photos taken under polarized light for the cracks in the red boxes in Figures 5.17a and d are shown in Figure 5.18. The symbols L_0 and L in Figure 5.17a show the lengths to calculate the tortuosity, L/L_0, shown in Figure 5.19e (Ishida et al., 2016).

Kusuda, 1994). We examined the crack patterns compared to the results of AE monitoring. Figure 5.17a(i), b(i), b(ii), c(i), and d(i) shows fluorescence images of the thin section taken under ultraviolet light. The bright and bluish-white regions correspond to cracks. Figure 5.17a(ii), b(iii), c(ii), and d(ii) shows sketches of the main crack induced by the injection. The crack propagates from right to left in all the figures. The periphery of the injection hole is shown with a red line on the right side of the sketch. The observed area corresponds to the area from the *Y*-coordinate of 4.105–114.10 mm in Figure 5.13, and the left side almost corresponds to the inner wall of the overcoring hole. The sketch was made by connecting segments of the main crack identified with the fluorescence method under ultraviolet light, considering the continuity

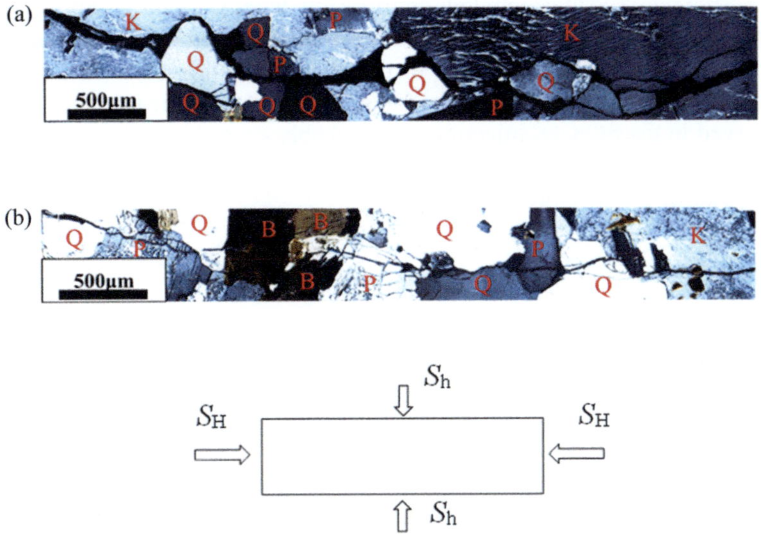

Figure 5.18 Observation on crack patterns and constituent mineral grains for (a) SC-CO_2 injection (red box in Figure 5.17a), and (b) oil injection (red box in Figure 5.17d), using a petrographic microscope in crossed-polarized light (crossed Nicols). (B: biotite, K: K-feldspar, P: plagioclase, and Q: quartz.) The bottom figure illustrates the stress condition when the cracks extended in the specimen (Ishida et al., 2016).

in the crack propagation direction. Although it was easy to identify the main cracks in the photos of SC-CO_2, water, and viscous oil (Figure 5.17a(i), c(i), and d(i)), it wasn't easy only in the figure of L-CO_2 (Figure 5.17b(i)); therefore, we used an enlarged photo (Figure 5.17b(ii)). The sketches of the main cracks in Figure 5.17a(ii), b(iii), c(ii), and d(ii) show that the cracks are tortuous and never propagate straight. However, they propagate along the direction of the maximum compressive stress and along the rift plane where the main preexisting microcracks align. Fluids with lower viscosity induce cracks with higher tortuosity because intermittent and stepwise crack extensions become more conspicuous.

To examine the correlations between crack patterns and the constituent mineral grains of the specimens, photomicrographs of the crack patterns on thin sections for SC-CO_2 and oil injection in the respective ranges shown in the boxes in Figure 5.17 were taken using a petrographic microscope in crossed-polarized light (crossed Nicols). The cracks induced by SC-CO_2 propagate mainly along the grain boundaries of the constituent minerals (Figure 5.18a). In contrast, Figure 5.18b shows that the cracks induced by oil cut through many mineral grains. This is most likely due to the reason why the crack pattern depends on the viscosity of the fracturing fluid.

The crack aperture shown in Figure 5.18a is larger than that in Figure 5.18b, although Ishida et al. (2004) and Stanchits et al. (2014) showed that the higher-viscosity fracturing fluid tends to induce cracks having a larger aperture. Ishida et al. (2004) injected water and viscous oil into blocks of Kurokami-jima granite while monitoring the pressure change of the flat jacks applying the confining pressure. Stanchits et al. (2014) injected two kinds of silicon oil having thousandfold different viscosities into low-permeability blocks of Colton sandstone while monitoring the volume of the injected fluid. Their results both indicated that the higher viscosity tends to induce larger aperture cracks, which is inconsistent with our results shown in Figure 5.18. Although we did not monitor changes in the flat jack pressure and the injected volume, the large aperture observed in the SC-CO_2 injection was probably caused by the expansion of CO_2 with the phase change from SC-CO_2 to gas due to the pressure decrease after the crack extends to the lateral free surface of the specimen.

5.4 Effect of fluid viscosity

In Figure 5.19, we show the dependency on the fracturing fluid viscosity of BD pressure, the AE distribution, and the fracturing mechanism based on ratios of the P wave first motion polarity of AE events and the crack features observed by the fluorescence method. To confirm the reproducibility of the results, we show data for our chosen specimens described in the previous section and for the alternative specimens G1202 for SC-CO_2, G1209 for L-CO_2, G1103 for water, and G1112 for viscous oil, respectively (Table 5.2). The viscosities of SC-CO_2 and L-CO_2 in Figure 5.19 were estimated from the temperature and pressure at the BD, following the equation of state for CO_2 implemented by Ohmori (http://hp.vector.co.jp/authors/VA030090/) based on the theory by Fenghour et al. (1998). The viscosities of the water were obtained from Chronological Scientific Tables (National Astronomical Observatory of Japan, 1995). The viscosities of the oil were obtained from the table for the transmission oil of a car (Mahha Super Hypoid Multi, GL-5 80W-90) provided by the oil manufacturer (Fuji Kosan Co. Ltd.), using the temperatures measured during the experiments.

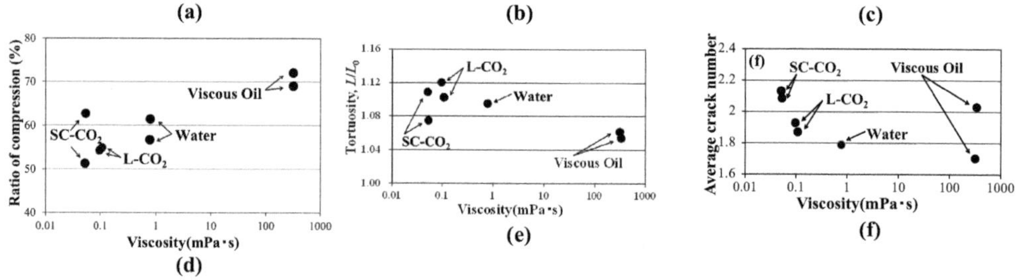

Figure 5.19 Effect of fracturing fluid viscosity on (a) BD pressure, (b) average distance L_{av} from a source to the estimated most likely plane, (c) FD, (d) the compression ratio to the total number of the polarities in P wave first motion, (e) crack tortuosity, and (f) average crack number. The results in Figures 5.19(b–d) were derived from AE data, whereas the results in Figures 5.19(e) and (f) were from microscopic crack observation using the fluorescence method (Ishida et al., 2016).

5.4.1 BD pressure and apparent tensile strength

Figure 5.19a shows the dependency of BD pressure on fracturing fluid viscosity. This figure demonstrates that BD pressure increases with viscosity. The tendency coincides with the experimental results by Morita et al. (1996) using water-based and oil-based muds and, in addition, the numerical simulation by Bunger et al. (2010).

Schmitt and Zoback (1992, 1993) discussed carefully the effect of pore pressure on crack initiation of HF through laboratory experiments using hollow cylinders of westerly granite and glass, where they changed the pressurization rate and the magnitude of the confining pressure by applying the outer surface of the cylinder. Here, we discuss the effect in a much simpler way than Schmitt and Zoback (1992, 1993), using the following equation proposed by Scheidegger (1962) for a vertical borehole:

$$P_b = 3S_{h\min} - S_{H\max} + T - P_p \tag{5.1}$$

where P_b is the BD pressure, P_p is the pore pressure, T is the tensile strength, $S_{h\min}$ is the minimum horizontal stress, and $S_{H\max}$ is the maximum horizontal stress. In our experiments, $S_{h\min}$ and $S_{H\max}$ correspond to S_h and S_H, respectively, as shown in Figure 5.1. By substituting that $S_{h\min}=S_h$, $S_{H\max}=S_H$, and $P_p=\alpha P_b$, the equation (5.1) becomes

$$T = S_H - 3S_h + \left(1+\alpha\right)P_b \tag{5.2}$$

The case that $\alpha=0$ corresponds to the state that no pore pressure applies without infiltration of fracturing fluid, whereas the case that $\alpha=1$ corresponds to the state that the pore pressure is equal to the fracturing fluid pressure applied on the wall of the injection hole. The magnitude of S_h fluctuated at the BD because a specimen expanded in the X direction, due to the intrusion of fracturing fluid into a crack induced along the Y direction, resulting in pushing the flat jacks to apply the confining pressure, S_h. However, because the fluctuation is not so large as shown in Figure 5.14, by substituting 3 MPa for S_h and 6 MPa for S_H, the equation (5.2) becomes

$$T = \left(1+\alpha\right)P_b - 3 \tag{5.3}$$

where the unit for T and P_b is MPa and α is a coefficient having no dimension.

The apparent tensile strength, T, obtained from equation (5.3), is shown in Table 5.2 when $\alpha=0$. Even when $\alpha=0$, the apparent tensile strength, T, ranges from 6 to 22 MPa, much larger than the average Brazilian tensile strength of the Kurokami-jima granite, 3.38 MPa. When $\alpha=1$, the apparent tensile strength, T, becomes much larger.

The tensile stress uniformly distributes along the fracture plane in the Brazilian test, whereas the tensile stress in the circumference direction is the largest on the wall of the injection hole and decreases with the square of the distance from the center of the hole in an HF experiment (Kirsch, 1898). When the difference of the stress distribution is taken into account by introducing a concept of "risk of rupture" proposed by Weibull (1939a,b) under the assumption of uniform distribution of Griffith cracks, the BD pressure, Pb, becomes 1.45 times larger than that of the Brazilian tensile strength (Ishida et al., 2005). The difference of 1.45 times showed good agreement with that of 1.47 obtained in their experiments (Ishida et al., 2005). However, the difference is still too small to explain the difference between the apparent and the average Brazilian tensile strength.

The cracks induced in our HF experiments most likely extend instantaneously to the lateral surface of the specimen. Discussion on the crack extension considering the stress intensity factors at a crack tip, shown in Figure 3.23 (Ishida et al., 1997) in Chapter 3, may help to interpret the difference in the apparent tensile strength depending on the viscosity of the fracturing fluid. The left column in Figure 3.23 depicts the case in which hydrostatic pressure, σ, applies throughout an ellipsoidal crack surface. In this case, $K_I = \sigma(\pi l)^{1/2}$, where K_I is the stress intensity factor of mode I at a crack tip and l is half of the crack length. According to this relation, the stress intensity factor at a crack tip increases with crack length; in other words, a crack never stops once it starts to extend. The right column in Figure 3.23 depicts the case in which a pair of point loads, F, is applied at the center of an ellipsoidal crack. In this case, $K_I = F(\pi l)^{-1/2}$, indicating that the stress intensity factor at a crack tip decreases with crack length; in other words, a crack never extends without an additional increase in applied loads. It can be expected that the closer the fracturing fluid pressure reaches to a crack tip, the lower the viscosity becomes. Thus, the low-viscosity fluid injection would make a state close to that in which hydrostatic pressure, σ, was applied throughout an ellipsoidal crack surface. On the other hand, the high-viscosity fracturing fluid injection would make a state close to that of a pair of point loads, F, applied at the center of an ellipsoidal crack. Consequently, the difference in the apparent tensile strength, T, that is, the difference in the BD pressure, P_b, among the fracturing fluids, is most likely caused by the differences in their viscosity. From the above discussion, in field operations for the same rock type stratum under the same in situ stress condition and flow rate, the BD pressure for injecting a low-viscosity fluid, such as SC-CO_2, is expected to be lower than that of water and much lower than that of high-viscosity fluid such as a gel.

5.4.2 AE distribution

Figure 5.19b shows the dependency on the viscosity of the average distance, L_{av}, from a source to the maximum likelihood plane. This figure indicates that L_{av} decreases with viscosity. This suggests that low-viscosity fracturing fluids induce cracks that propagate farther away from a flat plane, whereas viscous fluids, such as gels, tend to induce cracks along a flat plane. Figure 5.19c shows that the AE distributions' FD decreases with viscosity, suggesting that low-viscosity fracturing fluids induce cracks that propagate three-dimensionally, rather than two-dimensionally along a flat plane.

5.4.3 Fracturing mechanism deduced from P wave first motion polarity of AE

The differences in the crack features possibly arose from a difference in the fracture mechanism. Thus, we compared ratios of the P wave first motion polarity of approximately 30 AE events detected by at least seven sensors, allowing P wave arrivals to be determined for a source location. A calibration test, in which a small steel ball was dropped onto the upper surface of a thick steel plate and the resultant waveforms were recorded by sensors glued onto the plate's lower surface, showed that compression corresponded to an upward trace and dilation to a downward trace. In Figure 5.16, the closed triangles indicate compression in the P wave first motion, showing an upward trace, whereas the open triangles indicate dilatation. For example, in the case of Figure 5.16a, because the number of compressions is four and that of dilatation is four, the compression ratio in the P wave first motion for this AE event becomes 50%. Figure 5.19d shows that the P wave first motion compression ratio tends to increase with viscosity. The ratio would be 50% for a pure shear fracture and 100% for a pure tensile fracture. Thus, the results suggest that a low-viscosity fracturing fluid induces shear-dominant fractures, whereas viscous fluids induce tensile-dominant fractures.

As shown in Figure 5.18a, the cracks induced by SC-CO_2 propagate mainly along the grain boundaries of the constituent minerals, producing many small cracks inclined in the direction of the maximum compressive stress, S_H, which is the propagating direction of a main crack. As shown in Figure 5.15, the induced main cracks extended in planes almost parallel to the Z direction (the direction of the intermediate principal stress, S_V) in all the experiments, and we observed the cracks on the XY plane normal to the Z direction in Figure 5.18a. Thus, because shear stress develops on a plane inclined to the direction of the maximum compressive stress, S_H, by ignoring the effect of the intermediate principal stress and considering the two-dimensional stress condition, the inclined cracks are easily induced by shear fracture. In contrast, Figure 5.18b shows that the cracks induced by viscous oil injection cut through many mineral grains and propagate almost straight along the direction of the maximum compressive stress, S_{H}, without bending. Because shear stress does not develop on a plane parallel to the direction of the maximum compressive stress, S_H, the cracks are probably induced by tensile fracture rather than by shear. In addition, in the oil injection, from the fact that the cracks extended to both lateral surfaces of the specimens as shown in Figure 5.15d and the injected oil seeped out from the induced cracks on the specimen surfaces, the tensile cracks are most likely induced by the intrusion of the viscous oil.

5.4.4 *Crack features deduced from direct observation*

Figure 5.19e shows the dependency on the viscosity of the tortuosity, L/L_0, where L is the crack length measured by connecting 0.05 mm long line segments, namely, a divider, along the main crack, and L_0 is the length of the straight line from the initiation point of the crack on the wall of the injection hole to its endpoint on the inner wall of the overcoring (Figure 5.17a). The main cracks are traced in both directions from the fracturing hole, from Y coordinates of 75 to 51 mm and from 95 to 119 mm (see Figure 5.13). The tortuosity for each specimen is the average of the tortuosities obtained in the two opposite directions. Although the results for the two specimens for each fracturing fluid are plotted in Figure 5.19e, only one result is plotted for the water injection. This is because the cracks in specimen G1103 for water injection could not be filled with the fluorescent resin, and no observation was obtained. The tortuosity of the induced cracks tended to increase with the decrease of fracturing fluid viscosity, which is consistent with the crack propagation patterns in Figure 5.17.

Figure 5.19f shows the dependency on the viscosity of the average of the crack numbers, which were counted when the induced cracks crossed the scanning lines at intervals of 1 mm from the injection hole to the inner wall of the overcoring. The cracks were counted, excluding preexisting cracks, voids, and other defects, by checking the propagation direction and continuity of the cracks through careful observation of crack width, fluorescent resin brightness, and constituent mineral grains under polarized and ultraviolet light. We counted the cracks in the two opposite directions from the fracturing hole, and the numbers plotted in Figure 5.19f are the averages obtained in both directions. Thus, the number roughly corresponds to the average number of crack branches. The number of induced cracks increases with the decrease in fluid viscosity, although the number is large for oil injection in specimen G1112. This is an exceptional case because our observation of this specimen showed that one of the main crack propagations was obstructed by a large biotite grain near the injection hole; as a result, the crack was branched in many directions at the grain. The trends shown in Figure 5.19e–f are consistent with those obtained from the AE distribution, that is, the average distance, L_{av}, and the FDs, increased with the decrease of fluid viscosity, which were already shown in Figure 5.19b–c.

5.5 Discussion

Ishida et al. (2004) conducted HF experiments on the same kind of cubic granite as in the present study using water and viscous oil, which showed similar crack features and fracture mechanisms, as in Chapter 3 of this book. Ishida et al. (2012), Kizaki et al. (2013), Stanchits et al. (2014), and Bennour et al. (2015) also suggested the same dependency on the viscosity of the fracturing fluid, including SC-CO_2, based on their similar experiments using granite, tuff, sandstone, and shale. As described in the previous chapter, Chapter 4, Shimizu et al. (2011) simulated HF using their Distinct Element Method code and provided a persuasive interpretation for the dependency of the BD pressure, AE fracturing mechanism, and crack features on fluid viscosity.

In actual reservoirs, Baria and Green (1986) in the geothermal field and Warpinski et al. (2005) and Dinske et al. (2010) in tight gas reservoirs made comparative analyses of microseismicity induced by HF using viscous fluid and water. Baria and Green (1986) and Dinske et al. (2010) could not show a difference between the two fluids. In contrast, Warpinski et al. (2005) showed that slickwater fracturing extends a much larger area further in three dimensions than gel fracturing, which is consistent with the findings from our laboratory experiments. In actual field operations of HF, weak planes, such as preexisting cracks, sedimentary planes, and defects in strata, may behave similarly to grain boundaries in the granite specimens.

It could be concluded that HF using low-viscosity fracturing fluid tends to induce shear-dominant fractures, resulting in extensive three-dimensional cracking rather than the two-dimensional cracking observed for high-viscosity fluid. By HF using low-viscosity fracturing fluid, the three-dimensional cracks, which tend to be sinuous and have many secondary branches, could form pathways favorable for heat extraction from EGS, shale gas recovery, and other processes. From this point of view, CO_2 fracturing is better than conventional water fracturing for these purposes.

In actual field operations of HF, a solid material called proppant, typically treated sand or man-made ceramic materials, is usually injected into the induced cracks to prevent them from closing due to rock stresses. However, SC-CO_2 is hard to transport the proppant because of its low viscosity. In addition, because CO_2 is more corrosive than water to steel pipes and other units, this nature of CO_2 may cause trouble in actual HF operations. For the first problem, it may be better to inject L-CO_2, due to its larger viscosity than SC-CO_2, through a double pipe, having a vacuum between the inner and outer pipe to prevent it from heating underground. In another way, if we can make a new type of proppant to be dissolved in SC-CO_2 and deposited in the cracks due to the phase change of CO_2 caused by the decrease of pressure and temperature with the BD, the problem would be solved. For the second problem, since CO_2 injection is usually used to enhance oil recovery in the petroleum industry, the experiences would help to solve the problem.

CO_2 fracturing most likely forms pathways favorable for heat extraction, and large amounts of CO_2 would remain underground and distributed across many wells, even if only a small amount of CO_2 is used in each well injection. For this reason, by overcoming the problems for the field application of this approach, CO_2 fracturing for heat extraction from EGS could realize economically viable CO_2 sequestration.

5.6 Conclusions

We conducted HF experiments in 170 mm cubic granite specimens with a 20 mm diameter central hole using SC-CO_2, L-CO_2, water, and viscous oil with viscosities from 0.051 to 336.6 mPa·s. Based on the dependency on the fracturing fluid viscosity of the BD pressure, AE distribution, the fracturing mechanism deduced from ratios of the P wave first motion polarity of AE events, and crack features observed by the fluorescence method, the following conclusions can be drawn.

1 The BD pressure increases with viscosity, which is consistent with previous experiments and numerical simulations. From this result, in actual field operations, it can also be expected that the BD pressure with the injection of a low-viscosity fluid such as SC-CO_2 is lower than that of water and much lower than that of a viscous fluid such as gel.
2 AE source distributions clarified that the average distance, L_{av}, from a source to the maximum likelihood plane and the FDs, of the source distributions increased with the decrease in fluid viscosity. This indicates that HF using low-viscosity fluid tends to induce extensive three-dimensional cracking rather than the two-dimensional cracking observed for high-viscosity fluid.
3 Ratios of the compressional P wave first motion polarity elucidate that a low-viscosity fracturing fluid tends to induce shear-dominant fractures. In contrast, viscous fluids tend to induce tensile-dominant fractures.
4 The photomicrographs taken using a petrographic microscope in crossed-polarized light (crossed Nicols) show that the cracks induced by SC-CO_2 propagated mainly along the grain boundaries of the constituent minerals; as a result, many small cracks inclined to the direction of the maximum compressive stress were observed. In contrast, the cracks induced by viscous oil injection cut through the many mineral grains and propagated almost straight along the direction of the maximum compressive stress without bending. Because shear stress was developed on a plane inclined to the direction of the maximum compressive stress, the inclined cracks were probably induced by shear fracture. Consequently, the observation was consistent with the results obtained from AE monitoring; that is, shear-dominant fracturing occurred with low-viscosity fluid injection; in contrast, tensile-dominant fracturing occurred with high-viscosity fluid injection.
5 The microscopic crack observations showed that the tortuosity of a main crack and the average number of cracks increased with the decrease in fluid viscosity. This observation was consistent with the results of the AE source distributions, that is, the average distance, L_{av}, from a source to the maximum likelihood plane, and the FDs, of the source distributions increased with the decrease in fluid viscosity.
6 The observed dependency of crack features on the fracturing fluid viscosity was consistent with those reported in previous laboratory experiments, numerical simulations, and the comparison of the microseismic mappings between water and gel used in the field HF operations.
7 These results demonstrate that CO_2 fracturing induces three-dimensional sinuous cracks with many secondary branches, desirable pathways for heat extraction from EGS, shale gas recovery, and other processes. By overcoming the difficulty in transforming the proppant into induced cracks due to the low viscosity of SC-CO_2, CO_2 fracturing could realize economically viable CO_2 sequestration because large amounts of CO_2 would remain underground and distributed across many wells, even if only a small amount of CO_2 is used in each well injection.

References

Baria, R. & Green, A. S. P. (1986) Seismicity induced during a viscous stimulation at the Camborne School of Mines hot dry rock geothermal energy project in Cornwall, England. In *Progress in Acoustic Emission III*. The Jpn. Soc. of NDI, Tokyo, pp. 407–429.

Bennour, Z., Ishida, T., Nagaya, Y., Chen, Y., Nara, Y., Chen, Q., Sekine, K. & Nagano, Y. (2015) Crack extension in hydraulic fracturing of shale cores using viscous oil, water, and liquid carbon dioxide. *Rock Mech. Rock Eng.*, 48(4), 1463–1473. https://doi.org/10.1007/s00603-015-0774-2

Brown, D. W. (2000) *A Hot Dry Rock Geothermal Energy Concept Utilizing Supercritical CO_2 Instead of Water, Paper Presented at 25th Workshop on Geothermal Reservoir Engineering*. Stanford University, Stanford, California.

Bunger, A. P., Lakirouhani, A. & Detournay, E. (2010) Modelling the effect of injection system compressibility and viscous fluid flow on hydraulic fracture breakdown pressure. In *Proceedings of the 5th International Symposium on In Situ Rock Stress*, Int. Soc. For Rock Mech., Beijing pp. 59–67.

Dinske, C., Shapiro, S. A. & Rutledge, J. T. (2010) Interpretation of microseismicity resulting from gel and water fracturing of tight gas reservoirs. *Pure Appl. Geophys.*, 167, 169–182.

Elsworth, D., Gan, Q., Marone, C., Connolly, P., Alpern, J., Culp, B. & Im, K. J. (2014) Coupled processes related to gas-fracturing in unconventional reservoirs. In: *Proceedings of 8th Asian Rock Mechanics Symposium* October 2014, Sapporo, Japan.

Fenghour, A., Wakeham, W. A. & Vesovic, V. (1998) The viscosity of carbon dioxide. *J. Phys. Chem. Ref. Data*, 27(1), 31–44. https://doi.org/10.1063/1.556013

Ge, M. (2003a) Analysis of source location algorithms—Part I: overview and non-iterative methods. *J. Acoust. Emiss.*, 21, 21–28.

Ge, M. (2003b) Analysis of source location algorithms—Part II: iterative methods. *J. Acoust. Emiss.*, 21, 29–51.

Geiger, L. (1910) Herbsetimmung bei Erdbeben aus den Aukunfzeiten [German]. *K. Gessell. Will. Goett.*, 4, 331–349.

Geiger, L. (1912) Probability method for the determination of earthquake epicentres from the arrival time only. *Bull. St. Louis Univ.*, 8, 60–71.

Grassberger, P. (1983) Generalized dimensions of strange attractors. *Phys. Lett.*, 97A(6), 227–230.

Hirata, T., Satoh, T. & Ito, K. (1987) Fractal structure of spatial distribution of microfracturing in rock. *Geophys. J. R. Astron. Soc.*, 90, 369–374 https://doi.org/10.1111/j.1365-246X.1987.tb00732.x

International Society for Rock Mechanics, Commission on Standardization of Laboratory and Field Tests (1979) Suggested methods for determining water content, porosity, density, absorption and related properties, and swelling and slake-durability index properties. *Int. J. Rock Mech. Min. Sci. Geomech. Abstr.*, 16, 143–151.

Ishida, T., Aoyagi, K., Niwa, T., Chen, Y., Murata, S., Chen, Q. & Nakayama, Y. (2012) Acoustic emission monitoring of hydraulic fracturing laboratory experiment with supercritical and liquid CO_2, *Geophys. Res. Lett.*, 39, L16309. https://doi.org/10.1029/2012GL052788

Ishida, T., Chen, Q., Kaieda, H. & Mizuta, Y. (1998) Fractal analysis on a source distribution of acoustic emissions in a hot dry rock mass Reservoir. *J. JSCE*, 536(III-43), 49–63. (in Japanese with English Abstract).

Ishida, T., Chen, Q. & Mizuta, Y. (1997) Effect of injected water on hydraulic fracturing deduced from acoustic emission monitoring. *Pure Appl. Geophys.*, 150, 627–646.

Ishida, T., Chen, Q., Mizuta, Y. & Roegiers, J-C. (2004) Influence of fluid viscosity on the hydraulic fracturing mechanism. *Trans. ASME, J. Energy Res. Technol.*, 126, 190–200. https://doi.org/10.1115/1.1791651

Ishida, T., Chen, Y., Bennour, Z., Yamashita, H., Inui, S., Nagaya, Y., Naoi, M., Chen, Q., Nakayama, Y. & Nagano, Y. (2016) Features of CO_2 fracturing deduced from acoustic emission and microscopy in laboratory experiments. *J. Geophys. Res. Solid Earth,* 121(11), 8080–8098. https://doi.org/10.1002/2016JB013365

Ishida, T., Kitano, K., Kinoshita, N. & Wakabayashi, N. (1990) Acoustic emission and *P* wave velocity monitoring during thermal cracking in a locally heated granite block. *J. JSCE*, 418(III-13), 153–161. (in Japanese with English Abstract)

Ishida, T., Oka, H., Yonehiro, N., Tamura, Y. & Mizuta, Y. (2005) Examination on the tensile strengths of hydraulic fracturing using the probabilistic theory of fracture. *J. MMIJ (Shigen-to-Sizai)*, 121, 576–582. (in Japanese with English Abstract)

Ishida, T., Nagaya, Y., Inui, S., Bennour, Z., Chen, Y., Chen, Q. & Nakayama, Y. (2015) Laboratory-scale hydraulic fracturing experiments using super-critical state carbon dioxide, water and viscous oil. *J. of MMIJ*, 131, 115–121. (in Japanese with English Abstract)

Ishida, T. & Sasaki, S. (2011) Numerical simulation to examine accuracy of AE source location and its applications to in-situ rock monitoring. *J. Acoust. Emiss.*, 29, 260–272.

Jacobs, T. (2014) Shale revolution revisits the energized fracture. *J. Pet. Technol.*, June 2014, 48–56.

Jansen, D. P., Carlson, S. R., Young, R. P. & Hutchins, D. A. (1993) Ultrasonic imaging and acoustic emission monitoring of thermally induced microcracks in Lac du Bonnet granite. *J. Geophys. Res.*, 98(22), 231–243. https://doi.org/10.1029/93JB01816

Kirsch, C. (1898) Die theorie der elastizität und die bedürfnisse der festigkeitslehre [German]. *Z. Ver. Dtsch. Ing.*, 42, 797–807.

Kizaki, A., Ohashi, K., Tanaka, H. & Sakaguchi, K. (2013) Effects of vertical stress on fracture propagation using supercritical carbon dioxide. In *Proceedings of EUROCK, Int. Soc. for Rock Mech., Paper No. CH149*, Wroclaw, Poland.

Mohan, A. R., Turaga, U., Shembekar, V., Elsworth, D. & Pisupati, S. V. (2013) Utilization of carbon dioxide from coal-based power plants as a heat transfer fluid for electricity generation in enhanced geothermal system (EGS). *Energy*, 57, 505–512.

Morita, N., Black, A. D. and Fuh, G.-F. (1996) Borehole breakdown pressure with drilling fluids—1. Empirical results. *Int. J. Rock Mech. Min. Sci. Geomech. Abstr.*, 33, 39–51.

National Astronomical Observatory of Japan (1995) *Chronological Scientific Tables*, Maruzen, Tokyo, pp. 446–447.

Nishiyama, T. & Kusuda, H. (1994) Identification of pore spaces and microcracks using fluorescent resins. *Int. J. Rock Mech. Min. Sci. Geomech. Abstr.*, 31, 369–375.

Pruess, K. (2006) Enhanced geothermal systems (EGS) using CO_2 as working fluid—A novel approach for generating renewable energy with simultaneous sequestration of carbon. *Geothermics*, 35, 351–367.

Rothman, R. L., Greenfield, R. J. and Hardy Jr., H. R. (1974) Errors in hypocenter location due to velocity anisotropy. *Bull. Seismol. Soc. Am.*, 64, 1993–1996.

Sano, O., Kudo, K. & Mizuta, Y. (1992) Experimental determination of elastic constants of Oshima granite, Barre granite, and Chelmsford granite. *J. Geophys. Res.*, 97, 3367–3379. https://doi.org/10.1029/91JB02934

Scheidegger, A. E. (1962) Stress in Earth's crust as determined from hydraulic fracturing data. *Geol. Bauwes.*, 27, 45–53.

Schmitt, D. R. & Zoback, M. D. (1992) Diminished pore pressure in low-porosity crystalline rock under tensional failure: apparent strengthening by dilatancy. *J. Geophys. Res.*, 97, 273–288. https://doi.org/10.1029/91JB02256

Schmitt, D. R. & Zoback, M. D. (1993) Infiltration effects in the tensile rupture of thin walled cylinders of glass and granite: implications for the hydraulic fracturing breakdown equation. *Int. J. Rock Mech. Min. Sci. Geomech. Abstr.*, 30, 289–303. https://doi.org/10.1016/0148–9062(93)92731-5

Shimizu, H., Murata, S. & Ishida, T. (2011) The distinct element analysis for hydraulic fracturing in hard rock considering fluid viscosity and particle size distribution. *Int. J. Rock Mech. Min. Sci.*, 48(5), 712–727. https://doi.org/10.1016/j.ijrmms.2011.04.013

Stanchits, S., Surdi, A., Gathogo, P., Edelman, E. & Suarez-Rivera, R. (2014) Onset of hydraulic fracture initiation monitored by acoustic emission and volumetric deformation measurements. *Rock Mech. Rock Eng.*, 47, 1521–1532. https://doi.org/10.1007/s00603-014-0584-y

Ueda, A., Kato, K., Ohsumi, T., Yajima, T., Ito, H., Kaieda, H., Metcalfe, R. & Takase, H (2005) Experimental studies of CO_2-rock interaction at elevated temperatures under hydrothermal conditions. *Geochem. J.*, 39, 417–425.

Vutukuri, V. S. & Katsuyama, K. (1994) *Introduction to Rock Mechanics*. Ind. & Consult., Inc., Tokyo, pp. 125–129.

Warpinski, N. R., Kramm, R. C., Heinze, J. R. & Waltman, C. K. (2005) Comparison of single- and dual-array microseismic mapping techniques in the Barnett shale. *Paper Presented at 2005 Annual Technical Conference and Exhibition, Society of Pet. Eng.*, Dallas, Tex.

Weibull, W. (1939a) *A Statistical Theory of the Strength of Materials*. Ingeniörs Vetenskaps Akademiens, Handlingar, p. 151.

Weibull, W. (1939b) *The Phenomenon of Rapture in Solid*. Ingeniörs Vetenskaps Akademiens, Handlingar, p. 153.

Xu, C., Dowd, P. & Li, Q. (2016) Carbon sequestration potential of the Habanero reservoir when carbon dioxide is used at the heat exchange fluid. *J. Rock Mech. Geotech. Eng.*, 8, 50–59.

Zhou, X. & Burbey, T. J. (2014) Fluid effect on hydraulic fracture propagation behavior: a comparison between water and supercritical CO_2-like fluid. *Geofluids*, 14, 174–188. https://doi.org/10.1111/gfl.12061

Chapter 6

Small-scale field water refracturing experiments

6.1 Introduction

In laboratory experiments, we examined how rock properties and fracturing fluid viscosity affect the fracturing mechanism of AE (acoustic emission) and crack features in Chapters 2 and 3. In Chapter 4, we showed that the effects can be reproduced and explained well by the numerical simulation using the flow-coupled distinct element method. In this chapter, to bridge the gap between laboratory and field and to better understand the crack expansion mechanism and its fracturing mode, we introduce a small-scale field HF (hydraulic fracturing) experiment using water refracturing, which is a new crack extension associated with an increase in the flow rate of injected water after the first BD.

In a hot dry rock (HDR) project, water injection continues for as long as several weeks, and the region in which AE events are induced usually expands with time, as has been observed in Cooper Basin, Australia (Baisch et al., 2006, 2009, 2015), Soultz HDR, France (Evans et al., 2005), Hijiori, Japan (Sasaki, 1997, 1998), and Ogachi, Japan (Kaieda et al., 1995). However, when the injection flow rate increases during operations, AE activity often increases remarkably, suggesting crack expansion (e.g., Kaieda et al., 1995; Sasaki, 1997). By contrast, injections in oil and gas recovery from sandstone and shale reservoirs are much shorter than those in HDR; for example, the duration was around 5 h in Cotton Valley sandstone, Texas (Rutledge et al., 2004) and the Barnett Shale, Texas (Hummel and Shapiro, 2013). In addition, the refracturing made after the first HF has recently been focused on accelerating production rates and enhancing the ultimate recovery of depleted shale wells (Jacobs, 2014). Thus, it is necessary to clarify the mechanism of crack expansion to realize effective HF and associated refracturing. However, as indicated and discussed in the previous chapters of this book, many researchers have reported that shear events dominate in field monitoring (e.g., Talebi and Cornet, 1987), although elastic theory predicts that HF should induce tensile fracture (e.g., Hubbert and Willis, 1957). This paradox (e.g., Maxwell and Cipolla, 2011) means that the fracturing mechanism of HF remains ambiguous.

From the requirement, to better understand the crack expansion mechanism of HF and its fracturing mode, we conducted a small-scale field HF experiment of water refracturing, using a 10-m-long vertical hole to inject water in the test site in the Mizunami Underground Research Laboratory (MIU) in central Japan. In the chapter of this book, the method and results of the small-scale field HF experiment are shown following Ishida et al. (2019).

DOI: 10.1201/9781003282631-6

6.2 Site and experimental setup

6.2.1 Site and method of water injection

The site is located in a gallery situated 500 m below the surface in the MIU for an underground laboratory for research and development of the geological disposal of high-level radioactive waste of the Japan Atomic Energy Agency. The MIU is located in the Tono district of central Japan, and it consists of two main geological units, namely, the basement of late Cretaceous Toki granite and the overlying Akeyo formation of early Miocene mudstone and sandstone.

In two locations near the site on the 500 m level, the initial rock stress conditions were measured using the compact conical-ended borehole overcoring technique (Sugawara and Obara, 1999). One is location A, 80 m north, and the other is location B, 50 m south of our HF site. The existence of a sub-vertical fault having an north-northwest (NNW) strike is confirmed from the near-surface down to the 500 m level between locations A and B. The results measured at location A showed that $\sigma_1 = 16.8$ MPa (− 172°/59°), $\sigma_2 = 10.2$ MPa (5°/31°), and $\sigma_3 = 7.5$ MPa (96°/2°), and those at location B showed that $\sigma_1 = 15.1$ MPa (−173°/9°), $\sigma_2 = 10.9$ MPa (89°/41°), and $\sigma_3 = 10.0$ MPa (−74°/48°), where the numbers in the parentheses show an azimuth angle from north (positive to east) and an inclination angle from the horizontal of the respective principal stress directions (Kuwabara et al., 2014, 2015a). The magnitudes and the directions of the maximum and minimum horizontal stresses, as the tensor average, that are calculated from the respective three principal stresses and their directions are 11.9 MPa (N7°E) and 7.5 MPa (N97°E) at location A, and 15.0 MPa (N7°E) and 10.5 MPa (N97°E) at location B. Although the difference between locations A and B is likely due to the influence of the fault, both results indicate that the maximum horizontal stress almost lies in north-south (N7°E) and the minimum in east-west (N97°E). Young's modulus, Poisson's ratio, uniaxial compressive strength, and tensile strength of the Brazilian test for the cores obtained in the vicinity of our HF site were 52 GPa, 0.24, 160 MPa, and 5.5 MPa on average, respectively (Kuwabara et al., 2015b).

A schematic diagram of our HF site is shown in Figure 6.1, and the view of the test site is shown in Figure 6.2. An 86-mm diameter HF hole was drilled downward from the gallery floor, and four AE-monitoring holes of 66 mm diameter were drilled parallel to 1 m far from the HF hole. To inject water, we drilled a 36-mm-diameter pilot hole in the center of the bottom of the 86-mm-diameter HF hole. After sealing the upper end of the pressurizing section in the pilot hole with the upper O-ring attached to a packer unit, we poured cement paste above the upper O-ring. After the poured cement paste had hardened, we sealed the lower end of the pressurizing section with the lower O-ring and injected water into the 160-mm long pressurizing section at a depth of 5.34–5.49 m. The dimensioned diagram of the packer to seal the pressurizing section is shown in the lower left part of Figure 6.1. This sealing method is the same as that for our carbon dioxide injection, shown in Ishida et al. (2017, 2021, 2023) and Chapter 8 of this book. We used two syringe pumps, each with a 500-mL cylinder, to inject water at a constant flow rate of 10 or 30 mL/min; these pumps could be switched over to each other smoothly without interrupting the injection operation.

6.2.2 Methods for monitoring AE and injected pressure

We placed four AE sensors in each of the four AE holes (AE1-AE4) (see Figure 6.1). The AE sensor was a waterproof lead zirconate titanate type having a resonance frequency of 70 kHz (AE703SW-GAMP-0542; Fuji Ceramics Corp., Japan). We fixed each sensor to an aluminum

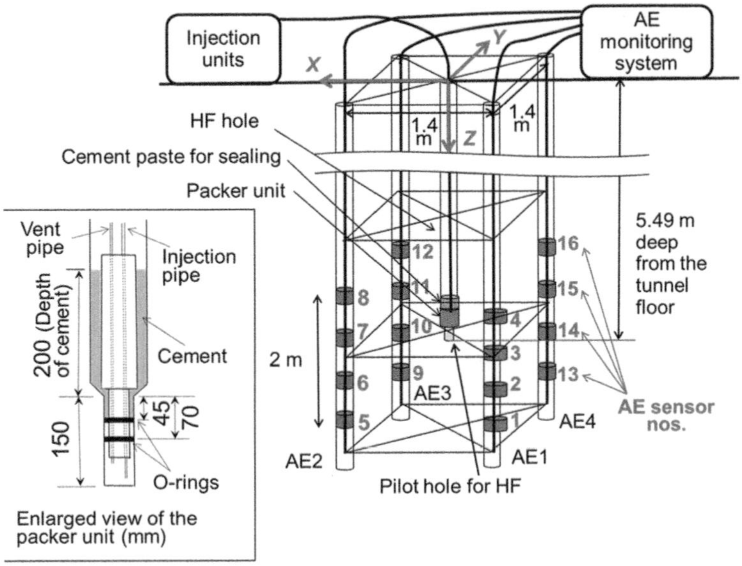

Figure 6.1 Bird's-eye view of the arrangement of AE sensors to enclose the pressurized section for HF under the test site. The Cartesian coordinate system used in the experiment is also shown. The dimensioned diagram of the packer to seal the pressurizing section is also shown in the lower left part of the figure (Ishida et al., 2019).

Figure 6.2 View of the test site.

rod with a pre-amplifier, inserting a thick rubber sheet between the sensor and the rod to block any vibration transmitted through the rod. After angling the sensitive direction of the sensor toward the HF hole, we pressed the sensor against the wall of the AE hole by applying an oil pressure of 1.5 MPa in a small hydraulic jack set behind the sensor. In each AE hole, we put the four sensors along a 2-m-long span with intervals of 0.7, 0.6, and 0.7 m, and centered the span at 5.40 m, which is very close to the central depth of the pressurizing section of the HF hole. AE signals detected on the 16 sensors were recorded continuously at a sampling time of 1 μs by exchanging a hard disk drive for a solid state drive in an analog-to-digital converter (PXI-5105; National Instruments Corp., USA). The AE signals were recorded after being amplified by 52 dB (40 dB preamplification and 12 dB main amplification) and applied through a 20–500-kHz bandpass filter.

The injection pressure was measured with a transducer (PW-50MPA; Tokyo Sokki Kenkyujo Co., Ltd., Japan) set on the injection pipe put on the gallery floor just outside of the HF hole, and recorded at a sampling time of 0.1 s through an analog-to-digital converter (PXI-6251; National Instruments Corp., USA). The injected flow rate was 30 mL/min at most, that is, 500 mm³/s. The inner diameter of the steel pipe to inject water was 2 mm, and the velocity of water in the pipe was $500/(1 \times 1 \times 3.14) = 160$ mm/s. Using these numbers, when we theoretically calculated the pressure drop along the pipe of 5 m, it was only 0.005 MPa. In addition, when we measured the pressure drop just outside of the HF hole on the gallery floor at 30 m horizontal distance from the pump outlet, the pressure drop was negligibly small, which is consistent with the theoretical calculation. From the results, we confirmed that the measured fluid pressure is the correct pressure in the pressurizing section at a depth of 5.34–5.49 m below the gallery floor.

6.3 Results

6.3.1 *Temporal changes of injected water pressure and located AE event rate during different constant flow rates*

Figure 6.3 shows the records of injected water pressure, flow rate, and rate of located AE events that satisfy the conditions described in the next section. We began injecting water at a flow rate of 10 mL/min at an elapsed time t of 22 s. The BD pressure, which is defined as a peak pressure immediately before it drops suddenly, was recorded as 9.20 MPa at $t = 630$ s. The abrupt pressure drop around $t = 200$ s in Figure 6.3 was due to trouble with the injection pumps, whereas the small one around $t = 400$ s was probably due to crack initiation because this was accompanied by AE events located around the pressurizing section, as shown later. We continued to inject at

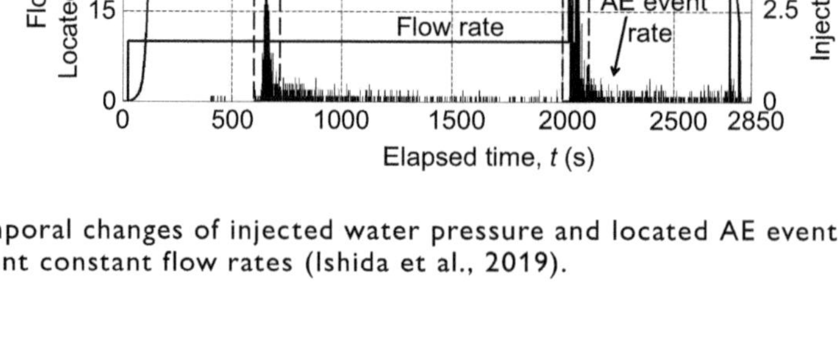

Figure 6.3 Temporal changes of injected water pressure and located AE event rate during different constant flow rates (Ishida et al., 2019).

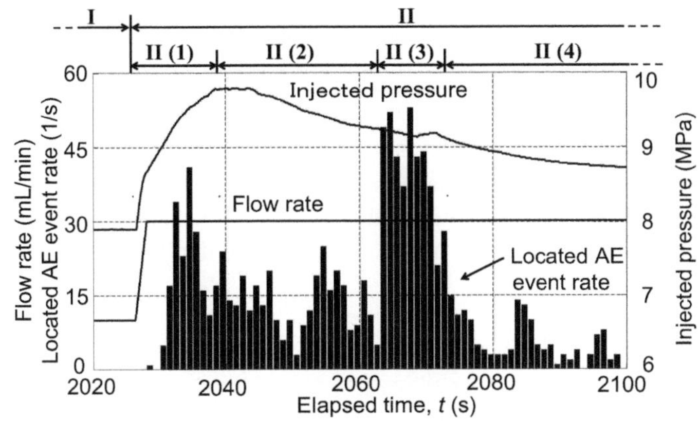

Figure 6.4 Temporal changes of injected water pressure, flow rate, and located AE event rate for the 80 s, $t = 2{,}020–2{,}100$ s, corresponding to the period enclosed by the right-hand broken rectangle in Figure 6.3 (Ishida et al., 2019).

a flow rate of 10 mL/min after the first BD. After we increased the flow rate to 30 mL/min at $t=2025$ s, the second BD pressure was recorded as 9.79 MPa at $t=2039$ s.

To discuss crack expansion and its fracturing mechanism induced at the second BD, we show the temporal changes of injected water pressure, flow rate, and located AE event rate for the 80 s from $t=2020–2100$ s in Figure 6.4; this period corresponds to the right-hand broken rectangle in Figure 6.3. In the upper parts of Figures 6.3 and 6.4, we marked the periods defined to discuss the distributions of located AE events. After the pressure increased linearly from 7.9 to 8.7 MPa with the increase in flow rate from 10 to 30 mL/min, in period II(1), the pressure continued to increase up to the secondary BD pressure of 9.79 MPa, and the located AE event rate increased with increasing pressure. Then, in period II(2), despite keeping the flow rate of the injected fluid at 30 mL/min, the AE activity decreased. The AE activity increased again immediately before a slight pressure increase in period II(3), before the AE activity decreased in period II(4).

6.3.2 *Temporal changes of AE hypocenter distributions*

To identify AE events in the waveforms recorded continuously by the 16 sensors in a sampling time of 1 μs, we searched for waveforms whose amplitudes exceeded $+0.125$ V on the full scale from -5 to $+5$ V. When we found waveforms meeting the amplitude criteria at one or more of the 16 sensors, we extracted 2,048 sample points in total (1,024 before and 1,024 after each time point) for all 16 channels. We located the hypocenter of each event using a least-squares method by reading the P-wave arrival times manually.

We measured the P-wave velocities between the HF hole and the 16 sensors just before the HF experiment using an emitter (AE703SWR-0840; Fuji Ceramics Corp.) attached just above the packer to seal the pressurizing section in the pilot hole to inject water. As a result of the measurements, we obtained an average velocity of 5.67 km/s with a standard deviation of 0.48 km/s. Because the scattering in the P-wave velocity due to the inhomogeneity was larger than the anisotropy, and in addition, the principal axes of the anisotropy could not be determined due to the limitation of our measuring paths, we used the average velocity for source location under the assumption of homogeneous rock.

Figure 6.5 shows the AE hypocenter distributions in periods I, II(1), II(2), II(3), and II(4) that were marked in the upper parts of Figures 6.3 and 6.4. Here, period I, t=0–2026 s, includes the period in which the flow rate was maintained at 10 mL/min, namely, t=22–2026 s, as shown in Figure 6.3. In Figure 6.5, we show only two projections on the *XY* horizontal plane and the *XZ* vertical plane; we omit the *YZ* plane projection because the AE hypocenter distribution is relatively narrow in the *Y* direction, as seen on the *XY* plane projection. Although we set the origin of the coordinate system at the center of the HF hole on the surface as shown in Figure 6.1, the central coordinates of the HF hole at the nearly central depth of the pressurizing section at

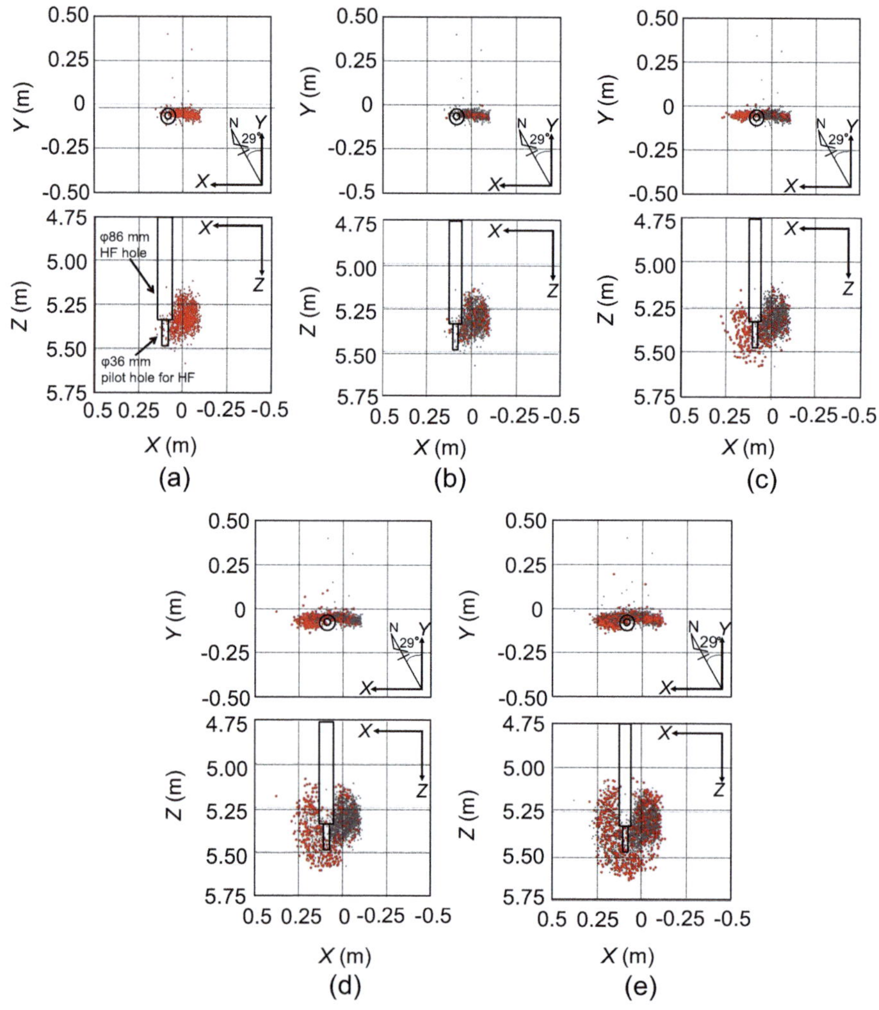

Figure 6.5 Projections on *XY* horizontal plane and *XZ* vertical plane distribution: (a) period I (0–2,026 s); (b) period II(1) (2,026–2,039 s); (c) period II(2) (2,039–2,063 s); (d) period III(3) (2,063–2,073 s); and (e) period III(4) (2,073–2,850 s). In Figure 6.5b–e, AE hypocenters newly located in the respective periods are shown with red points, whereas those located in the previous periods are shown with gray points (Ishida et al., 2019).

Z=5.415 m are X=0.080 and Y=−0.660 m as shown in Figure 6.5, respectively, because of a 1.1° tilt of the HF hole. Since the AE holes have similar (albeit small) tilts, we corrected the sensor coordinates for the AE source locations. The AE holes do not appear in Figure 6.5 because they are 1 m from the HF hole and on the outside of the figure frames.

The AE hypocenters shown in Figure 6.5 satisfy the following two conditions. First, P-wave arrival times had to be read at five or more sensors set in three or more AE holes to enclose the hypocenter three-dimensionally. Second, the maximum standard error, among the three principal axes of an error ellipsoid calculated from the variance-covariance matrix, had to be smaller than 50 mm for the hypocenter location. The number of located AE hypocenters satisfying these two conditions was 1098 (I), 189 (II(1)), 324(II(2)), 375(II(3)), and 638 (II(4)), respectively.

As shown in Figure 6.5a, in period I, while the flow rate remained at 10 mL/min, the AE hypocenters were distributed in the negative X direction from the pressurizing section along the pilot hole. In Figure 6.5b–e, the AE hypocenters newly observed in each period are shown with red points, whereas those in the previous periods are shown with gray points. In period II(1), just after the flow rate was increased to 30 mL/min, the AE hypocenters remained distributed within the region where they were distributed in the previous period I, as shown in Figure 6.5b, even though the pressure increased up to the secondary BD. In the following period II(2), just after the secondary BD, despite lower rates of located AE events with the decrease of injected pressure (see Figure 6.4), the AE hypocenters migrated into a new region in the positive X direction from the pressurizing section, as shown in Figure 6.5c. This suggests that new crack expansion occurred following the second BD pressure (9.79 MPa), which was larger than the first BD pressure (9.20 MPa). In period II(3), as shown in Figure 6.5d, AE events occurred throughout the regions where AE events actively occurred in the previous periods I–II(2), and some of them tended to migrate outward from their margins, suggesting slight crack expansions. In period II(4), as shown in Figure 6.5e, the AE events occurred throughout but only within the regions where they occurred in the previous periods, suggesting no crack expansion into any new region.

Because some of the located AE events lie along the wall of the 86 mm diameter hole, as shown in Figure 6.5a, the question may arise whether they reflect fracture growth along the packer or isolating cement early after the first BD. When we checked the locations of the six AE events immediately after the first BD, we found that five of the six are located only 10–40 mm away from the wall of the 86 mm diameter hole, and one of them is 90 mm away from it. Thus, we cannot rule out the possibility that the five events occurred on the interface, even if we consider an error in the location. However, the two cracks were observed in the upper part in the symmetrical positions (in the directions where the AE hypocenters were distributed) on the 86 mm hollow core recovered from the pressurizing section after the HF by drilling coaxially with the 36 mm pilot hole. From this fact, we guess that the crack was initiated from the upper part of the pressurized section and extended upward, and after that, the detectable events in the rock were caused close to the hole. In addition, even if an AE event occurred on the interface, it was probably small and undetectable because the interface between the hole wall and the cement was weaker than the matrix of intact rock.

6.3.3 Temporal changes of fracturing modes of AE events

We examined the focal mechanism of the AE events by analyzing the ratios of the P-wave initial motion polarities. In the analysis, we used only those located AE events whose P-wave polarities (i.e., compression or dilatation) could be read by 10 or more sensors. We checked the

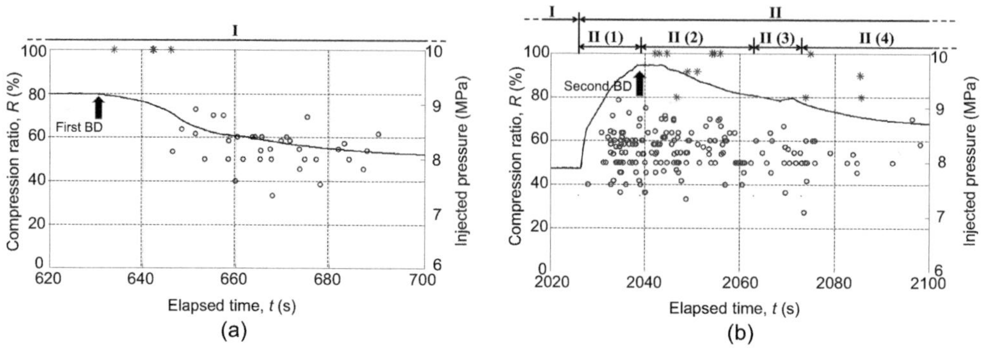

Figure 6.6 Percentage R of sensors that recorded a compressive wave out of all sensors at which the polarities could be read by 10 or more sensors for each AE event. Gray stars are AE events labeled as "tensile-dominant events" (TD events) ($80 \le R \le 100$); gray circles are those labeled as "shear dominant events" (SD) ($20 < R < 80$). (a–b) Show R for the 80 s from just before the first and second BD, respectively, corresponding to the periods shown with the two broken rectangles in Figure 6.3 (see the color figures online of Ishida et al., 2019).

polarity of the response of an AE sensor by dropping a small steel ball onto its surface and confirmed that an upward trace of P-wave initial motion corresponds to a compressive wave. For each AE event where the total number of sensors that could read the polarity was larger than 10, Figure 6.6 shows the percentage R of sensors that recorded the compressive wave. Figure 6.6a and b plots R for the 80 s time windows that begin just before the first and second BDs, respectively, corresponding to the periods enclosed by the two broken rectangles in Figure 6.3. Assuming that the AE sensors surround each AE hypocenter sufficiently, the ratio R should be 100% for pure tensile or explosion-type events, 50% for pure shear-type events, and 0% for pure crack closure or implosion-type events. Herein, we label those AE events with $80 \le R \le 100$ as "tensile-dominant events" (TD events; red stars) and those with $20 < R < 80$ as "shear-dominant events" (SD events; blue circles). Although we intended to label those with $0 \le R \le 20$ as "implosion dominant events", there were no such events. As shown in Figure 6.5a, only four TD events were recorded immediately after the first BD, followed by many SD events. This tendency is also seen in Figure 6.5b; that is, many TD events were recorded immediately after the second BD, whereas the frequency at which they were recorded decreased with time.

Figure 6.7 shows a typical example of TD events. As shown in Figure 6.7a, P-wave arrivals and their polarities could be read at the points indicated by the closed triangles at 13 out of the 16 sensors, and they all show upward traces corresponding to compression. As shown in Figure 6.7b, the polarities projected on a lower hemisphere Schmidt net show a well-constrained tensile fracturing mechanism.

Figure 6.8a and b plots the X, Y, and Z coordinates of the AE hypocenters along with the elapsed times corresponding to those of Figure 6.6a and b, respectively. In Figure 6.8a, the four TD events were induced immediately after the first BD, preceding SD events. In Figure 6.8b, if an AE event was induced on a frontier toward a new region into which no AE event had yet migrated, one of the X, Y, and Z coordinates of the event should be plotted on the frontier of a previous coordinate distribution. Out of the 14 TD events in Figure 6.8b, five are on the frontier of the X coordinate and five are on the frontier of the Z coordinate, as shown by the arrows. The arrow is attached to either the X or Z coordinate for an event, without duplication.

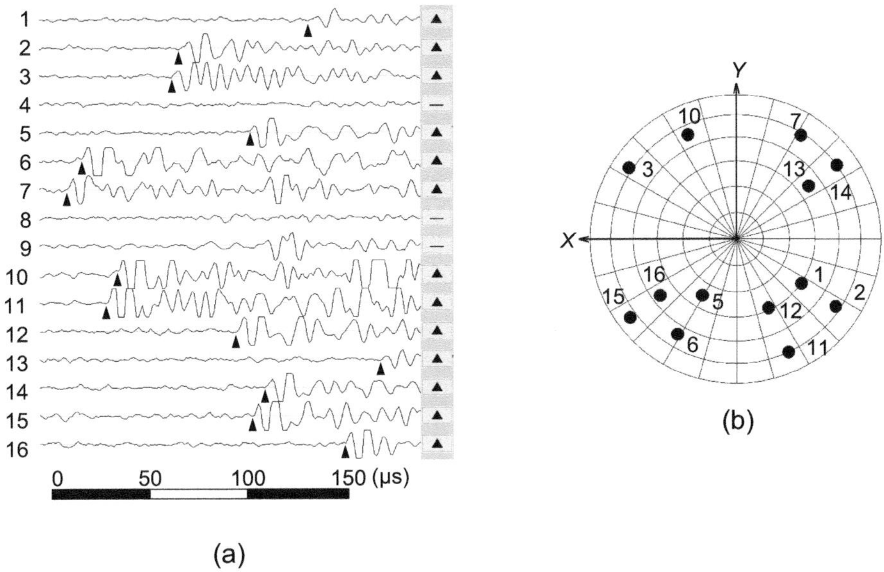

Figure 6.7 Typical example of TD AE events. (a) Shows recorded waveforms. A closed triangle indicates a P-wave arrival time. P-wave polarities were read by 13 sensors, all of which show upward traces corresponding to compression. (b) Polarities projected on the lower hemisphere projection of a Schmidt net. The occurrence time of the AE event was 2055.95 s, and the *X*, *Y*, and *Z* coordinates of its location were 0.266, −0.064, and 5.341 m, respectively (Ishida et al., 2019).

Because these 10 events represent 71% of the total number of events (i.e., 14 events), many were likely induced on frontiers toward new regions. At least, the ratio of TD events located on the frontiers is considerably larger than that of the SD events. This suggests that the frontier TD events are associated with the propagation of new cracks.

6.4 Discussion

6.4.1 *Direction of HF crack concerning rock stress and pre-existing cracks*

From the AE hypocenter distributions shown in Figure 6.5, the direction of the cracks induced by HF lay in N120°E. However, the direction is completely inconsistent with the rock stress conditions measured in the two locations near our HF site; that is, the maximum horizontal stress lies in north-south (N8°E) and the minimum in east-west (N98°E) as shown in Section 6.2.1. On the other hand, pre-existing cracks, which we avoided in the selection of the pressurizing section for our HF experiments, were observed at the rate of 1.4 cracks/m on average over around 50 m in total length along the 10 m long HF hole and the four 10 m long AE holes observed by using a borehole television system. The dominant strike and dip of the pre-existing cracks were N133°E/80°, close to the crack direction induced by HF lying in N120°E. From the facts, we consider that the crack direction extending in our HF experiment was more affected by the direction of inherent hidden weak planes corresponding to pre-existing crack directions

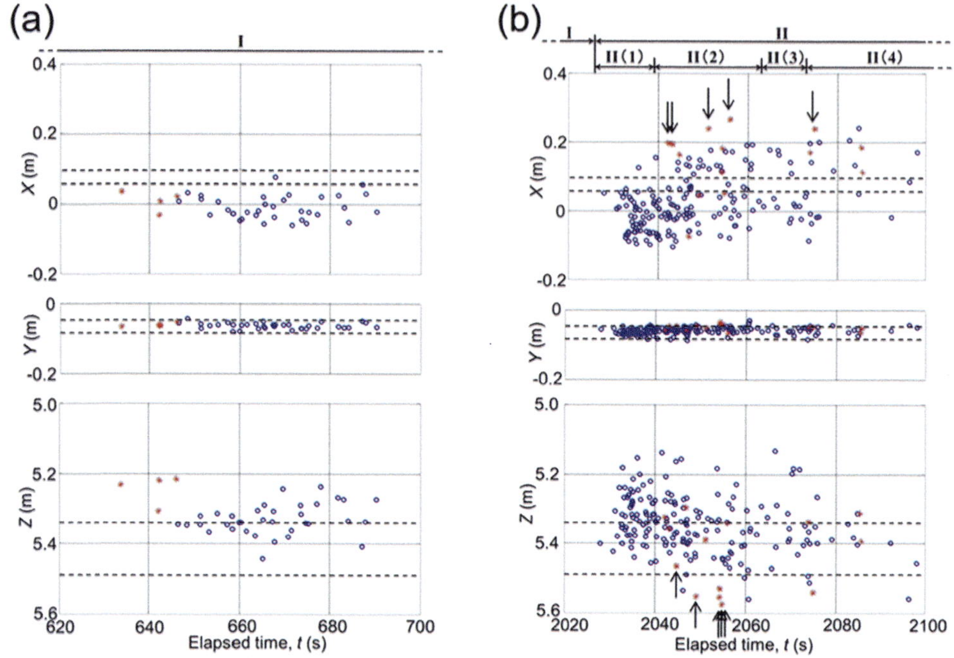

Figure 6.8 X, Y, and Z coordinates of AE hypocenters along with their elapsed times. Red stars are TD AE events ($80 \leq R \leq 100$); blue circles are SD AE events ($20 < R < 80$). Bands of broken lines indicate the span of the pressurizing section in each coordinate corresponding to the position of the pilot hole. (a–b) AE hypocenters for the periods corresponding to Figure 6.6a and b, respectively. The arrows in (b) indicate TD AE events located on the frontier of the previous distribution of each coordinate. An arrow is attached to either the X or Z coordinate of an event, without duplication (Ishida et al., 2019).

rather than the rock stress condition, although the HF was made in a small intact rock mass selected to avoid such pre-existing joints. In the larger-scale field experiments, the tendency was often observed in fracture generation around the HF hole, for example, at the Grimsel Test Site in Switzerland (Gischig et al., 2018), and in crack extensions farther from the HF hole, for example, at the Ogachi Site in Japan (Kaieda et al., 1993; Kondo, 1994; Ito, 2003).

However, we still cannot rule out the possibility that the fracture at the first BD was initiated under the control of stress in the intact rock, because TD events are unlikely to be induced by fracturing along a weak plane; in addition, the existence of a fault and many pre-existing cracks likely causes various scale stress inhomogeneity. Thus, due to the stress inhomogeneity, at the immediate locality around the borehole wall in the pressurizing section, the maximum horizontal compressive stress might apply in the direction where the HF crack propagated.

6.4.2 *Relationship between BD and crack expansion*

AE hypocenter distributions changed with elapsed time. In period I, with the first BD occurring at a flow rate of 10 mL/min, the AE hypocenters were distributed in the negative *X* direction from the pressurizing section in the pilot hole, as shown in Figure 6.5a. In period II(1), just before the second BD, as shown in Figure 6.5b, the AE hypocenters remained within the region in which they were distributed in the previous period I without migrating into any new region, despite the flow rate being increased to 30 mL/min. However, in period II(2), as shown in Figure 6.5c, the AE hypocenters started to migrate into a new region in the positive X direction from the pressurizing section immediately after the second BD at 9.79 MPa, which is larger than the first BD at 9.20 MPa. After that, in the periods II(3) and II(4), as shown in Figure 6.5d and e, respectively, the AE hypocenters were distributed in almost the same regions as those where they were distributed in the previous periods I–II(2), with slight migrations toward the positive X and negative Z directions, suggesting that crack extension during these periods was limited. From the temporal changes of the AE hypocenter distributions, crack expansions into new regions were likely induced only immediately after the first and second BDs.

In contrast to our results, AE events migrated into new regions and AE hypocenters distribution expanded with the duration of injection in actual field operations for HDR projects in Cooper Basin, Australia (Baisch et al., 2006, 2009, 2015), Soultz HDR, France (Evans et al., 2005), Hijiori, Japan (Sasaki, 1997; 1998), and Ogachi, Japan (Kaieda et al., 1995), and also in those for oil and gas recovery in CottonValley, Texas (Rutledge et al., 2004) and the Barnett Shale, Texas (Hummel and Shapiro, 2013). Sasaki (1998) examined the migration of AE hypocenters for 3 h at an injection flow rate of 6 m^3/min in Hijiori, Japan, using the CGDD (Christianovich, Geertsma, De Klerk, and Daneshy) model (Christianovich and Zheltov, 1955; Geertsma and De Klerk 1969; Daneshy, 1973) of a HF crack, named after their developers (Mendelsohn, 1984), which has been often used in the petroleum industry, having an ellipsoid shape on the horizontal section parallel to a crack extending direction with a rectangular shape on the vertical section. His findings using this model were that the distribution of AE hypocenters expanded as $t^{2/3}$ and as the square root of the injection flow rate. On the other hand, Hummel and Shapiro (2013) examined the migration for 5.4 h at an injection flow rate of 9 m^3/min in the Barnett Shale with fluid pressure diffusion. In our experiment, the pressurizing section of our HF experiment was selected in intact rock to avoid complications from pre-existing joints, and the expansion of AE migration was only around 0.5 m. In addition, the injection duration (48 min) was shorter and the injection flow rate (10 or 30 cm^3/min) was much smaller than those in the field operations. From the differences between our experiment and the one analyzed by Hummel and Shapiro (2013), the AE migration in our experiment was likely governed by new crack generations, whereas the AE migration in field operations seems to be controlled by pre-existing joints because of the long duration and large injection volume in a much larger rock mass having pre-existing joints. However, in some field operations, when the injection flow rate was temporarily increased during long-term injection, the AE activity increased remarkably (e.g., Kaieda et al. 1995; Sasaki 1997). When we consider the cases in fields where AE activity increases with the flow rate increase, the results of our experiment suggest that increasing the injection flow rate is an effective way to generate and expand new cracks if the same fracturing mechanism operates in large-volume injection in a field; in other words, new cracks also expand in intact rock masses with a pressure increase due to the flow rate increase.

Although refracturing performed after the first HF treatment has been proposed and used as a means to accelerate production and enhance the ultimate recovery of depleted shale wells as an economical alternative to drilling new wells (Jacobs, 2014; Foda, 2015; Malpani et al., 2015), the relationship between crack expansion and the injection flow rate has not been examined closely in actual reservoirs for reasons such as complicated injection histories and a time lag between injection and AE occurrence. Our results suggest that increasing the injection flow rate is an effective way to generate and expand new cracks, which may help understand and improve the refracturing.

6.4.3 *Different fracturing modes for crack expansion*

As shown in Figure 6.6, the ratios of the P-wave initial motion polarities indicate that TD events were induced immediately after the first and second BDs. The periods correspond to those when crack expansion was deduced from the migration of AE events. As for the fracturing mode induced by HF, elastic theory suggests tensile fracture (e.g., Hubbert and Willis, 1957; Zoback et al., 1977; Haimson, 1978; Schmitt and Zoback, 1993), whereas many researchers have reported that shear events are dominant in actual field monitoring (e.g., Talebi and Cornet, 1987; Cornet, 1992; Horalek et al., 2010; Maxwell and Cipolla, 2011). Recently, Ross et al. (1996), Šileny et al. (2009), and Julian et al. (2010) reported the existence of TD events induced by HF. However, Šileny et al. (2014) indicated that an insufficient number of AE sensors, their improper deployment, and waveform noise could result in spurious non-double couple components in the inverted moment tensor that would erroneously imply the TD mechanism. In laboratory experiments, while many SD events have been observed in water injection (e.g., Ishida et al., 2004, 2016), TD events have been observed in HF when viscous fracturing fluids were used (Matsunaga et al., 1993; Ishida et al., 2004, 2016; Rodriguez et al., 2017) or in HF for a hard intact granodiorite block consisting of small grains (Ishida et al., 2000). Thus, the fracturing mechanism active in HF operations using water is often ambiguous.

Although the fracturing mechanism induced by HF likely depends on factors such as the viscosity of the fracturing fluid, the nature of the rock matrixes, the density of pre-existing joints, and the rock stress conditions, Figure 6.6a shows that only TD events occurred immediately after the first BD, implying that they were induced by new crack expansion. After the second BD, although TD events occurred with SD events as shown in Figure 6.6b, 71% of these events were distributed on the frontiers of regions where AE events had already been distributed. This suggests that new crack expansions were induced in the parts through tensile fracturing, as was crack expansion after the first BD; this tendency has been observed in HF using viscous oil in an intact marble block (Matsunaga et al., 1993). The occurrence of TD events in our case was likely because our HF experiment was conducted in a small intact rock mass selected to avoid pre-existing joints.

6.4.4 *Fracturing mechanism of SD events*

Although we focused on TD events in the previous sections, most AE were SD events. Here, we discuss why these SD events were induced.

For example, in volcanic earthquake swarms, significant parts of seismic events show a shear mechanism, although magma intrusions or eruptions characterize many events. To explain the observation, Hill (1977) proposed a conceptual model that magma intrudes into the weak planes lying along the direction of the maximum compressive stress, among many weak planes prevailing in a volcanic region. The magma intrusion forming a dike would accompany some tensile fracturing, whereas shear fracture would form conjugate faults, connecting the tips of dikes, as indicated by symbols A and B in Figure 2.28, which has already been explained in Chapter 2.

As another example, the fact that SD events dominate even in a three-point-bend fracture test of a specimen helps us to understand how SD events were induced. Kao et al. (2011) conducted a three-point-bend fracture test on a granite specimen measuring $217 \times 73 \times 32\,mm^3$ (span × height × thickness) with a 4 mm notch as shown in Figure 2.25, and thus, AE events were located and their fracturing mechanism was analyzed. They found that all AE sources were shear-dominant due to tortuosity, reflecting the local deviation of the crack path due to grain-scale heterogeneity, although the macroscopic fracture was tensile, which has already been discussed in Section 2.5.2. From this experiment, we can say that SD events associated with macroscopic tensile fracture are a natural consequence of tortuosity, and the local mechanism, that is, the mechanism of the AE event, does not necessarily reveal the nature of the macro failure mode. The macroscopic observation in our laboratory HF experiments using very slick supercritical carbon dioxide (Ishida et al., 2016) revealed that the HF cracks propagate mainly along the grain boundaries of the constituent minerals, producing many small cracks inclined in the direction of the maximum compressive stress, σ_1, which is the propagating direction of a macroscopic crack. Because shear stress develops on a plane inclined to the direction of σ_1, the shear fracture can be easily induced along the plane, which was already pointed out in Section 5.4.3.

In our field experiment, when we consider that cracks had been newly induced in the rock mass after BD in addition to many pre-existing cracks, many SD events are most likely induced with the new crack extensions or slippages on the crack planes inclined to the direction of the macroscopic HF crack propagation. Suppose we can accept a concept like this regarding crack propagation and the origins of AE events. In that case, we can understand why SD events are dominant in HF in actual fields as reported by many researchers because rock masses in fields have various geological inhomogeneities, including pre-existing and newly generated HF cracks.

6.5 Conclusion

We conducted an HF experiment at a 500-m-level gallery in MIU in central Japan. We drilled an HF hole downward from the gallery floor and injected water into a section 36 mm in diameter and 160 mm in length that was selected to avoid pre-existing joints. We monitored AE events with 16 sensors set in four AE holes drilled 1 m away from the HF hole. From the experiment monitoring, we obtained the following results:

1. When we initially injected water at a flow rate of 10 mL/min, the first BD was induced at a pressure of 9.20 MPa. After that, when the flow rate was increased from 10 to 30 mL/min, the second BD associated with "refracturing" was induced at a pressure of 9.79 MPa, which is higher than the pressure of the first BD. Expansion of the regions where AE events were distributed was predominantly observed immediately after the first and second BDs. Many AE events in other periods occurred within the regions where AE events were already distributed.
2. The migration of AE events suggested that increasing the injection flow rate is an effective way to generate and expand new cracks, whereas AE events migrate step by step during injection without changing the flow rate in actual field operations. The differences can be interpreted as follows. Because our experiments were conducted in a small intact rock mass selected to avoid a pre-existing joint, the migration of AE events was controlled by the generation and expansion of new cracks. By contrast, in field operations, it is controlled by pre-existing joints because of the long duration and large injection volume in a much larger rock mass with pre-existing joints.
3. P-wave initial motion polarities indicate that TD AE events were induced immediately after the first and second BDs, corresponding to the periods when crack propagation was deduced from the migration of AE events. In addition, most of the TD AE events were distributed on

the frontiers of regions where AE events had already occurred. These results suggest that new crack propagation was induced by tensile fracturing.

4 Our results suggest that increasing the injection flow rate is an effective way to generate and expand new cracks in an intact rock, and the new crack expansions were associated with tensile fracturing, consistent with the elastic theory. These findings most likely help to understand and improve the refracturing in actual field operations for HDR projects and shale oil and gas recovery.

References

Baisch, S., Rothert, E., Stang, H., Voros, R., Koch, C. & McMahon, A. (2015) Continued geothermal reservoir stimulation experiments in the Cooper Basin (Australia). *Bull. Seismol. Soc. Am.*, 105(1), 198–209. https://doi.org/10.1785/0120140208

Baisch, S., Voros, R., Weidler, R. & Wyborn, D. (2009) Investigation of fault mechanisms during geothermal reservoir stimulation experiments in the Cooper Basin, Australia. *Bull. Seismol. Soc. Am.*, 99(1), 148–158. https://doi.org/10.1785/0120080055

Baisch, S., Weidler, R., Voros, R., Wyborn, D. & DeGraaf, L. (2006) Induced seismicity during the stimulation of a geothermal HFR reservoir in the Cooper Basin, Australia. *Bull. Seismol. Soc. Am.*, 96(6), 2242–2256. https://doi.org/10.1785/0120050255

Christianovich, A. & Zheltov, Y. P. (1955) Formation of vertical fractures by means of highly viscous liquid. In: *Proceedings of 4th World Petroleum Congress,* vol 2, pp. 579–586.

Cornet, F. H. (1992) Fracture processes induced by forced fluid percolation. In: Gasparini, P., Scarpa, R., Aki, K. (eds) *Volcanic Seismology. IAVCEI Proceedings in Volcanology*, vol 3. Springer, New York, pp. 407–431.

Daneshy, A. A. (1973) On the design of vertical hydraulic fractures. *J. Pet. Technol.*, 25, 83–97.

Evans, K. F., Moriya, H., Niitsuma, H., Jones, R., H., Phillips, W., S., Genter, A., Sausse, J., Jung, R. & Baria, R. (2005) Microseismicity and permeability enhancement of hydrogeologic structures during massive fluid injections into granite at 3 km depth at the Soultz HDR site. *Geophys. J. Int.*, 160, 388–412.

Foda, S. (2015) Refracturing: technology and reservoir understanding are giving new life to depleted unconventional assets. *J. Pet. Technol.* 67(7), 76–79.

Geertsma, J. & De Klerk, F. (1969) A rapid method of predicting width and extent of hydraulically induced fractures. *J. Pet. Technol.*, 21, 1571–1581.

Gischig, V. S., Doetsch, J., Maurer, H., Krietsch, H., Amann, F., Evans, K.F., Nejati, M., Jalali, M., Valley, B., Obermann, A.C., Wiemer, S. & Giardini, D. (2018) On the link between stress field and small-scale hydraulic fracture growth in anisotropic rock derived from microseismicity. *Solid Earth* 9, 39–61. https://doi.org/10.5194/se-9-39-2018

Haimson, B. C. (1978) The hydrofracturing stress measuring method and recent field results. *Int. J. Rock Mech. Min. Sci. Geomech. Abstr.,* 15, 167–178.

Hill, D. P. (1977) A model for earthquake swarms. *J. Geophys. Res.*, 82, 1347–1352.

Horalek, J., Jechumtalova, Z., Dorbath, L. & Šileny, J. (2010) Source mechanisms of micro-earthquakes induced in a fluid injection experiment at the HDR site Soultz-sous-Forets (Alsace) in 2003 and their temporal and spatial variations. *Geophys. J. Int.*, 181, 1547–1565.

Hubbert, M. K. & Willis, D. G. (1957) Mechanics of hydraulic fracturing. *Trans. Am. Inst. Min. Metall. Pet. Eng.*, 210, 153–168.

Hummel, N. & Shapiro, S. A. (2013) Nonlinear diffusion-based interpretation of induced microseismicity: a Barnett Shale hydraulic fracturing case study. *Geophysics*, 78(5), B211–B226. https://doi. org/10.1190/GEO2012-0242.1

Ishida, T., Chen, Q., Mizuta, Y. & Roegiers, J.-C. (2004) Influence of fluid viscosity on the hydraulic fracturing mechanism. *Trans. ASME J. Energy Resour. Technol.*, 126, 190–200.

Ishida, T., Chen, Y., Bennour, Z., Yamashita, H., Inui, S., Nagaya, Y., Naoi, M., Chen, Q., Nakayama, Y. & Nagano, Y. (2016) Features of CO_2 Fracturing deduced from acoustic emission and microscopy in laboratory experiments. *J. Geophys. Res. Solid Earth* 121(11), 8080–8098. https://doi.org/10.1002/2016JB01330

Ishida, T., Desaki, S., Yamashita, H., Inui, S., Naoi, M., Fujii, H. & Katayama, T. (2017) Injection of supercritical carbon dioxide into granitic rock and its acoustic emission monitoring. *Proc. Eng.*, 191, 476–482. https://doi.org/10.1016/j.proeng.2017.05.206

Ishida, T., Fujito, W., Yamashita, H., Naoi, M., Fuji, H., Suzuki, K. & Matsui, H. (2019) Crack expansion and fracturing mode of hydraulic refracturing from acoustic emission monitoring in a small-scale field experiment. *Rock Mech. Rock Eng.*, 52(2), 543–553. https://doi.org/10.1007/s00603-018-1697-5

Ishida. T., Desaki, S., Kishimoto, Y., Naoi, M. & Fujii, H. (2021) Acoustic emission monitoring of hydraulic fracturing using carbon dioxide in a small-scale field experiment. *Int. J. Rock Mech. Min. Sci.*, 141, Article 104712. https://doi.org/10.1016/j.ijrmms.2021.104712

Ishida, T., Kishimoto, Y, Desaki, S., Naoi, M., Suzuki, K. & Fujii, H. (2023) Characteristics of carbon dioxide fracturing in comparison to conventional water hydraulic fracturing: Evidence from acoustic emission monitoring of small-scale field experiments. *Int. J. Rock Mech. Min. Sci.*, 172, Article 105604. https://doi.org/10.1016/j.ijrmms.2023.105604

Ishida, T., Sasaki, S., Matsunaga, I., Chen, Q. & Mizuta, Y. (2000) Effect of grain size in granitic rocks on hydraulic fracturing mechanism. In: *Trends in Rock Mechanics (proceedings of Sessions of Geo-Denver 2000), Geotechnical Special Publication no. 102*, ASCE, pp. 128–139.

Ito, H. (2003) Inferred role of natural fractures, veins, and breccias in development of the artificial geothermal reservoir at the Ogachi hot dry rock site, Japan. *J. Geophys. Res.,* 108(B9), 2426. https://doi.org/10.1029/2001JB0016

Jacobs, T. (2014) Renewing mature shale wells through refracturing. *J. Pet. Technol.*, 66(4), 52–60.

Julian, B. R., Foulger, G. R., Monastero, F. C. & Bjornstad, S. (2010) Imaging hydraulic fracturing in a geothermal reservoir. *Geophys. Res. Lett.*, 37, L07305. https://doi.org/10.1029/2009GL0409

Kaieda, H., Kiho, K. & Motojima, I. (1993) Multiple fracture creation for hot dry rock development. *Trends Geophys. Res.,* 2, 127–139.

Kaieda, H., Fujimitsu, Y., Yamamoto, T., Mizunaga, H., Ushijima, K. & Sasaki, S. (1995) AE and mise-a-lamasse measurements during a 22 day water circulation test at Ogati HDR site, Japan. In: *Proceedings of World Geothermal Congress,* Florence, pp. 2695–2700.

Kao, C. S., Carvalho, F. C. S. & Labuz, J. F. (2011) Micromechanisms of fracture from acoustic emission. *Int. J. Rock Mech. Min. Sci.,* 48, 666–673.

Kondo, H. (1994) *Development of a method for prediction of the extending direction of fractures created by hydraulic fracturing for hot dry rock power generation—characterization of natural fractures in jointed rockmass*. CRIEPI (Central Research Institute of Electric Power Industry), Abiko, Chiba, Japan, Report no. U93039 (in Japanese with English abstract)

Kuwabara, K., Sato, T., Sanada, H. & Takayama, Y. (2015b) Mizunami underground research laboratory project—rock mechanical investigations at the—500 m stage, JAEA-Research 2015-005, Tono Geoscience Center. *Jpn. Atom Energy Agency*. https ://doi. org/10.11484 /jaea-research-2015-005 (in Japanese with English abstract)

Kuwabara, K., Sato, T., Takayama, Y., Tanno, T., Kato, H. & Itamoto, M. (2015a) Initial stress measurement by CCBO at GL. In: *500 m Reflood Gallery of the Mizunami Underground Research Laboratory, Proceedings of MMIJ Fall Meeting, Paper No. 3416*, Matsuyama, Japan (in Japanese)

Malpani, R., Sinha, S., Charry, L., Sinosic, B., Clark, B. & Gakhar, K. (2015) Improving hydrocarbon recovery of horizontal shale wells through refracturing, SPE-175920-MS. In: *Proceedings of the SPE/CSUR Unconventional Resources Conference*, Society of Petroleum Engineers, Calgary, Alberta, Canada.

Matsunaga, I., Kobayashi, H., Sasaki, S. & Ishida, T. (1993) Studying hydraulic fracturing mechanism by laboratory experiments with acoustic emission monitoring. *Int. J. Rock Mech. Min. Sci. Geomech. Abstr.,* 30, 909–912.

Maxwell, S. C. & Cipolla, C. (2011) What does microseismicity tell us about hydraulic fracturing? In: *SPE Annual Technical Conference and Exhibition*, Denver, Colorado, USA, SPE 146932.

Mendelsohn, D. A. (1984) A review of hydraulic fracture modeling— Part 1: General concepts, 2D models, motivation for 3D modeling. *J. Energy Resour. Technol.*, 106, 369–376.

Rodriguez, I. V., Stanchits, S. & Burghardt, J. (2017) Data-driven, in situ, relative calibration based on waveform fitting moment tensor inversion. *Rock Mech. Rock Eng.*, 50, 891–911. https://doi.org/10.1007/s00603-016-1144-4

Ross, A., Foulger, G. R. & Julian, B. R. (1996) Non-double-couple mechanisms at the Geysers geothermal area, California. *Geophys. Res. Lett.*, 23, 877–880.

Rutledge, J. T., Phillips, W. S. & Mayerhofer, M. J. (2004) Faulting induced by forced fluid injection and fluid flow forced by faulting: an interpretation of hydraulic-fracture microseismicity, Carthage Cotton Valley gas field, Texas. *Bull. Seismol. Soc. Am.*, 94(5), 1817–1830.

Sasaki, S. (1997) Microseismic activity induced during hydraulic fracturing experiment at the Hijiori hot dry rock geothermal energy site, Yamagata, Japan. In: *Proceedings of 4th International Symposium Rockburst and Seismicity in Mines*, Krakow, pp. 403–407.

Sasaki, S. (1998) Characteristics of microseismic events induced during hydraulic fracturing experiments at the Hijiori hot dry rock geothermal energy site, Yamagata, Japan. *Tectonophysics*, 289, 171–188.

Schmitt, D. R. & Zoback, M. D. (1993) Infiltration effects in the tensile rupture of thin walled cylinders of glass and granite: implications for the hydraulic fracturing breakdown equation. *Int. J. Rock Mech. Min. Sci. Geomech. Abstr.*, 30, 289–303. https://doi. org/10.1016/0148-9062(93)92731-5

Šileny, J., Hill, D. P., Eisner, L. & Cornet, F. H. (2009) Non-double-couple mechanisms of microearthquakes induced by hydraulic fracturing. *J. Geophys. Res.*, 114, B08307. https://doi.org/10.1029/2008JB0059

Šileny, J., Jechumtalova, Z. & Dorbath, C. (2014) Small scale earthquake mechanisms induced by fluid injection at the enhanced geothermal system reservoir Soultz (Alsace) in 2003 using alternative source models. *Pure Appl. Geophys.*, 171, 2783–2804. https://doi.org/10.10007/s00024-013-0750-2

Sugawara, K. & Obara, Y. (1999) Draft ISRM suggested method for insitu stress measurement using the compact conical-ended borehole overcoring (CCBO) technique. *Int. J. Rock Mech. Min. Sci.*, 36, 307–322.

Talebi, S. & Cornet, F. H. (1987) Analysis of the microseismicity induced by a fluid injection in a granitic rock mass. *Geophys. Res. Lett.*, 14, 227–230.

Zoback, M. D., Rummel, F., Jung, R. & Raleigh, C. B. (1977) Laboratory hydraulic fracturing experiments in intact and pre-fractured rock. *Int. J. Rock Mech. Min. Sci. Geomech. Abstr.*, 14, 49–58.

Chapter 7

Small-scale field carbon dioxide fracturing experiments

7.1 Introduction

Carbon capture and storage in underground reservoirs is a promising and feasible method for mitigating the greenhouse effect by decreasing the amount of carbon dioxide (CO_2) emitted into the air. If we could utilize CO_2 for energy production, carbon capture and storage would be much more eagerly developed and adopted by compensating for its cost. The technology is called carbon capture, utilization, and storage, and for examples of applications of carbon capture, utilization, and storage to underground resources, Xie et al. (2014) showed the recovery of enhanced geothermal systems, enhanced shale gas, enhanced natural gas, enhanced coalbed methane, enhanced oil, and others. Among these, for geothermal energy extraction in enhanced geothermal systems, Brown (2000) and Pruess (2006) stated that CO_2 is superior to water as a working fluid. Regarding the exploitation of shale gas in enhanced shale gas, Kalantari-Dahaghi (2010), Liu et al. (2013), Middleton et al. (2015), and Godec et al. (2013) suggested that it is feasible through numerical simulations because CO_2 has a higher affinity for shale than methane (CH_4) and CH_4 is desorbed with the absorption of CO_2 (Nuttall et al., 2006). For these projects, CO_2 is usually injected into rocks deeper than 1,000 m and sometimes deeper than 3,000 m, and the temperature and pressure at the depths cause the CO_2 to reach its supercritical (SC) state. SC-CO_2 has a much lower viscosity (0.01–0.1 mPa·s) than water (1 mPa·s). In Chapter 5, we clarified in laboratory experiments (Ishida et al., 2012, 2016) that hydraulic fracturing (HF) using SC-CO_2 tends to induce three-dimensional sinuous cracks with many secondary branches, which look like desirable pathways for such projects. In addition, more recently, Ranjith et al. (2019) and Zhang et al. (2020, 2021) showed that crack features induced by CO_2 HF (HF using CO_2) have many more advantages than those by water HF (HF using water) in their laboratory experiments on siltstone and shale samples.

Although laboratory experiments represent a powerful approach, as shown by these studies (Ishida et al., 2012, 2016; Ranjith et al., 2019; Zhang et al., 2020, 2021), they have limitations in their ability to simulate HF in actual production fields. For example, it is difficult to examine the effects of pre-existing cracks and the crack extending size because a specimen usually does not contain a macroscopic pre-existing crack, and the cracks extend to the surfaces of the specimen immediately after HF. Thus, we conducted HF experiments using CO_2 in the borehole arrangements similar to those shown in Chapter 6 from the tunnel floor of a hot rock mass, which had a pre-existing crack and satisfied the temperature necessary to form SC-CO_2. The experiments elucidated that the hypocenters of acoustic emission (AE) were distributed almost perpendicular to the direction of the initial HF crack after the injection pressure declined to zero.

DOI: 10.1201/9781003282631-7

This suggested that CO_2 can easily intrude and induce AE events along a pre-existing crack. In Section 7.2, we describe the results of the experiment following Ishida et al. (2021).

Two years after the previous experiments, we conducted new HF experiments at the same site in the same injection hole to compare the size of the AE hypocenter distribution and the AE focal mechanism between CO_2 HF and water HF (Ishida et al., 2023). The new experiments showed that CO_2 HF can stimulate a much larger region than water HF under the same constant injection flow rate. They also showed that CO_2 HF induces compression-dominant AE events never observed in water HF. These results indicate the remarkable differences between CO_2 HF and water HF, at least in our small field experiments, although Verdon et al. (2010) reported almost no difference between them at depths of 2,600–2,900 m in a productive oil-field in North America. In addition, by analyzing the occurrence times and locations of tensile and shear-dominant AE events, we found a tendency for tensile-dominant events to occur in intact rock, whereas shear-dominant events occurred with a linkage between cracks, that is, the interaction of cracks. This tendency is similar to that found in water HF in the Mizunami Underground Research Laboratory (MIU) site shown in Chapter 6. This finding suggests a solution for the paradox of the mechanism of HF; that is, the elastic theory expects tensile fractures, whereas shear events dominate in AE monitoring. In addition, to compare injectivity improvements between CO_2 HF and water HF, Injecting/Fall-Off Tests (IFOTs) were conducted using gaseous CO_2 before and after HF experiments using CO_2 and using water before and after HF experiments using water. The injectivity was improved around 400 times by CO_2 HF, whereas the improvement by water HF was only a few times. In Section 7.3, the experiments to compare the size of the AE hypocenter distribution and the AE focal mechanism between CO_2 HF and water HF, and the tests to compare injectivity improvements between CO_2 HF and water HF were described, following Ishida et al. (2023).

7.2 Small-scale field carbon dioxide fracturing experiments to examine the effects of a pre-existing crack

7.2.1 *Site and set up for experiments conducted in 2014*

7.2.1.1 *Site and experiments in 2014*

As shown in Figure 7.1, the experimental site was situated in a small railroad tunnel approximately 50 m below the surface of a mountainous area in central Japan, with a tunnel exit 40 m away from the site. The rock mass around the site consisted of granitic rock formed from the late Miocene to the Pliocene. Although the rock temperature was 165°C when this tunnel was excavated in 1940, it is now 35°C due to the cooling effect of air circulating through the tunnel for an extended time. An HF hole (86 mm in diameter) was drilled vertically downward from the tunnel floor, and four AE monitoring holes (66 mm in diameter each) were drilled parallel to and at 1 m from the HF hole as shown in Figure 7.2. To make a pressurizing section for HF, we drilled a pilot hole 36 mm in diameter at the center of the bottom of the HF hole. Figure 7.3 shows a view of the test site. The hole arrangement and the setup for HF were the same as those in the MIU site reported in Chapter 6. In this site, we conducted HF experiments using CO_2 and those using water, as shown in Table 7.1.

In November 2014, we conducted four HF experiments using CO_2 labeled KC1 to KC4 and two using water labeled KW1 and KW2 in depth from 6.28 to 7.72 m in the HF hole. Out of the six experiments, only in the experiments labeled KW1 and KC3, we observed breakdown

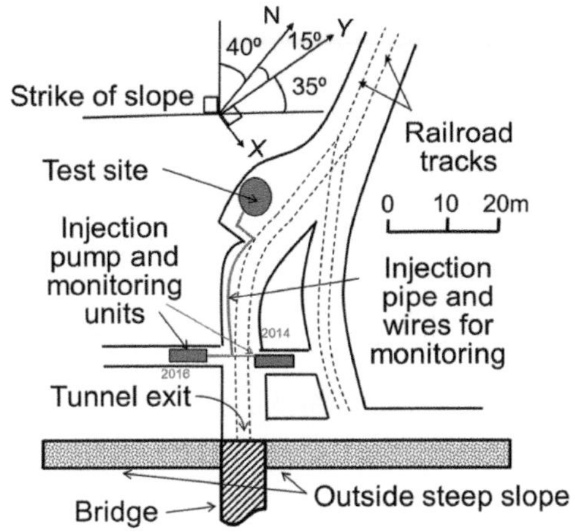

Figure 7.1 Horizontal layout of the test site and strike of the outside steep slope. The locations of the injection pump and monitoring units are slightly different in the experiments between 2014 and 2016. (Slightly revised from Ishida et al., 2023.)

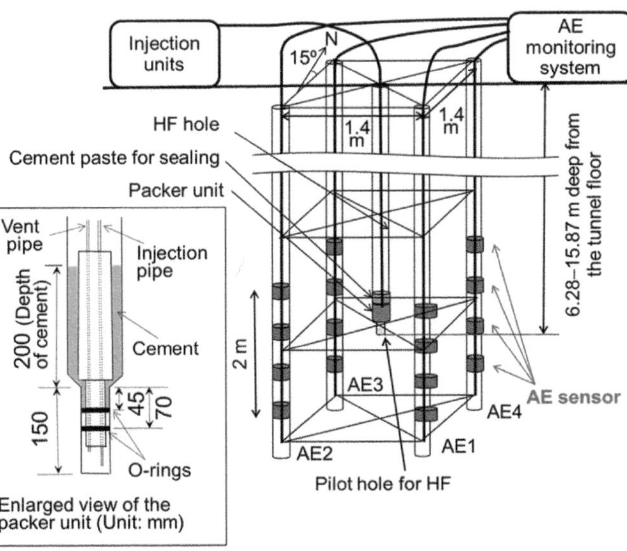

Figure 7.2 Arrangement of AE sensors surrounding a pressurized section under the test site. The experiments were conducted at depths from 6.28 to 9.73 m in 2014, whereas from 9.73 to 15.87 m in 2016. The inset shows a dimensioned diagram of the packer used to seal the pressurized section. (Slightly revised from Ishida et al., 2021.)

(BD) pressure defined as the peak pressure immediately before the large sudden drop, and recorded burst occurrences of AE associated with the BDs. In the other four HF experiments, we observed neither BD nor AE bursts, probably because the fracturing fluid leaked through cracks induced by the previous HF experiments. In the experiment, KW1, HF cracks extended horizontally along hair cracks that inherently existed in the pressurizing section before the experiment.

Figure 7.3 View of the test site.

The experiment is excluded because the crack extension was unlikely to be a typical HF crack extension; thus, we show and discuss only the case of the experiment KC3 out of the six experiments conducted in 2014. The experiment KC3 elucidated that the hypocenters of AE were distributed almost perpendicular to the direction of the initial HF crack after the injection pressure declined to zero, suggesting that CO_2 can easily intrude and induce AE events along a pre-existing crack.

7.2.1.2 *Method of CO_2 injection*

In the experiment KC3 that was conducted in 2014, we selected the pressurizing section of an intact rock at a depth of 7.24–7.40 m, where no visible crack was found by checking the recovered core of the pilot hole. We sealed the pilot hole's upper section with two O-rings attached to a packer unit and poured cement paste above the upper one of the two O-rings, as shown in the inset of Figure 7.2. CO_2 was injected into the pressurizing section under the lower O-ring. Figure 7.4 shows the injection system used in the experiment. To inject CO_2, we used two syringe pumps and alternated between them to achieve continuous injection during the experiment. We fed CO_2 from a bomb into the syringe-pump cylinders, each of which had a capacity of 500 mL. To fill the cylinders as fully as possible, we circulated coolant around them to cool the CO_2 and keep it in the liquid state. In the experiment, the liquid CO_2 was discharged from the cylinder at a constant flow rate of 50 mL/min by controlling the displacement of the syringe. The discharged CO_2 was warmed up to approximately 30°C of the ambient temperature, while flowing through a steel pipe along the tunnel, and it was injected into the sealed section of the HF hole at a temperature of 35°C. Referring to the phase diagram of Figure 5.3 that was shown for the laboratory CO_2 fracturing experiment in Chapter 5, CO_2 reached an SC state at this temperature of 35°C, which is larger than the critical temperature 31.1°C, and under a pressure higher than the critical pressure 7.38 MPa. Because the groundwater level was kept approximately 1 m below the tunnel floor, CO_2 and water were injected into a rock mass saturated with water.

7.2.1.3 *Measurement of AE, pressure, and temperature*

For AE monitoring, we installed four waterproof lead zirconate titanate sensors with a resonance frequency of 70 kHz (AE703SW-GAMP-0542; Fuji Ceramics Corp., Japan) in each of the four AE holes (AE1–4 in Figure 7.2). We attached each sensor to an aluminum rod with

Table 7.1 Table of experiments

Name of the experiment[a]	*Date*	*Fracturing fluid*	*Depth (m)*	*BD (MPa)*	*Number of well-located AE events*	*Pressurizing section*	*P-wave velocity (km/s)*	*Remarks*[b]
KW1	November 11, 2014	Water	6.28–6.53	4.77	583	Two hair cracks	5.62	Horizontal crack
KC1	November 12, 2014	CO_2	6.70–6.85	N.A.	0	N.A.	N.A.	No BD pressure
KC2	November 13, 2014	CO_2	7.00–7.15	N.A.	0	N.A.	N.A.	No BD pressure
KC3	**November 14, 2014**	**CO_2**	**7.24–7.40**	**15.62**	**1,478**	**Intact**	**5.54**	**in the text**
KW2	November 15, 2014	Water	7.49–7.65	N.A.	0	N.A.	N.A.	No BD pressure
KC4	November 16, 2014	CO_2	7.57– 7.72	N.A.	0	N.A.	N.A.	No BD pressure
KC5	October 8, 2016	CO_2	9.73–9.88	N.A.	2,595	Hair crack	5.51	No BD pressure
KC6	**October 9, 2016**	**CO_2**	**10.65–10.80**	**11.55**	**3,259**	**Intact**	**5.46**	**in the text**
KW3	**October 11, 2016**	**Water**	**13.32–13.49**	**12.78**	**465**	**Intact**	**5.51**	**in the text**
KW4	October 12, 2016	Water	14.09–14.24	N.A.	189	Hair crack	5.52	No BD pressure
KW5	**October 13, 2016**	**Water**	**15.20–15.35**	**14.15**	**247**	**Intact**	**5.54**	**in Appendix 7.A**
KC7	**October 14, 2016**	**CO_2**	**15.72–15.87**	**8.26**	**973**	**Hair crack**	**5.52**	**in Appendix 7.A**

[a] Name of the experiment. K: Head letter of site name, C: CO_2, W: Water.

[b] Experiments from KC1 to 4 and from KW1 to 2 were conducted in 2014. Out of them, only experiment KC3 was successful and described in the text. Out of the other six experiments conducted in 2016, we mainly discuss the differences in results between the KC6 and KW3 in the text. We briefly discuss those between KC7 and KW5 in Appendix 7.A to confirm the differences obtained between KC6 and KW3.

The bold values are explained and discussed in the experiments in the text or the appendixes as shown in the column, "Remarks", in the table.

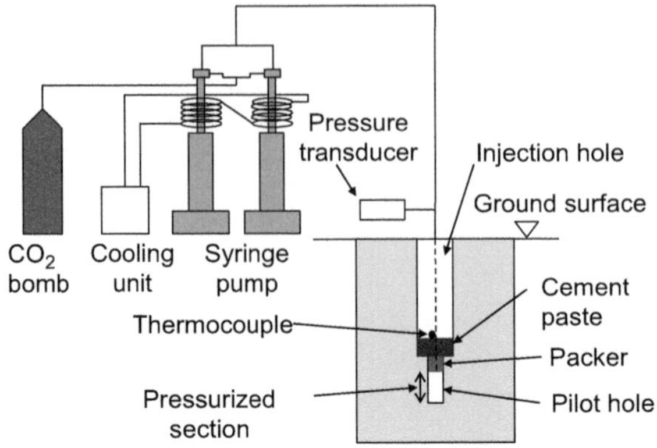

Figure 7.4 Injection system for SC-CO_2. (Slightly revised from Ishida et al., 2021.)

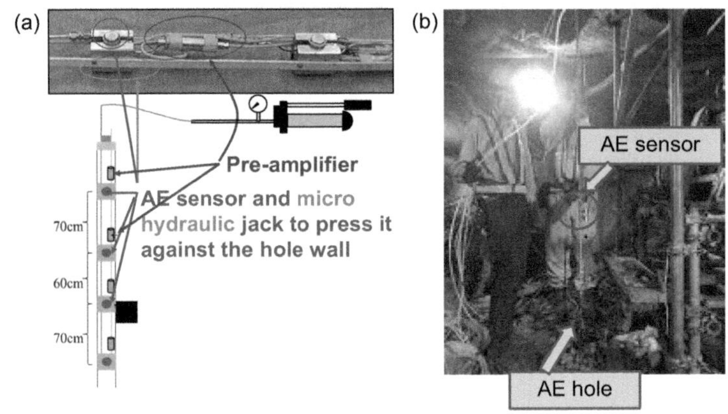

Figure 7.5 (a) Photo of AE sensor, pre-amplifier, and micro hydraulic jack fixed on a rod, and Illustration of positions of them in an AE monitoring hole. The micro hydraulic jack was set behind each AE sensor to press it against a wall of an AE monitoring hole. (b) A view of inserting the rod, on which AE sensor, pre-amplifier, and micro hydraulic jack are fixed, into an AE monitoring hole.

a pre-amplifier and inserted a thick rubber sheet between the sensor and the rod to block any vibrations transmitted through the rod, as shown in Figure 7.5. After orienting each sensor to face the HF hole, we pressed them against the walls of the AE holes by applying 1.5 MPa of oil pressure in a small hydraulic jack set behind each sensor. In each AE hole, we installed four sensors over a 2 m long span at intervals of 0.7, 0.6, and 0.7 m to center the 2 m section at the center depth of the pressurizing section of the HF hole. In the HF experiments conducted in 2014, the recording of an AE event was triggered when one of the signals from the 16 AE sensors (in total) set in the four holes exceeded 1 V. After the AE signals detected at the sensors were amplified by 70 dB in total (40 dB in a pre-amplifier and 30 dB in a main amplifier), they were applied through a band-pass filter (20–200 kHz) and recorded on a hard disk through an analog-to-digital (A/D) converter (PXI-5105; National Instruments Corp., USA). The AE signal of each sensor for each event was digitized into 2,048 samples with a 1 μs sampling time. We stopped recording for

10 ms, just after recording an event for 2.048 ms, to prevent the hard disk from recording too much noise caused by "ringing", which was the vibration following a large AE event.

We measured the injection pressure with a transducer (PW-50MPA; Tokyo Sokki Kenkyujo Co., Ltd., Japan) plugged into the pipe for injection on the floor just out of the HF hole. The temperature was measured with a T-type thermocouple glued onto the steel pipe for injection just above the cement paste, which sealed the pressurized section in the HF hole. We recorded the injection pressure and temperature on the hard disk through another A/D converter (PXI-6251; National Instruments Corp., USA) with a sampling time of 0.1 s.

7.2.1.4 AE data processing for hypocenter location

We adopted the least-squares iterative method using the P-wave arrival time for the hypocenter location. We measured the P-wave velocities between the HF hole and the 16 sensors immediately before and after the HF using an emitter (AE703SWR-0840; Fuji Ceramics Corp., Japan) attached to the upper part of the packer unit. In the case of the experiment KC3 that was conducted in 2014, after confirming that the difference in P-wave velocities before and after the HF was small, an average velocity of 5.4 km/s (standard deviation of 0.4 km/s) was used for the hypocenter locations under the assumption that the rock mass was isotropic and homogeneous.

In the discussion, we used well-located hypocenters whose accuracy is expected to be within 50 mm by satisfying the following two conditions (Ishida and Sasaki, 2011): (1) the P-wave arrival times could be read at six or more sensors set in three or more different AE holes to enclose the hypocenter three-dimensionally; (2) the standard deviation and the maximum of residuals of arrival times were within 10 and 20 μs, respectively. We located the hypocenters of 1,249 AE events satisfying these conditions during the CO_2 injection.

7.2.2 Results of the experiment KC3 to examine the effect of a pre-existing crack on CO_2 fracturing

7.2.2.1 Temporal changes in fluid pressure, temperature, AE rate, and flow rate

Figure 7.6 shows the temporal changes in fluid pressure, temperature, AE rate, and flow rate with the elapsed time, *t*. We took $t=0$ when the pressure reached the BD pressure. Here, the AE rate was counted for only well-located events (see the previous section, Section 7.2.3).

When we opened the valve of the injection system, the bomb pressure of the CO_2 (approximately 1 MPa) acted on the pressurized section. We injected CO_2 at a constant flow rate of 50 mL/min, with a sudden stop for 30 s in the elapsed time from −15 min 13 s to −14 min 43 s, due to a problem with switching the two syringe pumps. The injected pressure increased with the elapsed time, and the BD was induced at 15.62 MPa. Immediately after the BD, the injection was stopped. A few minutes after the BD, we found that bubbles of CO_2 were gushing out from the water that filled the AE1 and AE3 holes.

After starting the injection of CO_2 at a constant flow rate of 50 mL/min, the pressure increased gradually with time, and then, it increased steeply at approximately $t=-3$ min (indicated by the thick arrow in Figure 7.6a). This steep increase started at a pressure close to the critical pressure of 7.38 MPa. At that time, the temperature was approximately 35°C higher than the critical temperature of 31.1°C. Thus, the steep increase was most likely caused by the decrease in CO_2 compressibility, due to the phase change from the gaseous state to the SC state (refer to the

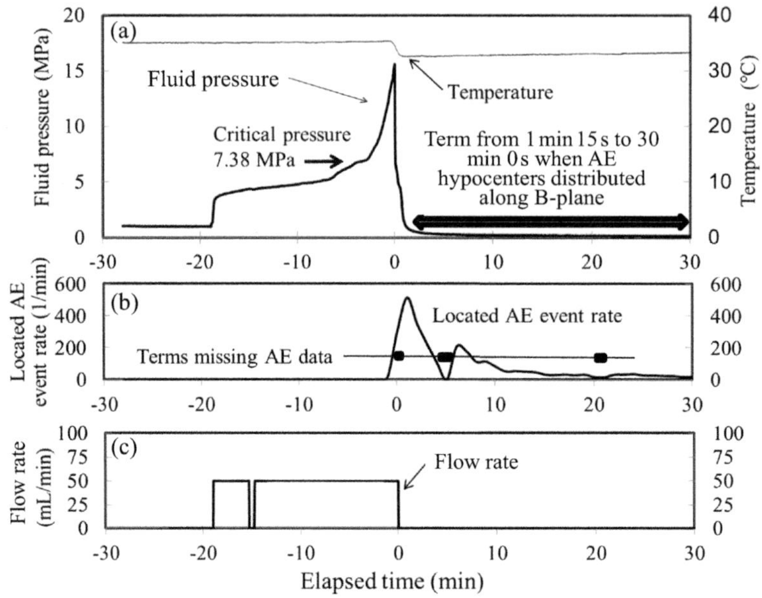

Figure 7.6 Temporal changes of fluid pressure, temperature, AE rate, and flow rate. Thick line segments on the fine line in Figure 7.6b indicate periods missing AE data. (Slightly revised from Ishida et al., 2018, 2021.)

phase diagram of CO_2 shown in Figure 5.3 in Chapter 5). The temperature decreased to 32.6°C at the BD, probably due to the adiabatic expansion of the CO_2. This adiabatic expansion of the CO_2 likely occurred because of the pressure decrease due to the leakage of SC-CO_2 into the induced cracks, which accompanied the phase change from the SC state to the gaseous state.

Three periods were missing the AE data, as shown by the thick solid line segments in Figure 7.6b. The first missing period from 1 to 26 s was due to the problem of an AE data acquisition because too many AE events were induced. The second and third periods, from 4 min 38 s to 6 min 12 s and from 20 min 31 s to 21 min 44 s, respectively, were due to replacing a file filled with AE data with a new file in the monitoring unit. The decrease in the located AE event rate of the second and third periods was caused by the missing data.

7.2.2.2 AE hypocenter locations and crack observations

Figure 7.7 shows the locations of the AE hypocenters projected onto the horizontal plane, *XY*, and the two vertical planes, *YZ* and *ZX*. In this figure, the dense cloud of AE hypocenters 1.3 m long in the vertical, 1.0 m long in the horizontal, and 0.4 m in thickness suggests that the HF induced the vertical crack of this size. Figure 7.8 shows the AE hypocenter distributions on the horizontal plane, *XY*, for the period from 0 (BD) to 1 min 15 s, comparing those from 1 min 15 s to 30 min 0 s. Figure 7.8 indicates AE hypocenters distributed along the A-direction from the HF hole to the AE1 and AE3 holes for the initial 1 min 15 s from BD. After that, they started to be distributed along the B-direction, almost perpendicular to the A-direction, from the portion close to the AE3 hole. From the vertical distributions on the *YZ* and *ZX* planes in Figure 7.7, the AE hypocenters are distributed on the two nearly vertical planes having strikes of the A- and B-directions.

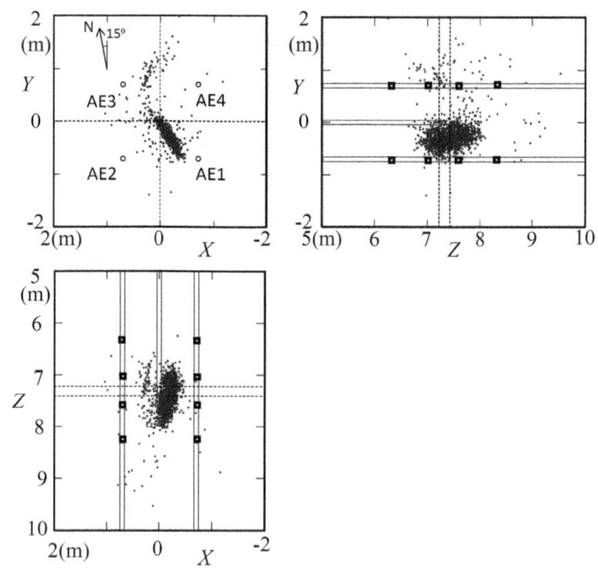

Figure 7.7 AE hypocenter distribution in the experiment KC3. Their projections onto the horizontal plane, *XY*, and two vertical planes, *YZ* and *ZX*, where the *X* axis is parallel to the direction from the hole AE1 to AE2 and that from AE4 to AE3, the *Y* axis is perpendicular to the *X* axis, and the *Z* axis is vertical. The origin is placed at the center of the HF hole on the horizontal plane, *XY*. The bars and open rectangles on the *YZ* and *ZX* planes correspond to HF, AE holes, and AE sensors, respectively. The broken lines on them indicate the pressurizing section from 7.24 to 7.40 m. Note that the +*X* direction is from right to left (Ishida et al., 2017, 2021).

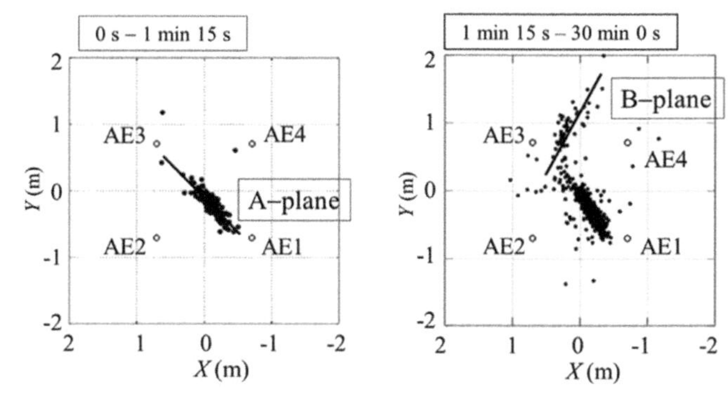

Figure 7.8 AE hypocenter distributions on the horizontal *XY* plane, for the period from 0 to 1 min 15 s, comparing those from 1 min 15 s to 30 min 0 s (Ishida et al., 2017, 2021).

After we analyzed and discussed the AE source distributions (Ishida et al. 2017), we observed closely a core drilled coaxially around the pressurizing section and obtained images on the wall of the HF hole by a borehole camera, to clarify the cracks induced by the HF.

Figure 7.9 shows a solid core of the 36 mm diameter pilot hole drilled to make the pressurizing section before the HF and a hollow core recovered by the 86 mm diameter coaxial drilling around the pressurizing section after the HF. We can observe a crack almost parallel

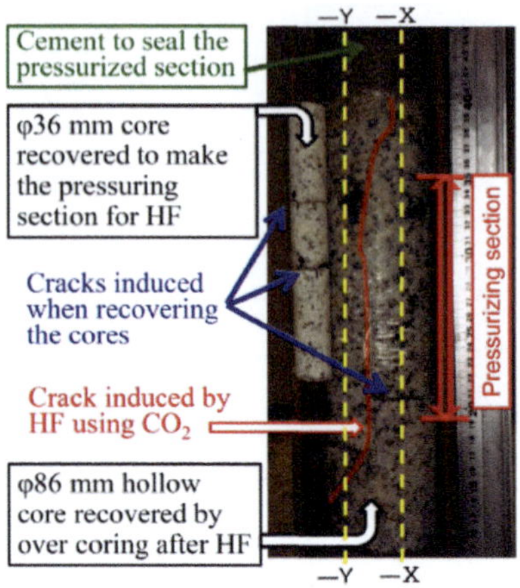

Figure 7.9 A solid core of 36 mm diameter pilot hole drilled to make a pressurizing section before HF and a hollow core recovered by 86 mm diameter coaxial drilling around the pressurizing section after HF. (Ishida et al., 2021).

to the drilling direction, as indicated by a red line on the surface of the hollow core recovered by the coaxial drilling. The crack direction was confirmed to face the AE1 hole from the HF hole. Also, on the opposite side of the core, a similar crack extending along the drilling direction faced the AE3 hole. These cracks indicated that the HF induced bi-wing vertical cracks in the A-direction. Because we confirmed that there was no pre-existing crack in the pressurizing section by inspecting the pilot hole core before the HF and the high BD pressure (15.62 MPa) recorded, it was most likely that the cracks were newly induced in the intact rock at the BD. They extended along the vertical plane having the strike of the A-direction for 1 min 15 s from the BD. The direction corresponded to the observation that the injected CO_2 gushed out from the AE1 and AE3 holes.

Figure 7.10 shows cracks observed on the wall of the HF hole by a borehole camera after the HF. From this, we can confirm that two parallel cracks were induced in the A-direction in the interval of the pressurizing section, and they extended 3.33 m in length from a depth of 6.15 m to 9.48 m, which resulted in an ellipsoidal crack. The vertical height of the crack extension almost corresponds to the AE hypocenter distribution in the *Z*-direction, as shown in Figure 7.7. On the other hand, the AE hypocenters horizontally distributed along the B-direction were almost perpendicular to the *A*-direction from a position of approximately 0.6 m away from the HF hole. However, no pressure acted in the pressurizing section for the period from 1 min 15 s to 30 min 0 s, as shown in Figure 7.6a. This suggests that a vertical crack existed in the B-direction before the HF, and the AE events were induced by the intrusion of gaseous state CO_2 into it.

7.2.2.3 Focal mechanism from P-wave first-motion polarities of AE

We examined the focal mechanisms using the AE's P-wave first-motion polarities. The polarity response of the sensors was checked by dropping a steel ball onto the surface of each AE sensor. We confirmed for each sensor that an upward trace corresponded to a compressive wave, and a downward trace corresponded to a dilatant wave. Figure 7.11 shows the percentage *R* (%) of

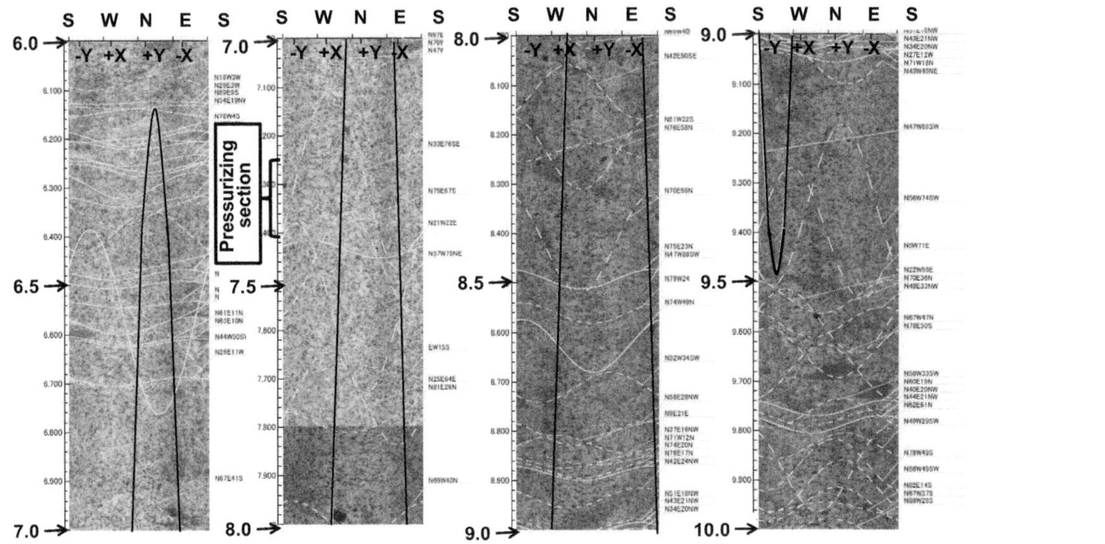

Figure 7.10 Cracks observed on the wall of the HF hole with a borehole camera after HF. CO_2 injection in the section from 7.24 to 7.40 m most likely induced the crack traced with the black lines. White thick lines indicate open cracks of the aperture within 0.5 mm, whereas white broken lines indicate closed adherent cracks. (Ishida et al., 2021).

Note: For interpretation of the references to color in this figure legend, the reader is referred to the Web version of Ishida et al. (2021).

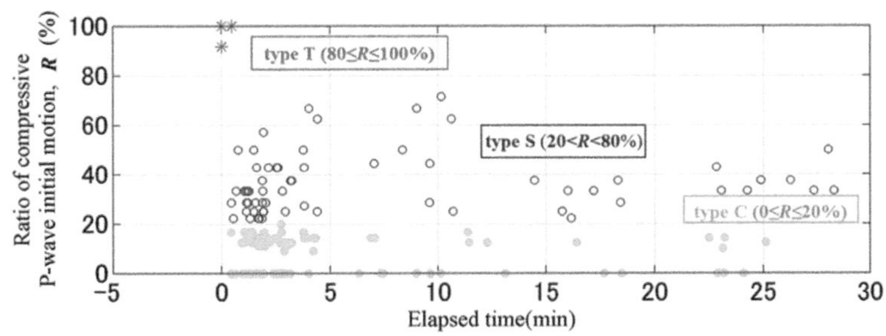

Figure 7.11 Compressive ratio R (%) of P-wave first-motion polarity of AE. In the present study, we labeled AE with $100 \geq R \geq 80$, $80 > R > 20$, and $20 \geq R \geq 0$ as tensile-dominant (type T: gray stars), shear-dominant (type S: open gray circles), and compression-dominant (type C: closed gray circles), respectively. (Ishida et al., 2018, 2021).

Note: For interpretation of the references to color in this figure legend, the reader is referred to the Web version of Ishida et al. (2021).

a number of the compression polarities to the total number of polarities we could read for each AE event, along with the elapsed time (Ishida et al., 2018). This analysis was conducted for AE whose polarities could be read at six or more sensors from the well-located AE events shown in Figures 7.7 and 7.8. The number of AEs that satisfied the condition was 161 out of the 1,249 well-located hypocenters.

The ratio R approaches 100% for a tensile or explosion source, 50% for a pure shear source, and 0% for a compression or implosion source when good coverage of AE sensors around a source is achieved. In the present study, we labeled AE with $100 \geq R \geq 80$ as tensile-dominant (type T), those with $80 > R > 20$ as shear-dominant (type S), and those with $20 \geq R \geq 0$ as compression-dominant (type C). In Figure 7.11, five events are labeled as type T within 30 s after the BD, although we can only see three symbols in the figure due to overlapping. However, no event labeled as type T was recorded after this time. Immediately after type T events were recorded, many type S events started to be recorded with type C events. They were recorded even after the injection was stopped and the injection pressure completely declined, as shown in Figure 7.6.

In addition to the temporal change of the ratio R, we show Figures 7.12 and 7.13 to examine the relation between the location of a source and its ratio R. Figure 7.12 plots the X, Y, and Z coordinates of the AE hypocenters along with the elapsed times for 30 min after BD, which corresponds to those in Figure 7.11. Figure 7.13 plots their distributions on the horizontal plane, XY, for the period from 0 (BD) to 1 min 15 s compared with those from 1 min 15 s to 30 min 0 s. The plotted AE events correspond to those shown in Figures 7.11 and 7.12; in other words, there are only 161 events satisfying the condition, although Figures 7.7 and 7.8 plotted all the 1,249 well-located AE hypocenters. Figures 7.12 and 7.13 elucidate that four out of the five type T events occurred near the pressurizing section of the HF hole immediately after the BD. Following that, as shown in Figure 7.12, the hypocenters of the type S and C events cluster in the range from −0.5 m to +0.1 m in both the X and Y coordinates, corresponding to the distribution along the A-plane in Figure 7.13a. As shown in Figure 7.13b, after 1 min 15 s, the type S and C events are not only located along the A-plane but also along the B-plane.

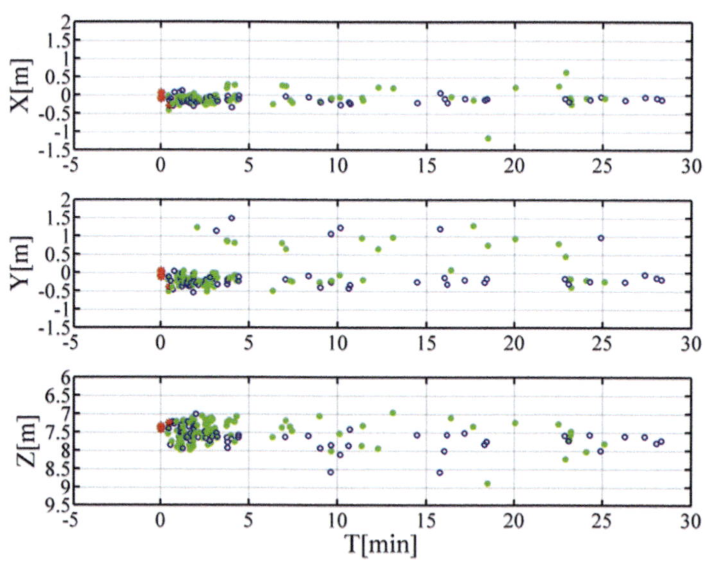

Figure 7.12 X, Y, and Z coordinates of AE hypocenters along with elapsed times for 30 min after BD, corresponding to those of Figure 7.8. Red stars, open blue circles, and closed green circles indicate the same AE types as those in Figure 7.11. A band of broken lines on the Z coordinate indicates the span of the pressurizing section in the pilot hole (Ishida et al., 2021).

Note: For interpretation of the references to color in this figure legend, the reader is referred to the Web version of Ishida et al. (2021).

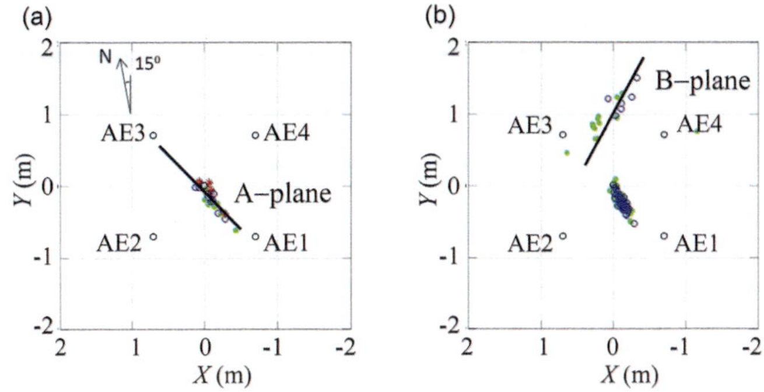

Figure 7.13 Distributions of AE hypocenters projected on the horizontal plane, *XY*. (a) From 0 s to 1 min 15 s. (b) From 1 min 15 s to 30 min 0 s. Figure 7.13a and b plots only 161 AEs out of the 1,249 well-located AE hypocenters plotted in Figures 7.7 and 7.8. The 161 AEs correspond to those shown in Figures 7.11 and 7.12 (Ishida et al., 2021).

Note: For interpretation of the references to color in this figure legend, the reader is referred to the Web version of Ishida et al. (2021).

7.2.2.4 *Crack extension with injection of SC-CO_2*

We injected SC-CO_2 into the pressurizing section after inspecting the recovered core and confirming that there were no pre-existing cracks in it. However, the 3.33 m long bi-wing crack that extended vertically along the boring axis was observed on images of the HF hole wall obtained by a borehole camera after HF. The horizontal direction of the crack observed on the outer surface of the hollow core recovered by overcoring after HF was consistent with the direction of the AE1 and AE3 holes, where the injected CO_2 was gushed out. Thus, the crack was most likely induced at the BD for the following reasons: (i) the BD pressure (15.62 MPa) was considerably high, (ii) the bi-wing crack extended vertically, and (iii) the horizontal directions of the crack extension corresponded to the AE hypocenter distributions just after the BD.

In a similar HF field experiment at MIU, water was injected into the same size pressurizing section of granite (36 mm in diameter and 160 mm long) (Ishida et al., 2019), as shown in the previous chapter, Chapter 6. In this case, new cracks extended 0.4 m vertically and 0.3 m horizontally under a flow rate of 10 mL/min. After the flow rate was increased to 30 mL/min, the cracks further extended to 0.7 m and 0.5 m, respectively. In the present experiment, SC-CO_2 was injected at a flow rate of 50 mL/min. Even considering that the flow rate was greater than that of water, the crack extension of 3.33 m vertical and 2 m horizontal was still notably larger than that of water. The difference could be attributed to the effects of expanding gaseous pressure with the phase change of the CO_2 from SC to gas and the lower viscosity of gaseous state CO_2 than that of liquid water. If so, SC-CO_2 can likely make a larger crack than water, even under the same flow rate, when both CO_2 and water are measured in a volume of the liquid state in the cylinder of the syringe pump.

After 1 min 15 s from BD, AE hypocenters started to distribute along the B-direction, almost perpendicular to the A-direction, from a position of 0.6 m distance from the HF hole. Because no pressure acted in the pressurizing section of the HF hole in the period as shown in Figure 7.6a, a pre-existing crack most likely existed along the B-direction in this position. If so, this indicates that CO_2 could easily intrude and enhance AE occurrence along a pre-existing crack without pressure.

7.3 Characteristics of carbon dioxide fracturing in comparison to conventional water fracturing: evidence from small-scale field experiments

7.3.1 *Site and set up for experiments conducted in 2016*

7.3.1.1 Site and experiments in 2016

Two years after the previous experiments described in Section 7.2, we conducted six new HF experiments at the same site in the same injection hole to compare the size of the AE hypocenter distribution and the AE focal mechanism between CO_2 HF and water HF. The horizontal layout of the site for the experiments is already shown in Figure 7.1. The location of the injection pump and monitoring units in 2016 is slightly different from that in 2014.

As already shown in Figure 7.2, from the tunnel floor, an HF hole with a diameter of 86 mm was drilled downward, and parallel to and 1 m away from the HF hole, four AE monitoring holes with a diameter of 66 mm were drilled. The HF hole and the four AE monitoring holes had already been drilled to 8.0 m and 12.3 m, respectively, two years before for the previous experiments described in Clause 7.2. The HF and four AE holes were additionally drilled from 8.0 to 16.2 m and 12.3 to 17.3 m in depth, respectively, for the new experiments in 2016. As the new experiments, we conducted HF experiments KC5, KC6, and KC7 using CO_2, and KW3, KW4, and KW5 using water, as shown in Table 7.1. Out of the six experiments, we mainly discuss the differences in results between the experiments KC6 and KW3, and briefly discuss those between KC7 and KW5 to confirm the differences obtained between KC6 and KW3. We do not mention the other two experiments, KC5 and KW4, because we observed neither BD nor AE burst for the two experiments, probably owing to leakage of the fracturing fluid through hair cracks inherently existing in the pressurizing sections before the experiments.

In addition, to compare the injectivity improvement, IFOTs were conducted using gaseous CO_2 before and after CO_2 HF and using water before and after water HF. The injectivity of rock around the HF hole was improved around 400 times by CO_2 HF in comparison to a few times larger by water HF.

7.3.1.2 Injection method of fracturing fluid

The system used for CO_2 injection in the experiments was the same as that already shown in Figure 7.4, except that the flow rate was kept constant at 10 mL/min. In all water HF experiments conducted in 2016, including that labeled KW3, water was injected using the same system and at the same constant flow rate of 10 mL/min as those for CO_2 injection conducted in 2016. Because the groundwater level was kept approximately 1 m below the tunnel floor, CO_2 and water were injected into a rock mass saturated with water, as well as in those conducted in 2014.

7.3.1.3 Measurement of AE, pressure, and temperature

Although AE monitoring was conducted by a triggering system in the experiments in 2014, the AE signals detected at the 16 sensors were recorded continuously with a sampling time of 1 μs in the experiments in 2016 by exchanging a hard disk drive for a solid state drive in the A/D converter (PXI-5105; National Instruments). In this case, the signals were recorded after being amplified by 52 dB (40 dB in a pre-amplifier and 12 dB in a main amplifier) and applied through a band-pass filter (20–500 kHz). The injection pressure and the temperature were measured with a sampling time of 0.1 s by another A/D converter (PXI-6251, National Instruments Corp.) in the same ways as those in 2014, as shown in Section 7.2.1.3.

7.3.1.4 AE data processing for hypocenter location

Because the AE signals were recorded continuously in the HF experiments in 2016, from the waveforms by the 16 AE sensors, we identified AE events by searching for the times at which the absolute amplitude exceeded 0.125 V (full scale: −5 to +5 V) as trigger points. When we found a record satisfying this criterion, we extracted 2,048 samples (1,024 before and 1,024 after the trigger point) from each of the 16 channels. A dead time of 2,048 μs (corresponding to 2,048 samples), during which trigger points were not searched for, was set after the end of the extracted waveform, to prevent extracting excessive noise due to "ringing", which was caused by the vibration following a large AE event. From the extracted waveforms, we manually identified the arrival times of the P-waves and located their hypocenters. Similarly, for the experiment KC3 conducted in 2014, the P-wave velocities were measured between the HF hole and the 16 sensors. Then, we located AE hypocenters by an iterative least-squares approach under the assumption of a homogeneous and isotropic medium. We used the average velocities of 5.46 km/s for the experiment KC6 and 5.51 km/s for KW3, respectively. From the located AE hypocenters, we selected "well-located events" that satisfied the following two conditions. 1) The P-wave arrival times could be read at six or more sensors, and in addition, the sensors were set in three or more different AE holes to enclose the hypocenters three-dimensionally. 2) The maximum principal axis of the error ellipsoid calculated from the variance-covariance matrix was smaller than 50 mm. The number of well-located AE events was 3,259 for the experiment KC6 and 465 for KW3, as shown in Table 7.1. We then relocated the hypocenters using the double difference (DD) method (Waldhauser and Ellsworth, 2000).

7.3.1.5 Methods of IFOTs

To compare the injectivity improvement between CO_2 HF and water HF, IFOTs were conducted before and after CO_2 HF and before and after water HF experiments. In each HF experiment, a pressure transient could be created by injecting fluid into the pressurizing section for HF at around 1 MPa, which is sufficiently less than the BD pressure. When the pressurizing section for HF was shut in after injection, the pressure in the pressurizing section fell off. We measured the pressure transient at intervals of 0.1 s for 10 min (Suzuki et al., 2022). To determine permeability by IFOTs using gaseous CO_2, we used the following formula proposed by Suenaga et al. (2004).

$$K = -\frac{\mu_g}{2\rho_{gs}}\frac{Mr^2}{\mathrm{RT}_s}\frac{1}{\alpha}\ln\left(\frac{L+\sqrt{L^2+4r^2}}{2r}\right) \tag{7.1}$$

where K is the permeability (m^2), μ_g is the viscosity of gas (1.53×10^{-5} Pa·s), ρ_{gs} is the standard-state density (1.977 kg/m^3), M is the molecular weight of gas (0.044 kg/mol for CO_2), R is the gas constant (8.31451 J/mol/K), T_s is the standard-state temperature (273.15 K), L is the length of the pressuring section (0.15 m), r is the borehole radius of the pressurizing section (0.018 m), P_{gs} is the pressure at the standard state (0 °C, 1 atm) (1.01325×10^5 Pa),

P_g is the measured pressure in the pressuring section at the elapsed time, t, and α is the coefficient of t in the following equation (7.2) that is obtained from an inclination on a graph of measured data.

$$\ln\left(\frac{P_g - P_{gs}}{P_g + P_{gs}}\right) = \alpha t + \frac{P_g - P_{gs}}{P_g + P_{gs}} \quad (7.2)$$

On the other hand, to determine hydraulic conductivity by IFOTs using water, we used the Hvorselev method (3) (Hirata, 1993), with a graph whose vertical axis is natural logarithms of the decrease of pressure and the horizontal axis is the elapsed time.

$$k = \frac{(2r_w)^2 \ln(\mathrm{mL}/r)}{8L(t_2 - t_1)} \ln(s_1/s_2) \quad (7.3)$$

where k is the hydraulic conductivity (m/s), r is the borehole radius of the pressurizing section (0.018 m), m is the ratio of hydraulic conductivity to the vertical direction ($m = 1$ in isotropy), L is the length of the sealed pressurizing section (0.15 m), s_1 and s_2 are differences (m) from the equilibrium water level at the elapsed time t_1 and t_2, respectively, and r_w is the radius of the piezometer (2.2×10^{-5}m). Although the number of r_w has some uncertainty because the number of r_w depends on Young's modulus of the packer to seal the pressurizing section and others, the uncertainty never affects when we discuss the effect of HF by the ratio of hydraulic conductivities, k, before and after HF.

7.3.2 Results of the CO_2 HF experiment KC6 in comparison to water HF experiment KW3

7.3.2.1 *Temporal changes in fluid pressure, temperature, AE rate, and flow rate*

7.3.2.1.1 CO_2 HF experiment KC6

Figure 7.14 shows temporal changes in fluid pressure, flow rate, temperature, and located AE event rate in the CO_2 HF experiment, KC6, with the elapsed time t when $t=0$ was taken as the time of BD pressure. Here, the AE rate is counted only for well-located events satisfying the two conditions (see Section 7.3.1.4). In the figure, for 7 min around -80 min on the elapsed time, the temperature and pressure records were missing due to an unexpected voltage drop in the power supply.

In the experiment KC6 (Figure 7.14), at the beginning of CO_2 injection, the pressure was gradually increased to 4 MPa by opening the valve of the injection system manually and slowly, so as not to apply the pressure in the CO_2 gas bomb (approximately 4 MPa) directly into the pressurizing section. Then, from $t=-72$ min 27 s, we started to inject CO_2 at a constant flow rate of 10 mL/min from the cylinder by controlling the displacement of the syringe pump. In the beginning, the pressure gradually increased, but it started to increase steeply approximately from $t=-10$ min (arrow A in Figure 7.14), which corresponds to the time when the pressure has almost reached the critical pressure of CO_2, 7.38 MPa. Thus, this steep increase was most likely

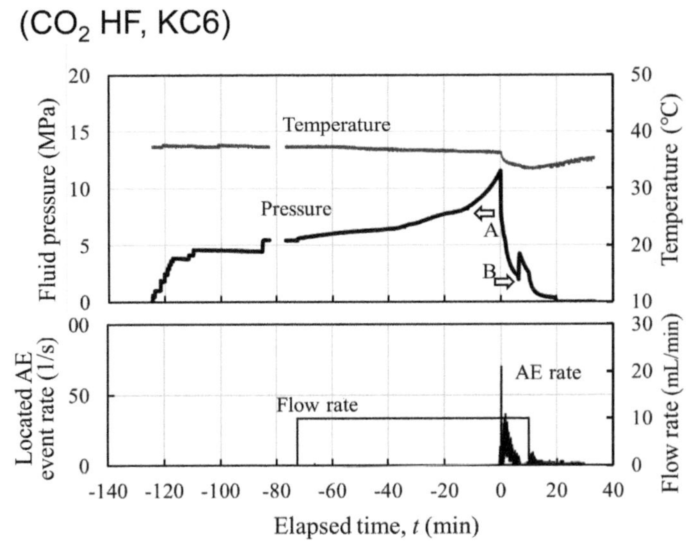

Figure 7.14 Temporal changes in fluid pressure, flow rate, temperature, and the located AE event rate for the CO_2 HF experiment, KC6 (Ishida et al., 2023).

due to the decrease in CO_2 compressibility, which was accompanied by the phase change from the gaseous state to the SC state, as well as the case of the experiment KC3 already shown in Figure 7.6a. When the pressure reached 11.55 MPa, BD occurred. The temperature decreased to 32.6°C at the same time, probably because of adiabatic expansion of CO_2 due to SC-CO_2 leaking into cracks induced at the BD, which likely accompanied the pressure decrease and resulted in a phase change from the SC state to the gaseous state. At t=6 min 26 s (arrow B in Figure 7.14), the pressure suddenly increased from 1.95 to 4.25 MPa, which was caused by switching the pump to the other to maintain the injection.

As shown in the bar diagram in Figure 7.14, the rate of the well-located AE events increased drastically at the BD. The rate decreased with time after the BD; however, the AE events continued to occur even after t= 10 min when the injection was stopped. They continued until approximately t=30 min, even after t=20 min, when the injection pressure decreased completely to zero.

7.3.2.1.2 Water HF experiment KW3

Figure 7.15 shows temporal changes in fluid pressure, flow rate, temperature, and located AE event rate with elapsed time t in the water HF experiment, KW3. In this experiment, the pressure increased over a much shorter time than that of the CO_2 HF experiment, KC6, under the same flow rate (10 mL/min), and BD occurred at 12.78 MPa, which is larger than 11.55 MPa of the CO_2 HF. These results also indicate that it took only 25 min for the water HF, whereas it took 3 h for the CO_2 HF, as shown in Figure 7.14. AE activity burst only at the BD and ceased completely immediately after BD. This is quite different from the AE activity observed in the CO_2 HF experiment, KC6.

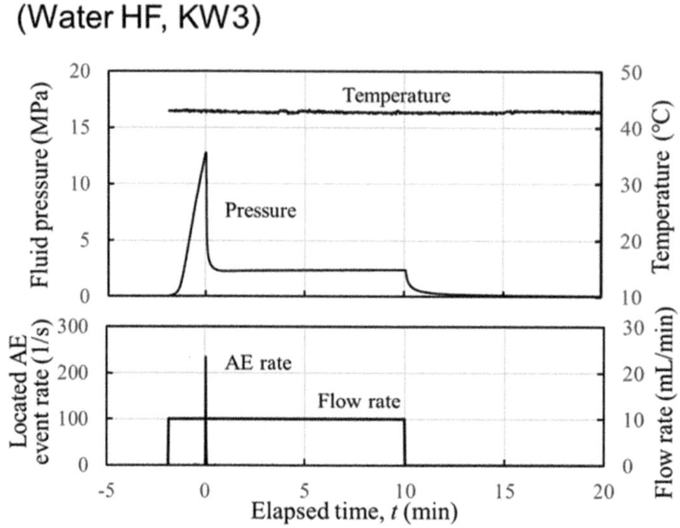

Figure 7.15 Temporal changes in fluid pressure, flow rate, temperature, and the located AE event rate for the water HF experiment, KW3 (Ishida et al., 2023).

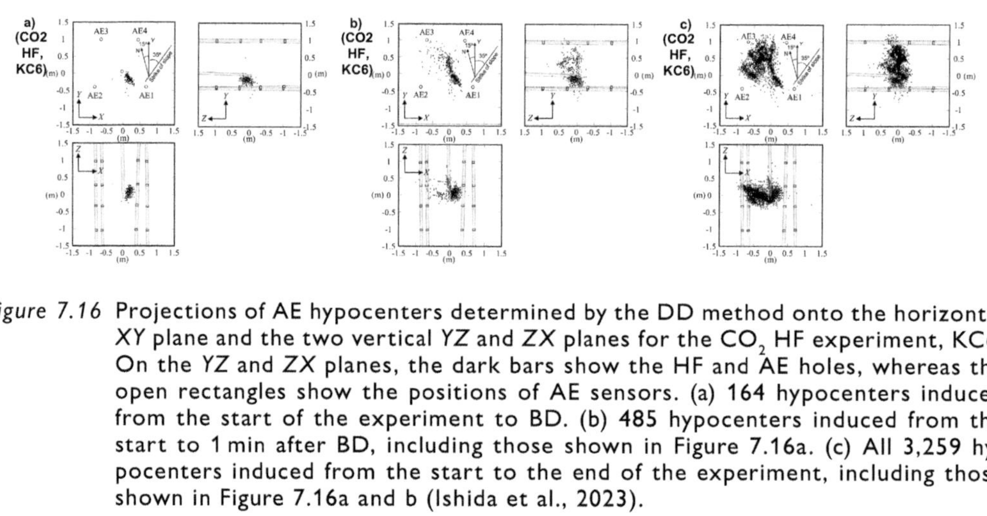

Figure 7.16 Projections of AE hypocenters determined by the DD method onto the horizontal *XY* plane and the two vertical *YZ* and *ZX* planes for the CO_2 HF experiment, KC6. On the *YZ* and *ZX* planes, the dark bars show the HF and AE holes, whereas the open rectangles show the positions of AE sensors. (a) 164 hypocenters induced from the start of the experiment to BD. (b) 485 hypocenters induced from the start to 1 min after BD, including those shown in Figure 7.16a. (c) All 3,259 hypocenters induced from the start to the end of the experiment, including those shown in Figure 7.16a and b (Ishida et al., 2023).

7.3.2.2 AE hypocenter distributions and cracks observed on recovered cores

7.3.2.2.1 CO_2 HF experiment KC6

Figure 7.16 shows the AE hypocenters projected onto the horizontal (XY) plane and the two vertical (*YZ* and *ZX*) planes for the CO_2 HF experiment, KC6. They were located following the procedure shown in the previous section, Section 7.3.1.4. Figure 7.16a shows those from the start of the experiment to the BD, whereas Figure 7.16b shows those from the start to 1 min after the BD, including those already shown in Figure 7.16a. Figure 7.16c shows all hypocenters from the start to the end of the experiment. Of all AE hypocenters shown in Figure 7.16c, the first AE event occurred 36.4 s before the BD, whereas the last AE event occurred 1785.0 s after the BD.

Although the HF hole was drilled at the center of the four AE holes on the tunnel floor (Figure 7.2), its position was not at the center of the injection depth due to the deviation of drilling. The HF hole tilted approximately 1.7°, and the AE holes tilted 0.4–1.0° from the vertical. The influence of the deviation on the sensor positions was corrected for the AE hypocenter locations in the experiments in 2016 because the influence is relatively larger in the deeper depths, although it was not corrected for those in 2014. In Figure 7.16, the origin of the *X* and *Y* coordinates is taken at the center of the HF hole, whereas that of the *Z* coordinate is taken at the central depth of the pressurizing section.

Figure 7.16a indicates that AE hypocenters are distributed in the direction from the HF to the AE1 hole on the horizontal *XY* plane, whereas they are distributed upward from the wall of the pilot hole having the pressurizing section on the vertical *ZX* and *ZY* planes. Because the difference between Figure 7.16a and b corresponds to hypocenters induced for one minute immediately after the BD, it indicates that AE hypocenters for the period are newly distributed from the HF hole to the direction between the AE3 hole and the +*Y* direction on the *XY* plane, which is almost opposite to that before the BD. On the *ZX* and *ZY* planes, they are also newly distributed in the opposite direction to that before the BD from the wall of the pilot hole, implying the generation of a bi-wing crack around the pressurizing section. The difference between Figure 7.2b and c indicates that many AE events were induced for approximately 30 min from one minute after the BD, and their hypocenters spread much more widely, reaching close to the four AE holes. The spread of AE hypocenters is consistent with our observation that bubbles of CO_2 gushed from all four AE holes a few minutes after the BD, indicating that the induced fractures reached the four AE holes.

Figure 7.17a and b shows the recovered solid core of the pilot hole for the CO_2 HF experiment, KC6. The drilled hole diameter was 36 mm, and the diameter of the recovered solid core was 22 mm. The two figures are photos taken from opposite sides of the core. Because we found

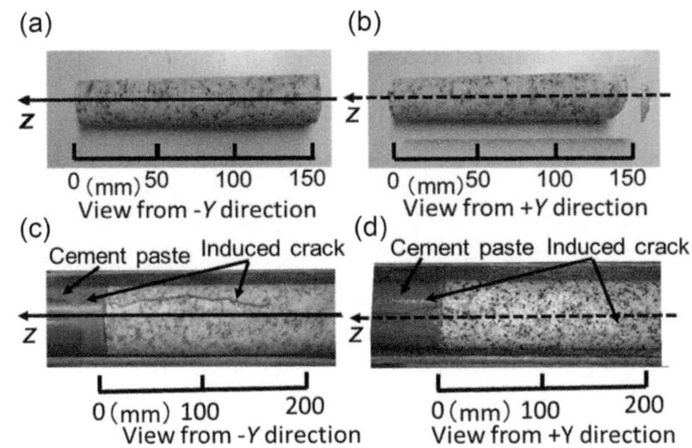

Figure 7.17 Photos of a solid core recovered from the pilot hole before HF (a and b) and a hollow core recovered by drilling coaxially (overcoring) around the pressurizing section after HF (c and d) from the pressurizing section in the CO_2 HF experiment, KC6. Photos a and b and Photos c and d were taken from opposite sides of the same cores, respectively. The straight solid lines and broken lines over the cores indicate the −Y and +Y directions, respectively. The gray lines drawn on the core of Photos c and d indicate cracks induced by the CO_2 HF. (Ishida et al., 2023).

Note: For interpretation of the references to color in this figure legend, the reader is referred to the Web version of Ishida et al. (2023).

no visible cracks on the core, we selected the interval as the pressurizing section for the HF experiment. Figure 7.17c and d shows a hollow core recovered by drilling coaxially (overcoring) around the pressurizing section after the CO_2 HF. The drilled hole diameter was 86 mm, and the diameter of the recovered core was 68 mm. The dark part at the left-hand end of the core was cement paste used to seal the pressurizing section. Two cracks along the borehole axis were observed on the core in almost symmetrical directions. The strike of the crack on Figure 7.17c is consistent with the AE hypocenter distribution 20–40° counterclockwise from the $-Y$ direction, projected on the XY plane in Figure 7.16a. The strike of the crack on Figure 7.17d is almost along the $+Y$ direction, which is consistent with the AE hypocenter distribution from the HF hole to the $+Y$ direction on the XY plane in Figure 7.16b. This suggests that the crack shown in Figure 7.17c was induced at the BD, whereas the crack shown in Figure 7.17d was induced for one minute after the BD. These cracks also confirm that the CO_2 HF induced almost vertical bi-wing cracks in the pressurizing section, and the bi-wing cracks extended outward from the HF hole.

7.3.2.2.2 Water HF experiment KW3

Figure 7.18 shows all the AE hypocenters during the water HF experiment, KW3. Out of the 465 AE events shown in Figure 7.18, the first AE event occurred 27.50 s before the BD, and the last AE event occurred 7.05 s after the BD; while only 4 AE events occurred before the BD, the remaining 461 events occurred only for 7.05 s immediately after the BD. The time length of 7.05 s is considerably smaller than that of 30 min 22 s for the CO_2 HF experiment, KC6. In addition to this, the 465 events of the located AE hypocenters shown in Figure 7.18 are only around 14% of the 3,259 events shown in Figure 7.16c. On the horizontal XY plane in Figure 7.18, the AE hypocenters are distributed in a short length of 0.1–0.2 m from the HF hole to the direction

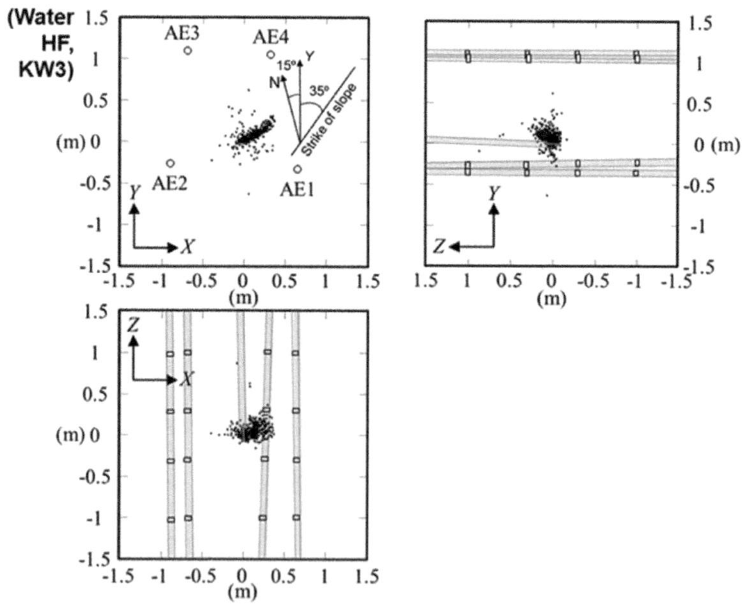

Figure 7.18 Projections of all 465 AE hypocenters determined by the DD method onto the three orthogonal planes for the water HF experiment, KW3 (Ishida et al., 2023).

of the AE2 hole, whereas they are distributed in a relatively long length of 0.4–0.5 m to the opposite direction. On the vertical planes, *ZX* and *ZY*, they distribute upward from both sides of the pilot hole, suggesting the generation of a bi-wing crack around the pressurizing section. The direction of the distribution is almost orthogonal to that at the BD in the experiment KC6 shown in Figure 7.16a and b. However, the much more remarkable difference between the experiment KC6 and KW3 is that the region of the AE hypocenter distributions induced by the CO_2 HF is much larger than that by the water HF, as discussed later.

Figure 7.19a and b shows the recovered solid core of the pilot hole for the water HF experiment, KW3. The two figures are photos taken from opposite sides of the core. Because we found no visible cracks on the core, we selected the interval as the pressurizing section for the HF experiment. Figure 7.19c and d shows a hollow core recovered by drilling coaxially around the pressurizing section after the water HF. All the core diameters shown in the figures are the same as those shown in Figure 7.17. On the core shown in Figure 7.19c, a crack extends along the borehole axis, whose strike is approximately 60° clockwise from the −*Y* direction. The strike corresponds to the direction of the AE hypocenter distribution in a short length of 0.1–0.2 m from the HF hole to the AE2 hole. On the core shown in Figure 7.19d, a crack extends along the borehole axis, whose strike is 0–30° clockwise from the +*Y* direction. Although there is an around 30° difference between the strike of the crack observed in Figure 7.19d and the horizontal direction of the AE hypocenter distribution in a relatively long length of 0.4–0.5 m from the HF hole in Figure 7.18, they are almost consistent if a crack was extended with slight bending outward from the HF hole. These cracks also confirm that the water HF induced almost vertical bi-wing cracks in the pressurizing section, and the bi-wing cracks extended outward from the HF hole, as well as in the CO_2 HF experiment, KC6.

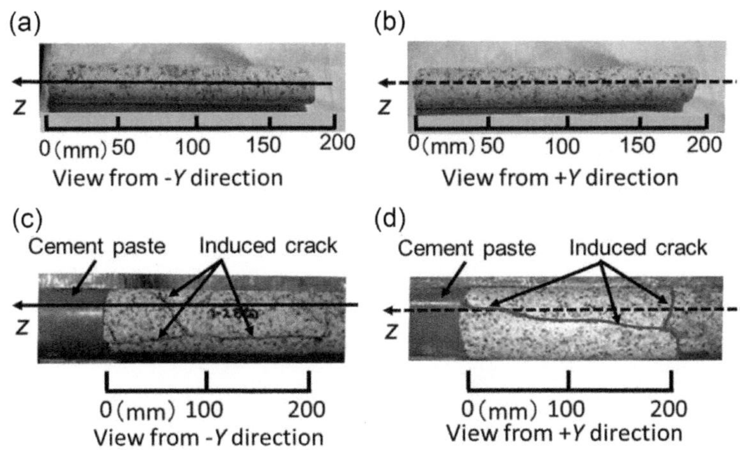

Figure 7.19 Photos of a solid core recovered from the pilot hole before HF (a and b) and a hollow core recovered by drilling coaxially (overcoring) around the pressurizing section after HF (c and d) from the pressurizing section in the water HF experiment, KW3. Photos a and b and Photos c and d were taken from opposite sides of the same cores, respectively. The straight solid and broken lines over the cores indicate the −*Y* and +*Y* directions, respectively. The gray lines drawn on the core of Photos c and d indicate cracks induced by the water HF. (Ishida et al., 2023).

Note: For interpretation of the references to color in this figure legend, the reader is referred to the Web version of Ishida et al. (2023).

7.3.2.3 *Fracturing mechanism deduced from P-wave first-motion polarities of AE events*

The focal mechanisms of AE events were examined by P-wave first-motion polarities. The polarity response of the sensors was checked by conducting a steel-ball drop test on an AE sensor surface in the same way as that shown in Section 7.2.2.3, and we confirmed that a downward trace corresponds to a dilatant wave and an upward trace corresponds to a compressive wave. We calculated the percentage R of the compression polarities of the P-wave first motions only for events whose polarities could be read at six or more sensors among the well-located hypocenters. As a result, the number of AE events whose percentage R was obtained by satisfying the condition was only 139 out of 3,259 well-located hypocenters for the experiment KC6 and 79 out of 465 for KW3. As the same shown in Section 7.2.2.3, we labeled AE events with $0 \leq R \leq 20$ as compression-dominant (type C), those with $20 \leq R \leq 80$ as shear-dominant (type S), and those with $80 \leq R \leq 100$ as tensile-dominant (type T), to discuss the HF mechanism of rock.

7.3.2.3.1 CO_2 HF experiment KC6

Figure 7.20 shows temporal changes in the percentage R in the experiment KC6. In the experiment, type T events were observed only at BD (with one exception at approximately 25 min), and type S events followed them. Following type S events, type C events occurred approximately 2 min after BD.

Figure 7.21 shows the X, Y, and Z coordinates of the AE hypocenters along the elapsed time in the experiment KC6, discriminating their three fracturing mechanisms, types T, S, and C, with the respective symbols. The figure indicates that type T events occurred near the HF hole, like the first AE event immediately after BD, whereas events of types S and C occurred away from the HF hole.

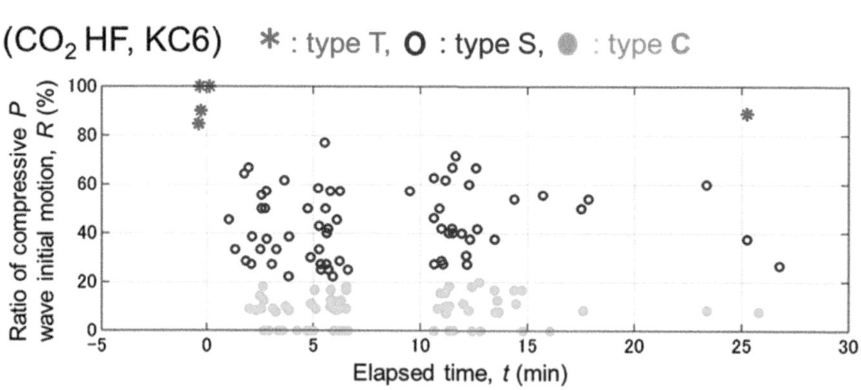

Figure 7.20 Temporal changes of the percentage, R, of the compression polarities of P-wave first motions of an AE event in the CO_2 HF experiment, KC6. AE events were labeled with $0 \leq R \leq 20$ as compression-dominant (type C), those with $20 \leq R \leq 80$ as shear-dominant (type S), and those with $80 \leq R \leq 100$ as tensile-dominant (type T), and they are shown with closed gray circles, open gray circles, and gray stars, respectively. (Ishida et al., 2023).

Note: For interpretation of the references to color in this figure legend, the reader is referred to the Web version of this Ishida et al. (2023).

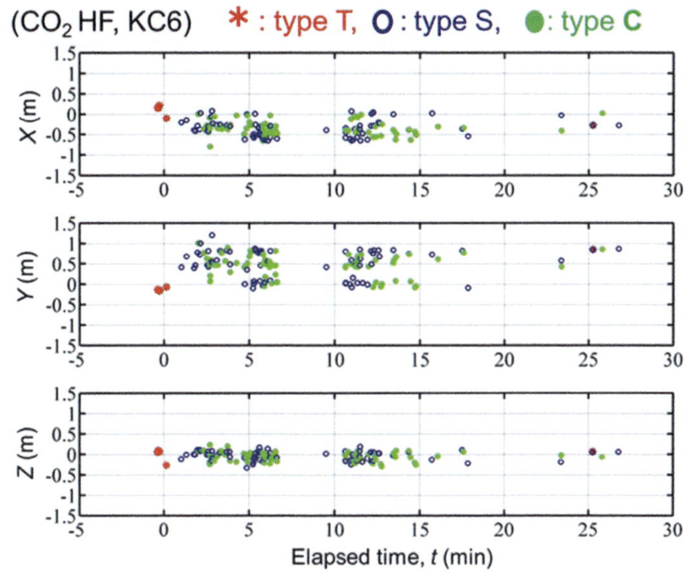

Figure 7.21 Temporal change of X, Y, and Z coordinates of AE hypocenters in the CO_2 HF experiment, KC6, discriminating their three fracturing mechanisms with the symbols of closed green circles (type C), open blue circles (type S), and red stars (type T), respectively. The origin of the X and Y coordinates is taken at the center of the HF hole, whereas that of the Z coordinate is taken at the center of the pressurizing section (Ishida et al., 2023).

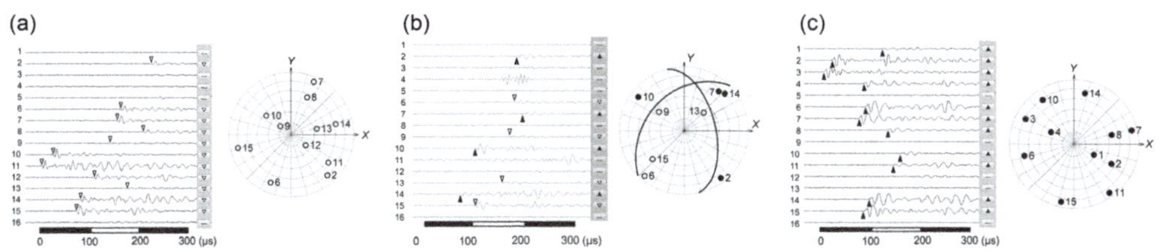

Figure 7.22 Respective typical examples of waveforms of AE events labeled as type C (compression-dominant), type S (shear-dominant), and type T (tensile-dominant) in the CO_2 HF experiment, KC6. Projections onto Schmidt nets (lower hemisphere) of the polarities of P-wave first motions are also shown to the right-hand side of the respective waveforms. The symbols of an upward closed and a downward open triangle along the waveforms indicate an upward and a downward trace of P-wave first motion, respectively. Closed and open circles on the Schmidt net indicate compressive and dilatational polarity, respectively. (a) Example of a type C event that $R = 100 \times 0/11$(number of compressive polarities/that of found total polarities) = 0%: the occurrence time is 3 min 23.2 s and the hypocenter coordinates are (−362.69, 773.91, 86.56). The unit of the coordinates is mm. (b) Example of a type S event that $R = 100 \times 4/8 = 50\%$: 2 min 46.7 s and (−50.07, 547.96, −157.72). (c) Example of a type T event that $R = 100 \times 11/11 = 100\%$: −19.6 s and (162.82, −135.23, 98.30) (Ishida et al., 2023).

Figure 7.22 shows respective typical examples of waveforms of AE events labeled as type C, S, and T, with projections of the polarities of P-wave first motions onto the Schmidt net. In Figure 7.22a and c, we can recognize with certainty that the P-wave first motions were exclusively downward or upward at all 11 stations with a considerably uniform distribution on a Schmidt net, respectively. In Figure 7.22b, which shows a typical example of type S events, respective strikes of the two nodal planes make an angle of −15° and +45° from the +*Y* direction. The AE event occurred at 2 min 46.7 s after BD at the location of the coordinates, (−50.07, 547.96, −157.72). Comparing the AE hypocenter distribution in Figure 7.16c, the direction of −15° from the +*Y* out of the two strike directions seems consistent with the direction of crack extension induced by the HF.

7.3.2.3.2 Water HF experiment KW3

Figure 7.23 shows temporal changes in the percentage R in the water HF experiment, KW3. In the experiment, type S events started immediately after the BD, and type T events also started around 1.4 s after the BD. A similar tendency was observed in the experiment KC6, as shown in Figure 7.20. The remarkable difference to that in the experiment KC6 is that type C events were never observed in the experiment KW3, whereas many type C events were observed in the CO_2 HF experiment, KC6.

Figure 7.24 shows the *X*, *Y*, and *Z* coordinates of the AE hypocenters along the elapsed time in the experiment, KW3. In the figures, if an AE event was induced on a frontier toward a new region where no AE had yet migrated, one of the *X*, *Y*, and *Z* coordinates of the event is plotted on the edge of a previous distribution. When we watch the figure from this point of view, we find that type T events tend to occur at the edges of regions where AE events have already occurred. We show arrows for such type T events at the coordinates to easily observe this tendency. From the 13 type T events, seven events (54%) shown with arrows were induced at the edge of a region where AE events were previously distributed.

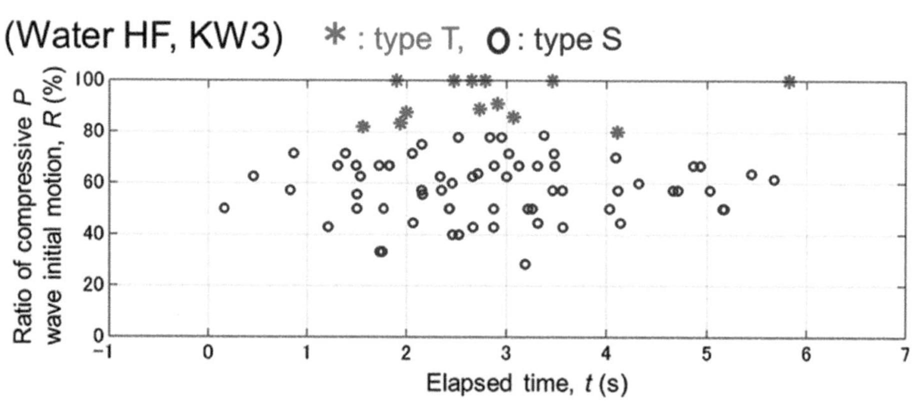

Figure 7.23 Temporal changes in the percentage, *R*, of the compression polarities of P-wave first motion of an AE event in the water HF experiment, KW3. The symbols of open gray circles and gray stars indicate the shear-dominant (type S) and the tensile-dominant (type T) AE events, respectively. (See their definitions in the text and the caption of Figure 7.20) Please note that the time span, 8 s, of this figure is only around 4/1,000 of 35 min of Figure 7.20. (Ishida et al., 2023).

Note: For interpretation of the references to color in this figure legend, the reader is referred to the Web version of Ishida et al. (2013).

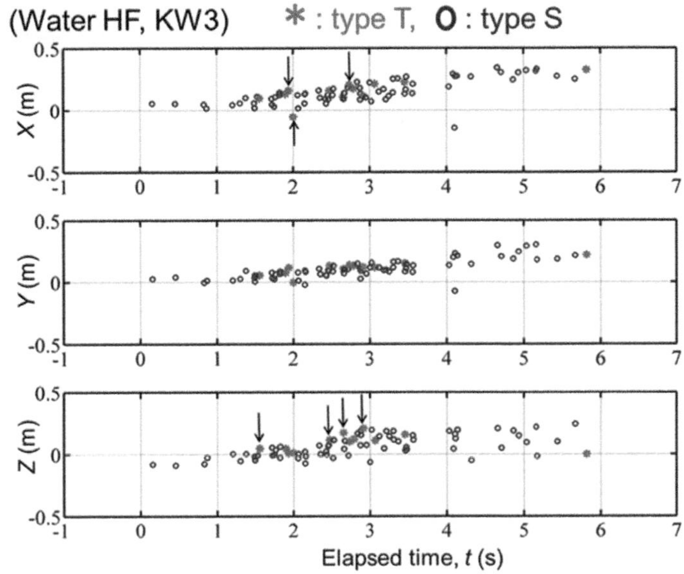

Figure 7.24 Temporal change of X, Y, and Z coordinates of AE hypocenters in the water HF experiment, KW3, discriminating the fracturing mechanisms with the symbols of open gray circles (type S) and gray stars (type T), respectively. The arrows in the figures of X and Z coordinates indicate the separate 7 type T events that were induced on a frontier toward a new region in which no AE had yet migrated, because one of the X, Y, and Z coordinates of such an event is plotted on the edge of a previous distribution. (Ishida et al., 2023).

Note: For interpretation of the references to color in this figure legend, the reader is referred to the Web version of Ishida et al. (2013).

7.3.3 *Injectivity improvement by HF*

Injectivity improvement by the HF experiments was originally shown and discussed by Suzuki et al. (2022). In this section, the testing methods and the results are explained following Suzuki et al. (2022) and Ishida et al. (2023).

7.3.3.1 *Injectivity improvement by CO_2 HF in the experiment KC6*

Figure 7.25 shows the results of IFOTs using gaseous CO_2 before and after the CO_2 HF in the experiment KC6. Gaseous CO_2 was injected at a pressure of around 1 MPa, and the pressure change was measured at intervals of 0.1 s for 10 min. Figure 7.25a shows the results before and after the CO_2 HF, respectively. We can easily find that the pressure decrease after the HF is faster than before the HF. Figure 7.25b shows graphs plotted from the data of Figure 7.25a. In Figure 7.25b, the vertical axis is ln[($PgPgs$)/(Pg+Pgs)] and the horizontal axis is the elapsed time, t. We calculated the numbers of α from the inclinations of the graphs in Figure 7.25b. We substituted the numbers into the formula (7.1) to obtain the permeabilities, K (before the HF)$=3.25\times10^{-19}$ (m^2) and K (after the HF)$=1.41\times10^{-16}$ (m^2), as shown in Table 7.2. Here, we define the permeability ratio K (after the HF)/K (before the HF) as an index of injectivity improvement. As a result, the injectivity improvement by the CO_2 HF was 433.

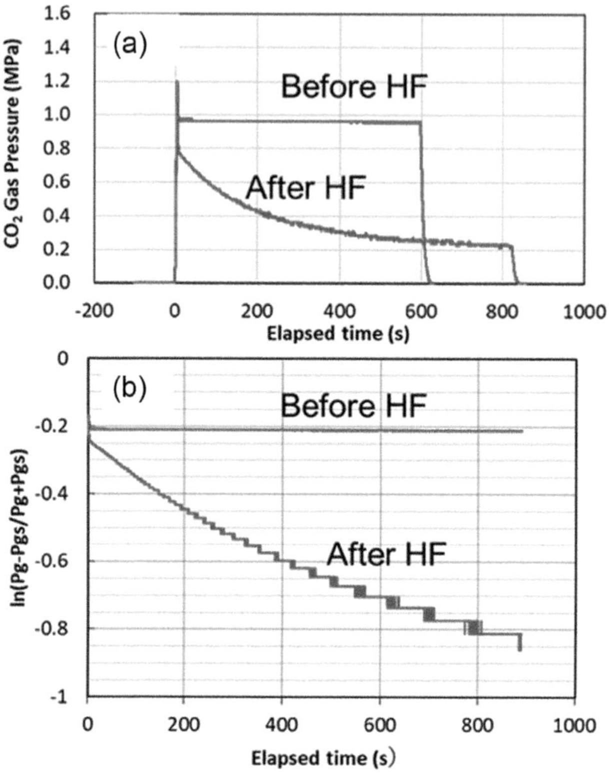

Figure 7.25 Results of IFOTs using gaseous CO_2 before and after the SC-CO_2 HF experiment KC6 (Ishida et al., 2023). (a) Pressure decreases with the elapsed time. (b) Graph plotted to obtain coefficient *a* in the formula (7.2) (Suzuki et al., 2022, Ishida et al., 2023).

Table 7.2 Results of IFOTs and ratios of injectivity improvements (Ishida et al., 2023)

Experiment	*Before HF*	*After HF*	*Ratio of injectivity improvement*	
CO_2 HF	*K* (m^2)	*K* (m^2)		
KC6	3.25×10^{-19}	1.41×10^{-16}	433	Text
KC7	1.68×10^{-19}	6.92×10^{-17}	411	Appendix
Water HF	*k* (m/s)	*k* (m/s)		
KW3	1.34×10^{-11}	9.41×10^{-11}	7.22	Text
KW5	2.45×10^{-11}	6.17×10^{-11}	2.52	Appendix

7.3.3.2 Injectivity improvement by water HF in experiment KW3

Figure 7.26 shows the results of IFOTs using water before and after the water HF in the experiment KW3. In the tests, water was injected at a pressure slightly larger than 1 MPa, and the pressure change was measured at intervals of 0.1 s for 10 min, as shown in Figure 7.26a. Figure 7.26b shows graphs plotted on a semi-log scale from the data of Figure 7.26a. In Figure 7.26b, the

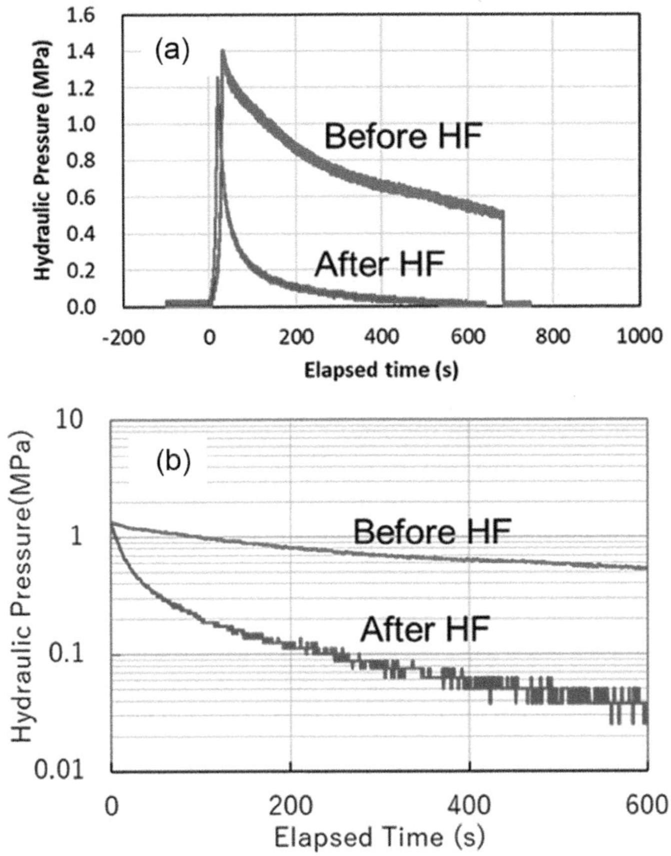

Figure 7.26 Results of IFOTs using water before and after the water HF experiment KW3. (a) Pressure decreases with the elapsed time (Ishida et al., 2023). (b) Graph plotted on a semi-log graph from the data of Figure 7.26a to obtain the hydraulic conductivity of the formula (7.3) (Suzuki et al., 2022, Ishida et al., 2023).

vertical axis is hydraulic pressure, s, sealed in the pressurized section, and the horizontal axis is the elapsed time, t. Because the relation between the logarithms of s and t becomes linear when the retention property is large, the hydraulic conductivity, k, is taken as the inclination of the curve at the starting point of the test. As shown in Table 7.2, the obtained numbers for the hydraulic conductivity are k (before the HF) $= 1.34 \times 10^{-11}$ (m/s) and k (after the HF) $= 9.41 \times 10^{-11}$ (m/s). As a result, the injectivity improvement by the water HF was 7.22.

7.4 Discussion

7.4.1 *Crack extension with CO_2 HF and the effects of a pre-existing crack observed in the experiment KC3 in 2014*

In the experiment KC3 conducted in 2014, we injected SC-CO_2 at a flow rate of 50 mL/min, which was 5 times larger than that in 2016, and the AE hypocenter distribution indicated that the injected CO_2 was most likely intruded into a pre-existing crack from newly generated HF cracks. From the results, we discuss the crack extension with CO_2 HF and the effects of a pre-existing crack on it.

We injected SC-CO_2 into the pressurizing section from 7.24 to 7.40 m in depth after confirming that there were no pre-existing cracks in the section by inspecting the recovered core. However, a 3.33 m long bi-wing crack that extended vertically along the boring axis was observed on images of the HF hole wall obtained by a borehole camera after HF. The bi-wing crack's horizontal direction was consistent with the crack observed on the outer surface of the hollow core recovered by overcoring after HF, and the direction of the AE1 and AE3 holes where the injected CO_2 was gushed out. The bi-wing crack was most likely induced at BD for the following reasons: (1) the BD pressure (15.62 MPa) was high, (2) the bi-wing crack extended vertically, and (3) the horizontal directions of the crack extension corresponded to the AE hypocenter distributions immediately after BD.

In the experiment, KC3, the dense clouds of AE hypocenters were distributed only over $1.0 \times 1.3 \times 0.4\,m^3$ ($0.52\,m^3$) as shown in Figure 7.7. This volume is not so large in comparison to the volumes in the experiments KC6 and KC7 that were conducted at a flow rate of 10 mL/min in 2016 whose AE hypocenter distributions were $1.1 \times 1.7 \times 0.8\,m^3$ ($1.496\,m^3$) and $1.75 \times 1.50 \times 1.25\,m^3$ ($3.281\,m^3$) as shown in Figure 7.16c and 7.A3, respectively. However, in the experiments, KC6 and KC7, any long crack extending along the boring axis was not found on the HF hole wall. Thus, a crack size of 3.3 m long extending along the boring axis with a large aperture observed on the HF hole wall is the only remarkable difference in the experiment KC3 compared to those in KC6 and KC7. Consequently, the difference was most likely caused by the flow rate difference between 50 mL/min in KC3 and 10 mL/min in KC6 and KC7.

After 1 min 15 s from the BD in the CO_2 HF experiment KC3, AE hypocenters started to distribute along the B-direction, almost perpendicular to the A-direction, from a position 0.6 m away from the HF hole, as shown in Figure 7.8. Because no pressure acted in the pressurizing section of the HF hole in the period as shown in Figure 7.6, a pre-existing crack most likely existed in the position along the B-direction. If so, this indicates that CO_2 could easily intrude and enhance AE occurrence along a pre-existing crack without injecting pressure.

7.4.2 Characteristics of CO_2 HF in comparison to those of water HF observed in the experiments KC6 and KW3 conducted in 2016

To compare the characteristics of CO_2 HF to those of water HF, we conducted the CO_2 HF experiment KC6 and water HF experiment KW3 in 2016, under the same flow rate of 10 mL/min. Among them, we discuss below the characteristic differences in BD pressure, duration times of AE occurrence, AE hypocenter distributions and crack extension, bi-wing cracks, and differences in their extending direction.

7.4.2.1 BD pressure

The BD occurred at 11.55 MPa for the CO_2 HF experiment KC6, whereas it occurred at 12.78 MPa for the water HF experiment KW3, under the same 10 mL/min flow rate. The tendency of the BD pressure for the CO_2 HF to be smaller than that for the water HF is consistent with the experimental results in the laboratory by Ishida et al. (2016) and Ranjith et al. (2019). This tendency is most likely attributed to the lower viscosity of CO_2 than that of water, because the BD pressure decreases with a decrease in the fracturing fluid viscosity, as suggested by Zoback et al. (1977), Morita et al. (1996), Ishida et al. (2004, 2016), Bunger and Detournay (2008), and Ranjith et al. (2019).

7.4.2.2 *Duration of AE occurrence*

The AE events continued to occur after BD and even after stopping injection in the CO_2 HF experiment KC6, as shown in Figure 7.14. In contrast, the AE events ceased immediately after BD in the water HF experiment KW3, as shown in Figure 7.15. In the experiment KC6, CO_2 was in the SC state at the BD because the temperature and pressure were larger than the critical points, 31.1°C, and 7.38 MPa, respectively. However, when a fracture extended after the BD, CO_2 was likely to become a gaseous state because the pressure decreased to below the critical pressure of 7.38 MPa, although the temperature remained higher than 31.1°C (please refer to Figure 7.14 in this chapter and the phase diagram of CO_2 shown in Figure 5.3 in Chapter 5). Therefore, the long duration of AE occurrence and the widespread distribution of AE hypocenters, which will be discussed in the next section, were probably induced by gaseous CO_2.

7.4.2.3 *AE hypocenter distributions and crack extension*

Although CO_2 and water were injected at the same flow rate, 10 mL/min, the AE hypocenters induced by the CO_2 HF were distributed more widely than those by the water HF. The AE hypocenters in the CO_2 HF experiment KC6 were distributed in a region of $1.1 \times 1.7 \times 0.8\,m^3$ ($1.496\,m^3$) as shown in Figure 7.16c, which is consistent with our observation that bubbles of CO_2 gushed from all four AE holes, indicating that the induced HF fractures had reached them. In contrast, they were distributed only in a region of $0.7 \times 0.2 \times 0.4\,m^3$ ($0.056\,m^3$) in the water HF experiment KW3, as shown in Figure 7.18. The difference in the volume was 27 times. In both experiments, the injections were performed under the same flow rate of 10 mL/min, measured in a cylinder of the syringe pump for the liquid state of CO_2 and water. Thus, the difference in the AE hypocenter distributions indicates that the CO_2 HF could stimulate a much larger region than the water HF; still, we cannot completely deny the possibility that the injected fluid leaked out of the surroundings. This tendency is also confirmed by the difference in the AE hypocenter distribution between the CO_2 HF experiment KC7 and the water HF experiment KW5, as shown in Figures 7.A3 and 7.A4, respectively, in Appendix 7.A.

7.4.2.4 *Bi-wing cracks and differences in their extending directions*

AE hypocenter distributions and observations of cracks on cores recovered by drilling coaxially around the pressuring section after the HF revealed that the HF in both the KC6 and KW3 experiments induced the bi-wing cracks along the pressuring section. However, the directions for crack extensions from the HF hole were different between the two experiments. The test site is situated in a small railroad tunnel, which is approximately 50 m below the surface of a mountainous area, and the tunnel exit was 40 m away from the site, as shown in Figure 7.1. The railroad exits from the tunnel to the outside steep slope, making a sidewall of the gorge, and the railroad continues on a bridge across a 50 m deep mountain stream. From the surface topography, the vertical stress, σ_Z, was estimated at 1.25 MPa as the overburden pressure in the experimental site. The horizontal stress in the direction along the railroad at the tunnel exit, that is, normal to the strike of the steep slope, N40°W (55° anticlockwise from the $+Y$ direction), should be almost zero. On the other hand, the horizontal stress in the direction along the strike of the steep slope, N50°E (35° clockwise from the $+Y$ direction), could be large (see Figure 7.1), although it is not certain because we could not measure rock stress at the site.

Although the crack extending direction from the HF hole in the water HF experiment KW3 is close to the direction along the strike of the steep slope, as shown in Figure 7.18, the crack direction in the CO_2 HF experiment KC6 cannot be explained by the stress condition expected from the surface topography, as shown in Figure 7.16. For this reason, we can consider two possibilities. One is the possibility that the stress in the direction along the strike of the steep slope was not so large; in other words, the stress condition around the HF hole was close to isotropic. The other possibility is the effect of a pre-existing crack. The borehole television observations at this site showed an average of 6.16 pre-existing cracks per meter in the HF and AE holes. One pre-existing crack or weak plane, which could not be found by inspecting the core recovered in the pressurizing section before the HF, might affect the crack extending directions in the experiment KC6.

7.4.2.5 Injectivity improvement by CO_2 HF in comparison to that by water HF

The results of the IFOTs for the experiments KC6 and KW3 are shown in Table 7.2. In the experiment KC6, as shown in Figure 7.16c, the AE hypocenters associated with the CO_2 HF extended from the HF hole to the surrounding four AE holes, and CO_2 gushed out from all the holes just after the BD. In addition, Figure 7.17c and d shows that HF cracks observed on the surfaces of the hollow core recovered by drilling coaxially around the pressurizing section extended to the part of the cement paste to seal the pressurizing section of the HF hole. From these facts, gaseous state CO_2 used for the IFOT after the HF likely leaked into the surrounding four AE holes and the upper part of the HF hole.

In the experiment KC7 shown in Appendix 7.A, although a hair crack was observed on a core recovered in the pressurizing section, gaseous CO_2 used for the IFOT unlikely leaked through the crack because the permeability measured before the HF is smaller than that measured before the HF in the experiment KC6 that had no crack in the pressurizing section. The distribution of AE hypocenters shown in Figure A7.3 indicates that a crack extending from the HF hole might reach the AE4 hole, but any crack is unlikely to reach the other three AE holes. In addition, the HF crack did not extend to the part of the cement paste on the hollow core recovered after the HF. These facts indicate that gaseous CO_2 used the IFOT after the HF was unlikely to have leaked into the surrounding AE holes and the upper part of the HF hole. In addition, there is no large difference in the permeabilities after HF between the experiments KC6 and KC7. From these, even if gaseous CO_2 injected after the HF leaks, its amount should be small, because the maximum pressure of 1 MPa was less than one-tenth of the BD pressure of the experiment KC6, 11.55 MPa. Thus, it is reasonable to consider that the measured permeability after the HF, even for the experiment KC6, is reliable, as well as that for the experiment KC7. Consequently, it can be concluded that the injectivity of rock around the HF hole was improved by around 400 times by the CO_2 HF in the experiments KC6 and KC7.

On the other hand, in the experiment KW3, as shown in Figure 7.18, the AE hypocenter distribution associated with the water HF concentrated around the HF hole without extending to the surrounding AE holes. In Figure 7.19c and d, we find that the HF crack did not extend into the part of the cement paste. From these facts, the measured hydraulic conductivity is most likely reliable because the water used for the IFOT after the HF was unlikely to have leaked into the surrounding AE holes and the upper part of the HF hole.

In the experiment KW5 shown in Appendix 7.A, because the AE hypocenter distribution concentrated around the HF hole without extending to the surrounding AE holes as shown in Figure 7.A4, the water used for the IFOT after the HF was unlikely to have leaked into the surrounding AE holes and the upper part of HF hole, as well as in the experiment KW3. Thus, the measured hydraulic conductivities after the HF for both experiments, KW5 and KW3, seem reliable. In addition, from the fact that the measured hydraulic conductivities not only after but also before HF are not so different between the experiments KW5 and KW3, it can be concluded that the water HF in the experiments KW3 and KW5 improved the injectivity of the rock around the HF hole by only a few times.

From the above discussion, we can conclude that the injectivity of rock around the HF hole was improved around 400 times by the CO_2 HF in comparison to only a few times larger by the water HF, because the reproductivity is shown in the measuring results of the two experiments KC5 and KC7 for the CO_2 HF and KW3 and KW5 for the water HF.

However, this is probably the first trial in the world to measure the difference between CO_2 and water in injectivity improvement induced by HF. In the measurements, we did not examine the effects of crack shape generated by the HF, the leak into the surrounding AE monitoring holes, the difference in nature of the fluids used for the IFOTs (gaseous state CO_2 and water), and others. Thus, further investigations are necessary to evaluate the injectivity improvement more accurately, including checking the validity of the assumption in the IFOTs that rock is a continuum.

7.4.3 *Focal mechanism deduced from P-wave first-motion polarities of AE*

The focal mechanisms of AE events were examined using P-wave first-motion polarities. We calculated the percentage *R* of compression polarities in the first motion of the P-wave for each AE event, and classified the events as follows: those with $0 \leq R \leq 20$ were labeled as compression-dominant (type C), those with $20 \leq R \leq 80$ as shear-dominant (type S), and those with $80 \leq R \leq 100$ as tensile-dominant (type T). The notable difference between the CO_2 HF experiment KC6 and the water HF experiment KW3 is that many type C events were observed in the CO_2 HF experiment, whereas no type C events were observed in the water HF experiment. In the following, we discuss the characteristics of the three types of AE events –type C, type T, and type S, adding new findings to those already discussed in the previous chapters.

7.4.3.1 *Characteristics of type C events*

As shown in Figure 7.20, type C events began to occur approximately 2 min after the BD in the CO_2 HF experiment KC6. In contrast, no type C events were observed in the water HF experiment, KW3. This difference was also confirmed in the experiments KC7 and KW5, as shown in Figures 7.A5 and 7.A6 in Appendix 7.A. Furthermore, in our previous similar HF experiments, many type C events were observed in the CO_2 HF experiment, KC3, conducted two years prior in the same HF hole. In contrast, none were observed in the water HF experiments in intact granite in the MIU. Thus, type C events are most likely a distinctive feature associated with CO_2 HF.

In volcanic regions, compressive events are often observed. Shimizu et al. (1987) and Foulger et al. (1989) suggested that these events are induced by crack closure due to degassing from magma. Additionally, Martínez-Garzón et al. (2017) recorded compressive events associated with the removal of large volumes of steam at the Geysers geothermal reservoir, in line with the expectations of Ross et al. (1996). On the other hand, Bohnhoff and Zoback (2010) observed non-shear-mechanism events linked to the leakage of injected CO_2 from a depth of 900 m underground. Since this depth was close to the pressure at which CO_2 transitions from SC to a gaseous state, they proposed that mass advection of CO_2 caused the non-shear-mechanism events. In a laboratory experiment on hydraulics, Warjito and Ishikawa (2002) demonstrated that a bubble breaking into smaller bubbles induced acoustic waves, while Kolaini (1999) pointed out that the acoustic radiation released by bubble breakage varied with the concentration of salt. Thus, the type C events observed in our present experiment were likely induced by crack closure due to degassing of injected CO_2, mass advection, or bubble bursting. Although CO_2 injection could induce many type C and S events, at least the type C events are unlikely to trigger natural earthquakes, as they do not create new cracks.

7.4.3.2 Characteristics of type T events

Type T events tended to occur near the HF hole as the first AE events immediately after the BD in the CO_2 HF experiment KC6, as shown in Figure 7.21. This tendency was also observed in the CO_2 HF experiment KC3, conducted two years prior to KC6 at the same hole, as shown in Figures 7.11–7.13. On the other hand, type T events also occurred following type S events after the BD, but primarily at the edges of regions where AEs had previously been distributed, as in the water HF experiment KW3 shown in Figure 7.24. This trend was also observed in the water HF experiments in the MIU, as shown in Figures 6.5 and 6.7 in the previous chapter.

Regarding the mechanism induced by HF, the conventional elastic theory predicts tensile fractures (e.g., Hubbert and Willis, 1957; Zoback et al., 1977; Haimson, 1978; Schmitt and Zoback, 1993). In contrast, many previous studies involving actual field monitoring have shown that shear events dominate (e.g., Talebi and Cornet, 1987; Cornet, 1992; Horálek et al., 2010). This discrepancy is one of the paradoxes between the modes of HF and observed microseismicity, as Maxwell and Cipolla (2011) noted. However, recent studies (Haimson, 1978; Šílený et al., 2009; Julian et al., 2010) have reported the existence of tensile events induced by HF in the field.

In laboratory HF experiments, Ishida et al. (2000a,b) demonstrated that tensile events become dominant as the grain size decreases in granitic rock samples of four different grain sizes. Additionally, Sasaki et al. (1988) performed a laboratory HF experiment on a block of impermeable acrylic resin and found that all 10 AEs observed were tensile events, without exception. Since the Brazilian tensile strengths and BD pressures of the granitic rock specimens increased as grain size decreased (Ishida et al. 2000a,b), and the tensile strengths and BD pressures of the acrylic resin were much higher than those of the granitic rock specimens (Sasaki et al., 1988), we can consider the acrylic resin block as an extremely fine-grained specimen. This suggests that adherence between grain boundaries increases as grain size decreases, including in the case of the acrylic resin.

Consequently, the results of these laboratory experiments indicate that tensile fracture becomes dominant when crack propagation along grain boundaries is restricted due to the increased adherence between boundaries as grain size decreases. Moreover, Rodriguez et al. (2017) demonstrated that tensile events are concentrated near the edges of the propagating fracture before BD.

As discussed in Chapter 2 of this book, a granitic rock generally has two sets of preferred orientations of microcracks (Dale, 1923; Sano et al., 1992). Most microcracks are preferentially oriented along the "rift plane", and a secondary orientation is known as the "grain plane". The "hardway plane" is defined as the plane perpendicular to the rift plane and the grain plane. The three planes were named by quarrymen and are used to split granitic rock along the rift plane, making it easier to break. Also in HF experiments, there is a tendency for cracks to extend more easily along the rift plane (Ishida, 2001). Recently, using this characteristic of granitic rock, Yamamoto et al. (2019) conducted HF experiments on samples of two types of granite under uniaxial loading: one with the rift plane perpendicular to the expected direction of macroscopic fracture propagation, and the other with the rift plane parallel to it (i.e., along the loading axis). They found that tensile events were dominant when the fracture propagated perpendicular to the rift plane, whereas shear events were dominant when the fracture propagated parallel to the rift plane, as shown in Chapter 2. These laboratory experiments suggest that new crack propagation in intact rock is induced by tensile fracturing. This observation is consistent with the fact that shear events become dominant when crack extensions occur easily along grain boundaries, as discussed in the section on type S events below.

7.4.3.3 Characteristics of type S events

Most of the monitored AE events were type S in both CO_2 and water HF, as shown in Figures 7.20 and 7.23. Many previous studies have reported that shear events are associated with HF, although tensile events were theoretically expected, as discussed in the previous section, Section 7.4.3.2.

Such shear-dominant activity has been reported even in volcanic earthquake swarms and three-point bending fracture laboratory tests, where tensile fractures are expected. Hill (1977) proposed a conceptual model (Figure 2.29) for the volcanic earthquake swarm, suggesting that tensile-dominant fracturing occurs when magma intrudes into a pre-existing crack along the direction of maximum compressive stress in a volcanic region. In contrast, shear-dominant fracturing is induced by the formation of conjugate faults connecting the dike tips. In a three-point bending laboratory fracture test using a granite specimen with a notch at the center of its span, Kao et al. (2011) showed that all AE events were shear-dominant, despite the expectation of a tensile fracture, which was macroscopically observed. They explained that the dominance of shear events resulted from the tortuosity of the induced fracture, caused by the local deviation in the crack path due to grain-scale heterogeneity. This also implies that the local fracturing mechanism – i.e., the mechanism inducing an AE event – does not always reflect the global failure mode.

The HF experiments by Ishida et al. (2016) using very slick SC-CO_2 in a granite block revealed that the HF cracks propagated mainly along the grain boundaries of the constituent minerals. This produced many small cracks inclined to the direction of the maximum compressive stress, σ_1, which is also the direction of propagation for a macroscopic crack. Since significant shear stress acts on a plane inclined to the direction of σ_1, shear fractures are likely to occur on the plane of the inclined small cracks.

On the other hand, Lei et al. (2000) stated that tensile fracture dominates in the process zone (the area at the front of the crack), whereas shear fracture dominates in crack extension within the damage zone, where cracks are linked (crack interaction), even during fault nucleation and its quasi-static growth of hornblende schist in a triaxial compression experiment. From this, it can be inferred that, not only in crack extension with HF but also in fault growth under triaxial compression, the general mechanism is that tensile fracture dominates at crack initiation in intact rock, while shear fracture dominates in crack extension with crack linkage.

In our current field experiments, there were likely many pre-existing cracks around the HF hole, as the borehole television observations revealed an average of 6.16 cracks per meter before the HF experiments in both the HF and AE holes. Thus, we can infer that type S events occurred due to the interaction between pre-existing cracks and cracks newly induced by HF, some of which were inclined either to σ_1 or the propagating direction of a macroscopic crack. Accepting this concept makes it clear why shear events are associated with HF in actual production fields, where various discontinuities exist, including pre-existing and newly generated HF cracks.

By analyzing the occurrence times and locations of tensile- and shear-dominant AE events in our current field experiments, we found that tensile-dominant events occur in intact rock, while shear-dominant events are associated with crack linkages, that is, the interaction of cracks. This finding provides a solution to the paradox in the mechanism of HF, where elastic theory predicts tensile fractures, but shear events dominate in AE monitoring.

7.5 Conclusions

We conducted HF experiments using CO_2 and water in a hole drilled into hot granitic rock at a depth of 7–16 m below the tunnel floor, which represents a more realistic scenario than laboratory conditions. All HF experiments using CO_2 confirmed that the injected CO_2 was in a SC state, at least at BD, based on its pressure and temperature. In the experiment KC3, conducted at a depth of 7.24–7.40 m, we examined the effects of a pre-existing crack under CO_2 injection at a 50 mL/min flow rate.

In the experiments KC6, KC7, KW3, and KW5 at a depth of 10–14 m, we examined differences in the size of the stimulated regions and the fracturing mechanisms between CO_2 and water HF under the same constant injection rate of 10 mL/min.

We monitored AE events, as well as the pressure and temperature of the injected CO_2 and water. After the experiments, we observed cracks induced by HF on the surface of the cores recovered through overcoring, as well as on images of the HF hole wall obtained with a borehole camera. Based on the hypocenter locations, the focal mechanisms of the monitored AE events, and the crack observations, we can draw the following conclusions.

1 The images of the HF hole wall obtained with a borehole camera after HF showed that the 3.33 m long vertical bi-wing crack with a large aperture along the HF hole was induced by the SC-CO_2 injection at a flow rate of 50 mL/min in the experiment KC3. The orientation of the crack, as observed with the borehole camera and on the surface of the core recovered through overcoring after HF, was consistent with the directions from the HF hole to the AE1 and AE3 holes, where the injected CO_2 was expelled. Since the BD pressure of 15.62 MPa was sufficiently high and the crack extension in both the vertical and horizontal directions corresponded to the AE hypocenter distributions immediately after BD, it was confirmed that the crack was induced by the BD.

2 In the CO_2 HF experiment KC3, the dense clouds of AE hypocenters are distributed over $1.0 \times 1.3 \times 0.4\,m^3$ ($0.52\,m^3$), which is relatively small compared to the volumes in the experiments KC6 ($1.1 \times 1.7 \times 0.8\,m^3$, or $1.496\,m^3$) and KC7 ($1.75 \times 1.50 \times 1.25\,m^3$, or $3.281\,m^3$). However, a long vertical bi-wing crack with a large aperture was induced only in KC3, while it was not induced in either KC6 or KC7. This is most likely due to the difference in the injection rates: 50 mL/min in KC3 compared to 10 mL/min in KC6 and KC7.

3 In the experiment KC3, after 1 min and 15 s after the BD, AE hypocenters began to distribute along a direction almost perpendicular to that of the initial crack induced by the HF, starting from a position 0.6 m from the HF hole. Since no pressure was applied in the pressurizing section of the HF hole during this period, a pre-existing crack likely existed along that direction. If this is the case, it suggests that CO_2 can easily intrude and enhance AE occurrence along a pre-existing crack without the need for pressure.

4 The BD occurred at 11.55 MPa in the CO_2 HF experiment KC6, whereas it occurred at 12.78 in the water HF experiment KW3. The tendency for the BD pressure in the CO_2 HF experiment to be smaller than in the water HF experiment is consistent with the results of previous laboratory experiments. This tendency is most likely attributed to the lower viscosity of CO_2 compared to that of water.

5 In the CO_2 HF experiment, KC6, AE events continued to occur after BD and even after stopping the injection, whereas in the water HF experiment KW3, AE events ceased immediately after BD.
6 The AE hypocenters in the CO_2 HF experiment KC6 were distributed over a region of $1.1 \times 1.7 \times 0.8\,m^3$ ($1.496\,m^3$), whereas in the water HF experiment KW3, they were distributed over a region of only $0.7 \times 0.2 \times 0.4\,m^3$ ($0.056\,m^3$). The difference in volume was 27 times. This result suggests that CO_2 HF can stimulate a much larger region than water HF. This tendency is also confirmed by the difference in the AE hypocenter distribution between the CO_2 HF experiment KC7 and the water HF experiment KW5, as shown in Appendix 7.A.
7 In both CO_2 and water HF experiments, AE hypocenter distributions and observations of cracks on cores recovered by drilling coaxially around the pressurizing section after HF indicate that bi-wing cracks were induced. However, the directions of crack extensions were almost orthogonal, and the reason for this could not be sufficiently clarified.
8 Many compression-dominant (type C) events were observed in the CO_2 HF experiments, but never in the water HF experiments. Our previous similar experiments also observed this distinctive difference between CO_2 and water HF. Discussions by other researchers, based on field observations, suggest that the crack closure due to the degassing of injected CO_2 and/or mass advection most likely induced the type C events observed in our present experiments. Since these events never create new cracks, type C events are unlikely to trigger natural earthquakes, unlike those classified as shear-dominant (type S) and tensile-dominant (type T).
9 The CO_2 HF tended to induce tensile-dominant (type T) events near the HF hole immediately after BD, whereas the water HF induced type T events following type S events after BD and primarily at the edges of regions where AE events had occurred previously. This strongly suggests that these type T events were induced by crack initiation in intact rock, based on evidence obtained from laboratory experiments and recent studies in the field.
10 The majority of monitored AE events were shear-dominant (type S) events in both CO_2 and water HF. The results of AE monitoring for volcanic earthquake swarms and laboratory experiments, where type T events are expected, suggest that many shear events could be induced by crack extension with crack linkage, that is, the interaction of cracks. This suggests why shear events dominate in HF in actual production fields, as various discontinuities in the fields are likely to interact with the crack extensions of HF.
11 The two conclusions regarding type T and S events suggest a solution to the paradox in the mechanism of HF: while elastic theory predicts tensile fractures, shear events dominate in AE monitoring in actual production fields.
12 To compare injectivity improvement, IFOTs were conducted before and after HF, using gaseous CO_2 for CO_2 HF and water for water HF. The injectivity of rock around the HF hole improved by about 400 times with CO_2 HF, compared to a few times with water HF. The results are likely reliable, as we obtained almost the same results for the other two experiments: KC7 for CO_2 HF and KW5 for water HF. It is hoped that further investigations will verify the quantitative reliability of the improvement effect, including checking the validity of the assumption in the permeability tests that rock behaves as a continuum.

7.A Appendix

To help readers better understand, we briefly present the results of the CO_2 HF experiment KC7 and the water HF experiment KW5. However, we could not obtain typical pressure records showing a clear single BD pressure in these experiments.

Figures 7.A1 and 7.A2 show the temporal changes in fluid pressure, flow rate, temperature, and AE event rate for the CO_2 HF experiment KC7 and the water HF experiment KW5, respectively. As shown in Figure 7.A1, in the CO_2 HF experiment KC7, the pressure gradually increased over time, and then it began to steeply increase near the critical pressure of 7.38 MPa, exhibiting the same trend as in the experiment KC6. However, its BD pressure of 8.26 MPa was considerably lower than the 11.55, 12.78, and 14.15 MPa observed in the experiments KC6, KW3, and KW5, respectively, as shown in Figures 7.14, 7.15, and 7.A2. The gradual decrease after BD contrasts with the sharp reductions observed in these three experiments. Because a hairline crack was observed on the core recovered from the pressurizing section before HF, CO_2 likely leaked from the crack at a lower pressure than in the other three experiments.

In the experiment KW5 shown in Figure 7.A2, with water injection, two peak pressures were observed. Since the vent pipe for the pressurizing section (see the inset in Figure 7.2) was

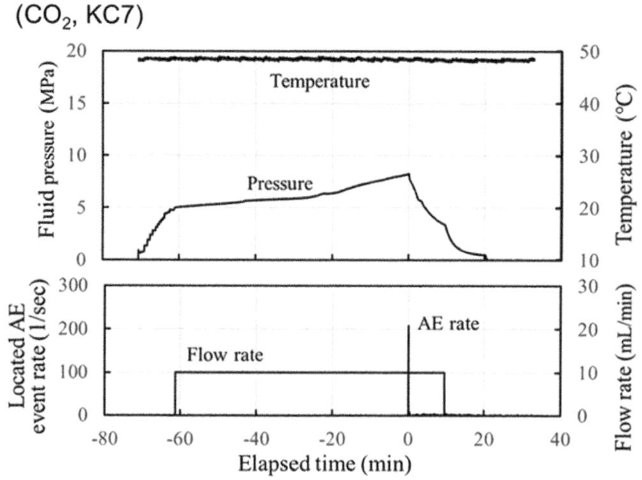

Figure 7.A1 Temporal changes in fluid pressure, flow rate, temperature, and the located AE event rate for the CO_2 HF experiment, KC7 (Appendix of Ishida et al., 2023).

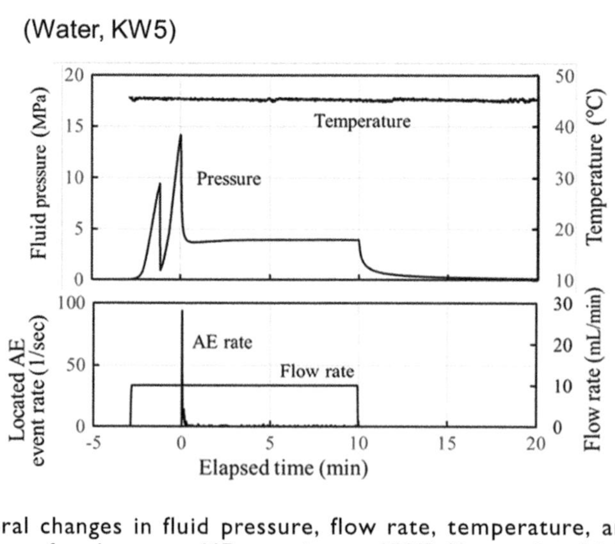

Figure 7.A2 Temporal changes in fluid pressure, flow rate, temperature, and the located AE event rate for the water HF experiment, KW5 (Appendix of Ishida et al., 2023).

found to be blocked with boring slime before the experiment, the first peak pressure was most likely recorded just before the slime was expelled. Therefore, the second peak pressure, namely 14.15 MPa, represents the BD pressure in the experiment. This is further supported by the fact that an AE burst occurred immediately after the second peak pressure.

Figures 7.A3 and 7.A4 show projections of the AE hypocenters on the *XY*, *YZ*, and *ZX* planes for the experiments KC7 and KW5, respectively. We observe that the AE hypocenters with CO_2

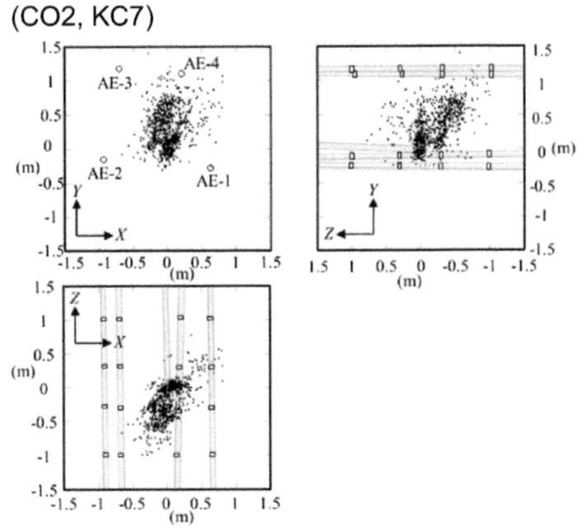

Figure 7.A3 Projections of all 973 AE hypocenters determined by the DD method onto the three orthogonal planes for the CO_2 HF experiment, KC7 (Appendix of Ishida et al., 2023).

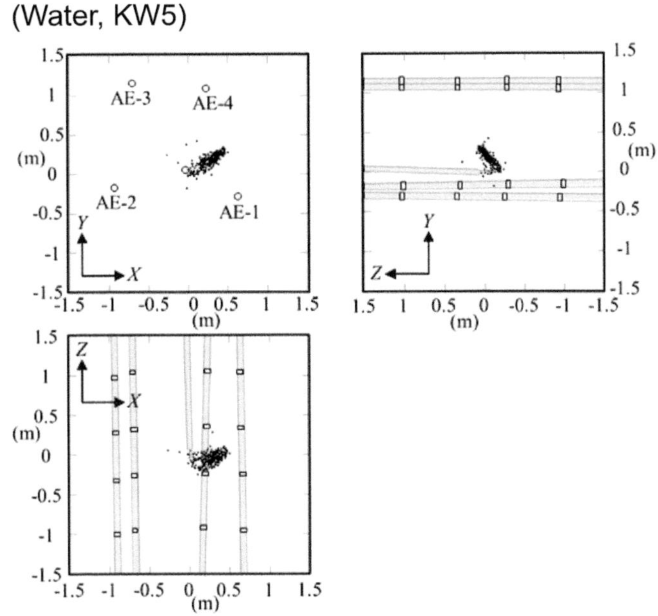

Figure 7.A4 Projections of all 247 AE hypocenters determined by the DD method onto the three orthogonal planes for the water HF experiment, KW5 (Appendix of Ishida et al., 2023).

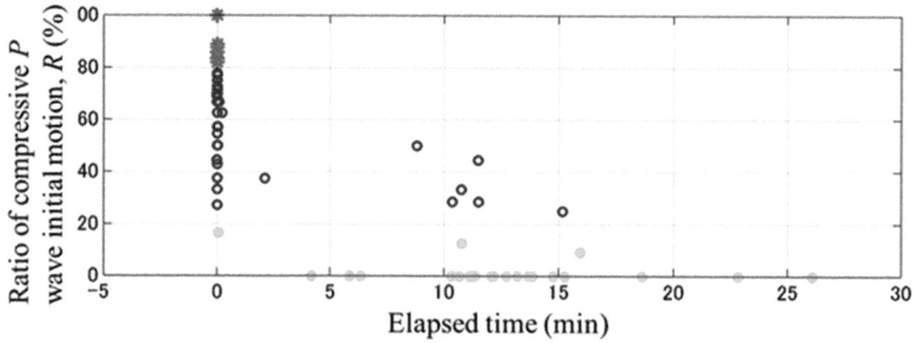

Figure 7.A5 Temporal changes of the percentage, *R*, of the compression polarities of P-wave first motions of an AE event in the CO_2 HF experiment, KC7. The symbols of open gray circles and gray stars indicate the shear-dominant (type S) and the tensile-dominant (type T) AE events, respectively. See their definitions in the caption of Figure 7.20. (Appendix of Ishida et al., 2023).

Note: For interpretation of the 1,044 references to color in this figure legend, the reader is referred to the Appendix of the Web version of Ishida et al. (2023).

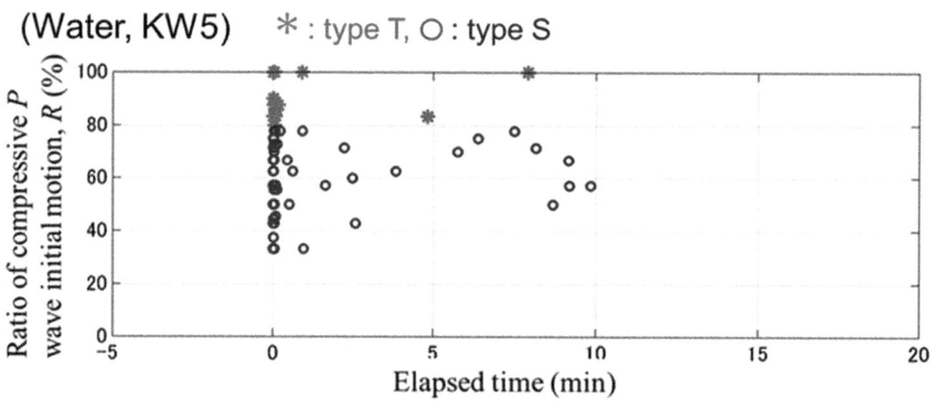

Figure 7.A6 Temporal changes in the percentage, *R*, of the compression polarities of P-wave first motion of an AE event in the water HF experiment, KW5. The symbols of open gray circles and gray stars indicate the shear-dominant (type S) and the tensile-dominant (type T) AE events, respectively. See their definitions in the caption of Figure 7.20. (Appendix of Ishida et al., 2023).

Note: For interpretation of the references to color in this figure legend, the reader is referred to the Appendix of the Web version of Ishida et al. (2023).

HF were distributed more widely than those with water HF, a tendency also observed in the experiments KC6 and KW3, as shown in Figures 7.16c and 7.18, respectively.

Figures 7.A5 and 7.A6 show the temporal changes in the compressive ratio *R* [%] of the P-wave first-motion polarity of the AE events for the experiments KC7 and KW5, respectively. By comparing Figures 7.A5 and 7.A6, we observe the same trend seen in experiments KC6 and KW3 (Figures 7.20 and 7.23): many type C events were observed with CO_2 HF, whereas none were observed with water HF.

References

Bohnhoff, M. & Zoback, M. D. (2010) Oscillation of fluid-filled cracks triggered by degassing of CO_2 due to leakage along wellbores. *J. Geophys. Res.*, 115, B11305. https://doi:10.1029/2010JB000848

Brown, D. W. (2000) A hot dry rock geothermal energy concept utilizing supercritical CO_2 instead of water. *Paper Presented at 25th Workshop on Geothermal Reservoir Engineering*. Stanford Univ., Stanford, Calif.

Bunger, A. P. & Detournay, E. (2008) Experimental validation of the tip asymptotics for a fluid driven fracture. *J. Mech. Phys. Solids.* 56, 3101–3115.

Cornet, F. H. (1992) Fracture processes induced by forced fluid percolation. In: Gasparini, P, Scarpa, R., Aki, K. (eds) *Volcanic Seismology, IAVCEI Proceedings Volcanolgy*, vol 3. , Springer Verlag, Berlin, pp. 407–431.

Dale, T. N. (1923) The commercial granites of New England(I). *U.S. Geol. Surv. Bull.*, 738, 1–97.

Foulger, G. R., Long, R. E., Einarsson, P. & Bjornsson, A. (1989) Implosive earthquakes at the active accretionary plate boundary in northern Iceland. *Nature*, 337, 640–642. hpps://doi:10.1038/337640a0

Godec, M., Koperna, G., Petrusak, R. & Oudinot, A. (2013) Potential for enhanced gas recovery and CO_2 storage in the Marcellus shale in the Eastern United States. *Int. J. Coal Geol.*, 118, 95–104.

Haimson, B. C. (1978) The hydrofracturing stress measuring method and recent field results. *Int. J. Rock Mech. Min. Sci. Geomech. Abstr.*, 15, 167–178.

Hill, D. P. (1977) A model for earthquake swarms. *J. Geophys. Res.*, 82, 1347–1352.

Hirata, Y. (1993) Lecture series, Recent research method and measuring technology for underground water, 5. Measurement for characteristics of a stratum of water, 5.1 In-situ permeability test using a single hole. *J. Groundw. Hydrol.*, 35(3), 247–254. (in Japanese)

Horálek, J., Jechumtálová, Z., Dorbath, L. & Šílený, J. (2010) Source mechanisms of microearthquakes induced in a fluid injection experiment at the HDR site Soultz-sous Forêts (Alsace) in 2003 and their temporal and spatial variations. *Geophys. J. Int.*, 181, 1547–1565. https://doi.org/10.1111/j.1365-246X.2010.04506.x

Hubbert, M. K. & Willis, D. G. (1957) Mechanics of hydraulic fracturing. *Petrol. Trans. Am. Soc. Min. Eng.*, 210, 153–168.

Ishida, T. (2001) Acoustic emission monitoring of hydraulic fracturing in laboratory and field. *Construct. Build. Mater.*, 15, 283–295. https://doi.org/10.1016/S0950-0618 (00)00077-5

Ishida, T., Aoyagi, K., Niwa, T., Chen, Y., Murata, S., Chen, Q. & Nakayama, Y. (2012) Acoustic emission monitoring of hydraulic fracturing laboratory experiment with supercritical and liquid CO_2. *Gephys. Res. Lett.*, 39, L16309.

Ishida, T., Chen, Y., Bennour, Z., Yamashita, H., Inui, S., Nagaya, Y., Naoi, M., Chen, Q., Nakayama, Y. & Nagano, Y. (2016) Features of CO_2 fracturing deduced from acoustic emission and microscopy in laboratory experiments. *J. Geophys. Res. Solid Earth.*, 121(11), 8080–8098. https://doi.org/10.1002/2016JB013365

Ishida, T., Chen, Q., Mizuta, Y. & Roegiers, J-C. (2004) Influence of fluid viscosity on the hydraulic fracturing mechanism. *Trans. ASME, J. Energy Res. Technol.*, 126, 190–200.

Ishida, T., Desaki, S., Yamashita, H., Inui, S., Naoi, M., Fujii, H. & Katayama, T. (2017) Injection of supercritical carbon dioxide into granitic rock and its acoustic emission monitoring. *Proc. Eng.*, 191, 476–482. https://doi.org/10.1016/j

Ishida, T., Desaki, S., Kishimoto, Y., Naoi, M. & Fujii H. (2018) Implosive acoustic emissions induced by injection of supercritical carbon dioxide into a hot granitic rock mass. In: *Geomechanics and Geodynamics of Rock Masses.* Taylor & Francis Group, London, pp. 1369–1374.

Ishida, T., Desaki, S., Kishimoto, Y., Naoi, M. & Fujii, H. (2021) Acoustic emission monitoring of hydraulic fracturing using carbon dioxide in a small-scale field experiment. *Int. J. Rock Mech. Min. Sci.*, 141, 104712. https://doi.org/10.1016/j.ijrmms.2021.104712

Ishida, T., Fujito, W., Yamashita, H., Naoi, M., Fuji, H., Suzuki, K. & Matsui, H. (2019) Crack expansion and fracturing mode of hydraulic refracturing from acoustic emission monitoring in a small-scale field experiment. *Rock Mech. Rock Eng.*, 52(2), 543–553. https://doi.org/10.1007/s00603-018-1697-5

Ishida, T., Kishimoto, Y, Desaki, S., Naoi, M., Suzuki, K. & Fujii, H. (2023) Characteristics of carbon dioxide fracturing in comparison to conventional water hydraulic fracturing: Evidence from acoustic emission monitoring of small-scale field experiments. *Int. J. Rock Mech. Min. Sci.*, 172, 105604. https://doi.org/10.1016/j.ijrmms.2023.105604

Ishida, T., Mizuta, Y., Matsunaga, I., Sasaki, S. & Chen, Q. (2000a). *Effect of grain size in granitic rock on crack extension in hydraulic fracturing. In: Proceedings of 4th North American Rock Mechanics Symposium*, Seattle, Washington, USA, 2000, pp. 1105–1111.

Ishida, T. & Sasaki, S. (2011) Numerical simulation to examine accuracy of AE source location and its applications to in-situ rock monitoring. *J. Acoust. Emiss.*, 29, 260–272.

Ishida, T., Sasaki, S., Matsunaga, I., Chen, Q. & Mizuta Y. (2000b) Effect of grain size in granitic rocks on hydraulic fracturing mechanism. In: *Trends in Rock Mechanics (Proceedings of 923 Sessions of Geo-Denver 2000), geotechnical Special Publication no. 102*, ASCE, vol 924, 2000, pp. 128–139.

Julian, B. R., Foulger, G. R., Monastero, F. C. & Bjornstad, S. (2010) Imaging hydraulic fracturing in a geothermal reservoir. *Gephys. Res. Lett.*, 37, L07305.https://doi:10.1029/2009GL04093

Kalantari-Dahaghi, A. (2010) numerical simulation and modeling of enhanced gas recovery and CO_2 sequestration in shale gas reservoirs: a feasibility study. *Paper Presented at International Conference on CO_2 Capture, Storage, and Utilization*. Soc. of Pet. Eng., New Orleans, La.

Kao, C. S., Carvalho, F. C. S. & Labuz J. F. (2011) Micromechanisms of fracture from acoustic emission. *Int. J. Rock Mech. Min. Sci.*, 48, 666–673.

Kolaini, A. R. (1999) Effects of salt on babble acoustic radiation in water. *J. Acoust. Soc. Am.*, 105(4), 2181–2186.

Lei, X., Kusunose, K., Rao, M. V. M. S., Nishizawa, O. & Satoh, T. (2000) Quasi-static fault growth and cracking in homogeneous brittle rock under triaxial compression using acoustic emission monitoring. *J. Geophys. Res.*, 105(B3), 6127–6139.

Liu, F., Ellett, K., Xiao, Y. & Rupp, J.A. (2013) Assessing the feasibility of CO_2 storage in the New Albany Shale (Devonian-Mississippian) with potential enhanced gas recovery using reservoir simulation. *Int. J. Greenh. Gas Control*, 17, 111–126.

Martínez-Garzón, P., Kwiatek, G., Bohnhoff, M. & Dresen, G. (2017) Volumetric components in the earthquake source related to fluid injection and stress state. *Gephys. Res. Lett.*, 44, 800–809. https://doi:10.1002/2016GL071963

Maxwell, S. C. & Cipolla, C. (2011) What does microseismicity tell us about hydraulic fracturing? In: *SPE Annual Technical Conference and Exhibition,* Denver, Colorado, USA, SPE 146932.

Middleton, R. S., Carey, J. W., Currier, R. P., Hyman, J. D., Kang, Q., Karra, S., Jiménez -Martínez, J., Porter, M. L. & Viswanathan, H. S. (2015) Shale gas and non-aqueous fracturing fluids: opportunities and challenges for supercritical CO_2. *Appl. Energy*, 147, 500–509.

Morita, N., Black, A. D. & Fuh, G-F. (1996) Borehole breakdown pressure with drilling fluids—1. Empirical results. *Int. J. Rock Mech. Min. Sci. Geomech. Abstr.*, 33, 39–51.

Nuttall, B. C., Drahovzal, J. A., Eble, C. F., & Bustin, R. M. (2006) *CO_2 Sequestration in Gas Shales of Kentucky, Poster No. 106 at the 5th Annual Conference on Carbon Capture and Sequestration.* Alexandria, VA, USA.

Pruess, K. (2006) Enhanced geothermal systems (EGS) using CO_2 as working fluid—a novel approach for generating renewable energy with simultaneous sequestration of carbon. *Geothermics*, 35, 351–367.

Ranjith, P. G., Zhang, C. P. & Zhang, Z. Y. (2019) Experimental study of fracturing behaviour in ultralow permeability formations: a comparison between CO_2 and water fracturing. *Eng. Fract. Mech.*, 217, 106541. https://doi.org/10.1016/j.engfracmech.2019.106541

Rodriguez, I. V., Stanchits, S. & Burghardt, J. (2017) Data-driven, in situ, relative sensor calibration based on waveform fitting moment tensor inversion. *Rock Mech. Rock Eng.*, 50, 891–911. https://doi.org/10.1007/s00603-016-1144-4

Ross, A., Foulger, G. R. & Julian, B. R. (1996) Non-double-couple mechanisms at The Geysers geothermal area, California. *Gephys. Res. Lett.*, 23, 877–880.

Sano, S., Kudo, Y. & Mizuta, Y. (1992) Experimental determination of elastic constants of Oshima granite, Barre granite, and Chelmsford granite. *J. Geophys. Res.*, 97(B3), 3367–3379.

Sasaki, S., Matsunaga, I., Kobayashi, H. & Ishida, T. (1988) *Development of a Fracture Evaluation Technique for HDR Geothermal Energy Extraction — Mechanism of Hydraulic Fracturing Inferred from*

Acoustic Emission. CRIEPI (Central Research Institute of Electric Power Industry), Abiko Research Laboratory Report, Abiko, Chiba, Japan; U88034 (in Japanese with English abstract and figure captions)
Schmitt, D. R. & Zoback M. D. (1993) Infiltration effects in the tensile rupture of thin walled cylinders of glass and granite: Implications for the hydraulic fracturing breakdown equation. *Int. J. Rock Mech. Min. Sci. Geomech. Abstr.*, 30, 289–303.
Shimizu, H., Ueki, S. & Koyama, J. (1987) A tensile-shear crack model for the mechanism of volcanic earthquakes. *Tectonophysics*, 144, 287–300.
Šílený, J., Hill, D. P., Eisner, L. & Cornet, F. H. (2009) Non-double-couple mechanisms of microearthquakes induced by hydraulic fracturing. *J. Geophys. Res.,* 114, 913. https://doi:10.1029/2008JB005987
Suenaga, H., Nakagawa, K., Nakata, E. & Ichihara, Y. (2004) *Study on a Method of In-Situ Pneumatic Test in Borehole for Rock of Low Permeability.* Research Report of Central Research Institute of Electric Power Industry, U03057. (in Japanese with English abstract)
Suzuki, K., Desaki, S., Fujito, W., Ishida, T., Naoi, M. & Fujii, H. (2022) On injectivity improvements measured in a small-scale field experiment of hydraulic fracturing using supercritical carbon dioxide. In: *Proceedings of 2022 Annual Meeting of Japan Society of Civil Engineers,* Kyoto, Japan, September 12–14, 2022. (in Japanese)
Talebi, S. & Cornet, F. H. (1987) Analysis of the microseismicity induced by a fluid injection in a granitic rock mass. *Gephys. Res. Lett.,* 14: 227–2.30
Verdon, J. P., Kendall. J. M. & Maxwell, S. C. (2010) A comparison of passive seismic monitoring of fracture stimulation from water and CO_2 injection. *Geophysics*, 75(3). https://doi.10.1190/1.3377789
Warjito, M. O. & Ishikawa, H. (2002) Breakup of a single babble and its sound. *Nagare*, 21, 165–172 (in Japanese with English abstract)
Waldhauser, F. & Ellsworth, L. W. (2000) A double difference earthquake location algorithm: method and application to the Northern Hayward fault, California. *Bull. Seism. Soc. Am.,* 90, 1353–1368.
Xie, H., Li, X., Fang, Z., Wang, Y., Li, Q., Shi, L., Bai, B., Wei, N. & Hou, Z. (2014) Carbon geological utilization and storage in China: current status and perspectives. *Acta Geotech.*, 9, 7–27.
Yamamoto, K., Naoi, M., Chen, Y., Nishihara, K., Yano, S., Kawakata, H., Akai, T., Kurosawa, I. & Ishida T. (2019) Moment tensor analysis of acoustic emissions induced by laboratory-based hydraulic fracturing in granite. *Geophys. J. Int.*, 216(3).1507–1516. https://doi. org/10.1093/gji/ggy493
Zhang, C. P. Chen, P. Ranjith, P. G., Lu, Y. Y. & Zhou, J. P. (2020) A comparative study of fracture surface roughness and flow characteristics between CO_2 and water fracturing. *J. Nat. Gas Sci. Eng.,* 76, 103188. https://doi.org/10.1016/j.jngse.2020.103188
Zhang, C. P., Chen, P., Ma, Z. Y., Ranjith, P. G. & Zhou, J. P. (2021) Comparison of fracturing unconventional gas reservoirs using CO_2 and water: an experimental study. *J. Pet. Sci. Eng.,* 203, 108598. https://doi.org/10.1016/j.petrol.2021.108598
Zoback, M. D., Rummel, F., Jung, R. & Raleigh C. B. (1977) Laboratory hydraulic fracturing experiments in intact and pre-fractured rock. *Int. J. Rock Mech. Min. Sci. Geomech. Abstr.,* 14, 49–58.

Chapter 8

Conclusions

This book focused on two subjects: the first was solving the paradox of the fracturing mode of hydraulic fracturing (HF), and the second was clarifying the characteristics of HF using carbon dioxide (CO_2) as a fracturing fluid compared to conventional HF using water. This chapter summarizes how the book addresses, discusses, and concludes these two topics.

8.1 Paradox of the fracturing mode of HF

The first topic addressed in this book was the paradox of the fracturing mode of HF. Specifically, the stress conditions around an HF hole, derived from traditional elastic theory, suggest that HF is induced by tensile fracturing. However, the focal mechanisms of microseismicity and acoustic emission (AE) associated with HF point to shear fracturing. Chapter 1 explained this paradox by examining the stress conditions around an HF hole and the analytical methods used to interpret the focal mechanism of microseismicity and AE. Chapter 2 presented the results of the various laboratory experiments in which the author has been involved, demonstrating that tensile-dominant fractures tend to form in intact homogeneous rock, without the influence of inherent pre-existing cracks. In contrast, shear-dominant fractures are induced when cracks extend and link, i.e., through the interaction of inherent pre-existing cracks.

Chapter 4 discussed the influence of grain size distribution on the fracturing mode of HF through numerical simulations using the distinct element method. The simulation results suggested that shear fracturing occurs along grain boundaries diagonally across the direction of the maximum confining stress, in the inhomogeneous rock model with a wide particle size distribution and large particles. In contrast, in the homogeneous rock model, where all particles are small, hydraulic fractures develop straight in the direction of the maximum confining stress through tensile fracturing. These simulation results were consistent with the experimental findings presented in Chapter 2. Additionally, the simulation results align with the stress conditions on the fracture plane: shear stress acts on a plane inclined relative to the direction of the maximum compressive stress, making inclined cracks more likely to form through shear fracturing. On the other hand, shear stress does not act on a plane parallel to the direction of the maximum compressive stress, making tensile fracturing more likely than shear fracture.

Chapter 6 presented a small-scale field experiment designed to extend an HF crack by increasing the flow rate of injected water, mimicking the refracturing process commonly used in shale oil and gas fields. Analyses of the monitored AE events indicated that most were shear-dominant. However, tensile-dominant AE events occurred near the HF hole immediately after the first breakdown (BD). Following the second BD, with an increased flow rate, tensile-dominant AE events primarily appeared at the frontiers of regions where AE events had already

DOI: 10.1201/9781003282631-8

been induced. The results suggest that tensile-dominant AE events occurred in the intact rock near the HF hole during the initiation stage of HF in the first BD. In contrast, the new crack expansion associated with the second BD was induced by tensile fracturing at the frontiers of the regions where fractures had already formed, accompanied by the tensile-dominant AE events. These findings suggest that tensile fracturing occurs in intact rock without the influence of pre-existing cracks, while shear fracturing is induced through crack extension with the interaction of pre-existing cracks.

Based on these results, it is concluded that tensile fracturing occurs during crack extension in the direction of maximum compressive stress in intact rock. In contrast, shear fracturing occurs on a plane inclined relative to the direction of the maximum compressive stress due to shear stress acting on the inclined plane when the influence of pre-existing cracks deflects the crack propagation direction. This insight helps to explain why shear events dominate in HF in actual production fields, where various discontinuities are expected to interact with HF crack extensions. Consequently, this understanding contributes to solving the paradox of the fracturing mechanism of HF, reconciling the elastic theory with the focal mechanism of AE.

8.2 Features of CO_2 fracturing in comparison to water fracturing

Various technologies of carbon dioxide capture, utilization, and storage have been investigated and proposed to mitigate global warming. HF using CO_2 has also gained attention for applications such as extracting heat from hot dry rock and exploiting shale gas and oil, as profits from resource production offset the costs. Therefore, as the second topic of this book, the author focused on the features of CO_2-based HF in comparison to traditional water-based HF, based on the results of laboratory and small-scale field experiments conducted by the author's research group.

HF to extract heat from hot dry rock and to exploit shale gas and oil is conducted at depths of a few thousand meters, where the temperature and pressure are such that CO_2 exists in the supercritical (SC) state. In this state, SC-CO_2 has a much lower viscosity than water. Thus, Chapter 3 examined the effects of the injected fluid's viscosity on HF through laboratory experiments using water and viscous oil as injected fluids. The observation of cracks induced by HF revealed that viscous oil tends to generate thicker, more planar cracks with fewer branches and wider apertures compared to water.

Chapter 5 presented the results of laboratory HF experiments using SC-CO_2, liquid CO_2 (L-state), water, and viscous oil. The results revealed that HF using SC-CO_2 tends to induce three-dimensional sinuous cracks with numerous secondary branches, which appear to be desirable pathways for heat recovery from Hot Dry Rock (HDR) and oil and gas extraction from shale reservoirs. Regarding the fracturing mechanism, the ratios of the P wave first motion polarity of AE events suggested that low-viscosity fracturing fluids, such as SC-CO_2, tend to induce shear-dominant fractures, while viscous fluids tend to induce tensile-dominant fractures. The results derived from AE data were also consistent with microscopic observations of the HF cracks.

On the other hand, Chapter 4 examined the effects of the fracturing fluid viscosity on induced crack features through numerical simulations using the distinct element method. The simulation results showed that low-viscosity fluid can easily infiltrate into the fracture, and the fluid pressure acts on the fracture tip. When the fluid pressure increased sufficiently, shear cracks formed to connect pre-existing microcracks, resulting in a hydraulic fracture with tortuous paths that follow the boundaries of the mineral grains. In contrast, when the high-viscosity fluid was used, the fluid pressure was applied only at the fracture inlet, causing the fracture to open due to wedging. Once the fracture was opened sufficiently by the fluid pressure, a tensile crack formed in the direction of the maximum confining pressure. As a result, a thick, planar hydraulic fracture with few branches is generated. These numerical simulation results were consistent with the experimental findings presented in Chapters 3 and 5.

Chapter 7 presented the results of small-scale field HF experiments that compared the effects of a pre-existing crack and induced crack size on SC-CO_2 fracturing versus conventional water fracturing, under more realistic conditions than laboratory experiments. The fracturing fluids were injected into a 10-m-long HF hole drilled in a hot rock mass, which met the temperature requirements to convert CO_2 into its SC state. The AE source distribution during HF indicated that CO_2 can easily intrude and increase AE occurrence along a pre-existing crack. Additionally, CO_2 fracturing in intact rock stimulated a much larger region and improved injectivity to a greater extent than conventional water fracturing, even under the same flow rate. Regarding the fracturing mechanism, the ratios of the P wave first motion polarity of AE events suggested that CO_2 fracturing induces a higher number of compression-dominant AE events, which were never observed in our similar water fracturing experiments. These compression-dominant events were likely caused by crack closure due to the degassing of injected CO_2. Even if CO_2 injection induces many AE events, the compressive-dominant events are unlikely to trigger natural earthquakes, as they do not create new cracks.

Through the above discussions, it can be concluded that CO_2 fracturing induces three-dimensional sinuous cracks with numerous secondary branches, which appear to be desirable pathways for the recovery of energy resources. In addition, CO_2 fracturing stimulates a much larger region and improves injectivity to a greater extent than conventional water fracturing. However, since the small-scale field experiment demonstrated that CO_2 can easily intrude and enhance AE occurrence along pre-existing cracks, it is important to determine the CO_2 injection site by ensuring no faults or large fractures that could act as leakage pathways.

Index

Note: **Bold** page numbers refer to tables.

For Product Safety Concerns and Information please contact our EU representative GPSR@taylorandfrancis.com Taylor & Francis Verlag GmbH, Kaufingerstraße 24, 80331 München, Germany